LIQUID CRYSTALS

LIQUID CRYSTALS

THE FOURTH STATE OF MATTER

Edited by Franklin D. Saeva

Xerox Corporation
Webster Research Center
Webster, New York

MARCEL DEKKER, INC.   New York and Basel

Library of Congress Cataloging in Publication Data
Main entry under title:

Liquid crystals.

Includes indexes.
1. Liquid crystals. I. Saeva, Franklin D.
QD923.L57 548'.9 79-18501
ISBN 0-8247-6813-2

MARCEL DEKKER, INC.
270 Madison Avenue, New York, New York 10016

Current printing (last digit):
10 9 8 7 6 5 4 3 2 1

PRINTED IN THE UNITED STATES OF AMERICA

To my wife Irene, and daughters Sandra, Maureen, and Rebecca

CONTRIBUTORS

EDWARD M. BARRALL, II, IBM General Products Division, San Jose, California

CHARLES L. BEATTY,[1] Webster Research Center, Xerox Corporation, Webster, New York

BERNARD J. BULKIN,[2] Department of Chemistry, Hunter College of the City University of New York, New York, New York

JOSEPH A. CASTELLANO,[3] Princeton Materials Science, Inc., Princeton, New Jersey

DENNIS CHAPMAN, Department of Biochemistry and Chemistry, Royal Free Hospital School of Medicine, University of London, London, England

ADRIAAN DE VRIES, Liquid Crystal Institute, Kent State University, Kent, Ohio

DONALD B. DUPRÉ, Department of Chemistry, University of Louisville, Louisville, Kentucky

MICHAEL F. FROIX,[4] Webster Research Center, Xerox Corporation, Webster, New York

---

[1] Department of Materials Science and Engineering, University of Florida, Gainesville, Florida

[2] School of Arts & Sciences, Polytechnic Institute of New York, Brooklyn, New York

[3] Stanford Resources, Inc., San Jose, California.

[4] Celanese Research Company, Summit, New Jersey

HARRY W. GIBSON, Webster Research Center, Xerox Corporation, Webster, New York

ANDREAS O. GOEDDE,[5] Webster Research Center, Xerox Corporation, Webster, New York

S.E.B. PETRIE, Chemistry Division, Eastman Kodak Company Research Laboratories, Rochester, New York

DARLYN F. POCHAN, Webster Research Center, Xerox Corporation, Webster, New York

JOHN M. POCHAN, Joseph C. Wilson Center for Technology, Xerox Corporation, Webster, New York

FRANKLIN D. SAEVA,[6] Webster Research Center, Xerox Corporation, Webster, New York

EDWARD T. SAMULSKI, Department of Chemistry and Institute of Materials Science, University of Connecticut, Storrs, Connecticut

[5] Corporate Audit and Operational Analysis Division, Xerox Corporation, Rochester, New York

[6] Research Laboratories, Eastman Kodak Company, Rochester, New York

PREFACE

Since the first detailed description of the nature of liquid crystals by Reinitzer in 1888 there has been considerable research activity designed to provide a better understanding of these intriguing compounds. With approximately 5% of all organic compounds exhibiting liquid crystalline behavior, several researchers have been concerned with the relationship between chemical structure and mesomorphic behavior. From these investigations the importance of molecular shape and group polarizability to the existence of mesomorphic behavior has evolved. Subtle alterations in chemical structure have also been utilized to modify the electric and magnetic field behavior of liquid crystals without significantly altering mesomorphic properties for their many applications in optical displays, watches, etc. In the editor's opinion, the field of liquid crystals is an exciting and challenging area of research, especially as it relates to biological processes, the effect of molecular organization on chemical reactivity, and commercial applications.

This monograph presents a series of contributions from leading scientists in the area of liquid crystals and is designed primarily to be an introductory text. The reader is encouraged to proceed beyond the material presented and go into the scientific literature for a more current view of the scientific advances in this relatively novel area of research.

I would like to thank the contributors of the chapters in this monograph for sharing their scientific expertise. A special thanks goes to Michael L. Hair who played a crucial role in making this monograph possible and to the Xerox Corporation for making its facilities available to us. I also want to thank Joan J. Preston for her assistance in preparing this monograph as well as Harry W. Gibson and Wolfgang H. H. Gunther for their several valuable suggestions.

Franklin D. Saeva

CONTENTS

Chapter 1

STRUCTURE AND CLASSIFICATION OF THERMOTROPIC LIQUID CRYSTALS

Adriaan de Vries

Liquid Crystal Institute
Kent State University
Kent, Ohio

I. INTRODUCTION

The title of this chapter mentions classification as well as structure. We have combined these two subjects, not because classification would be impossible without knowledge of structure, but because one cannot discuss structure very well without classification.

The earliest classification system was given by Friedel [1]. He distinguished between nematics, cholesterics, and smectics, mainly on the basis of optical studies. He also argued, however, that cholesterics should be regarded as a special kind of nematics, and that there are only two basically different types of thermotropic liquid crystals: nematics and smectics. The distinction between nematics and smectics is indeed the major distinction among thermotropic liquid crystals, but even so it is advantageous to differentiate between "subgroups" within these two main groups. Within the nematic category, e.g., it remains useful to distinguish between "twisted nematic" (cholesteric) and "nontwisted nematic" (usually designated simply as "nematic"). With respect to smectic phases, it has been found that many compounds show first-order phase transitions within the smectic range, suggesting the existence of different types of smectics.

In the following we shall first review and comment upon what is at present generally accepted with regard to the classification of thermotropic liquid crystals and with respect to the structure of the various types of phases (Sec. II). After that we shall approach the question of classification more specifically from the point of view of structure, and we shall discuss a number of structural aspects which are relevant to classification (Sec. III). Section IV summarizes some of the results from the discussions in Sec. III and gives some tentative conclusions. It includes a table listing the relationships between the structural aspects considered in Sec. III and all known thermotropic phase types.

II. THE CURRENT STATE OF AFFAIRS

A. Smectic Phases

1. Introduction

A systematic study of the classification of smectic liquid crystals has been undertaken by Sackmann, Arnold, and Demus and their colleagues in Halle. In the course of their work they developed, as their main tool, the "miscibility rule," which says that liquid crystals are of the same type if they are miscible in all proportions ("all liquid crystalline modifications which exhibit an uninterrupted series of mixed crystals in binary systems without contradiction can be marked with the same symbol" [2]). Other criteria used in their classification work were: microscopic textures, the order in which different phases appear in compounds with more than one smectic phase, and x-ray diffraction patterns.

The Halle group introduced the now-common identification of different smectic phase types by capital letters: A, B, C, etc. These letters were assigned to the phase types in the order in which they were discovered. At present the Halle group has identified phases A through G [2], and we had added to this the H phase [3]. In the following we shall briefly describe the structures of these phase types according to general current understanding. A summary of these structural data is given in Table 1. For a discussion of the x-ray diffraction patterns of the various phase types we refer the reader to Ref. 6.

2. Smectic A

Smectic A phases are the least ordered of all smectic phases: If a system has a smectic A phase in addition to other smectic phases, the smectic A phase is always found at the high-temperature end of the smectic range. The molecules in a smectic A phase are arranged in layers (as is true for all smectic phases except perhaps smectic D) and have their long axis on the average perpendicular to the layer plane.

TABLE 1. Data on the structure of smectic phases

Phase type	Molecular direction[a]	Arrangement of the molecular axes within the smectic layer
A	Normal	Disordered
B_n	Normal	Ordered; hexagonal lattice in the layer plane
B_t	Tilted	Ordered; pseudohexagonal lattice in the plane perpendicular to the molecules
C	Tilted	Disordered
D	--	Cubic arrangement of spherical aggregates of molecules
E_n	Normal	Ordered; nonhexagonal lattice in the layer plane
E_t	Tilted	Ordered
F	Tilted	Disordered
G(H)	Tilted	Ordered; lattice as in B_t

[a]Direction of the long molecular axis with respect to the smectic plane.

Within each layer the molecules are arranged in a rather random manner and have considerable freedom of translation and of rotation around their long axis. This rather loose organization of the molecules makes the layers rather flexible, and they often adopt complex curved arrangements (Dupin cyclides) [4]. Also, flat layers have been observed to slide very easily over each other, but the viscosity perpendicular to the layers is much higher due to a strong tendency to maintain the integrity of the layers.

The perpendicular arrangement of the molecules with respect to the layers is most directly demonstrated by x-ray diffraction photographs of monodomains (Fig. 1), taken with the x-ray beam parallel to the smectic layers (by monodomain we mean a region within a liquid crystal sample in which there is uniform orientation; for a smectic A this implies that the smectic layers in this region are all flat and parallel to each other). In such

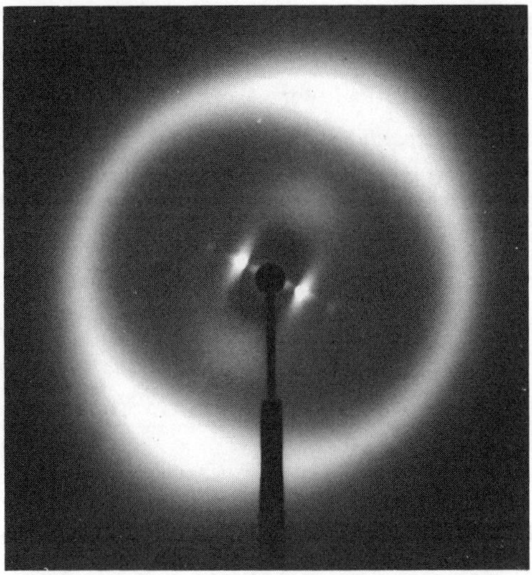

FIG. 1. X-ray diffraction photograph of a monodomain of the A phase of 4-n-heptyloxybenzal-4'-ethylaniline, taken with the incident beam approximately perpendicular to the mean direction of the long molecular axis and parallel to the smectic planes.

photographs, the angle between the molecular axes and the smectic planes can be directly measured as the angle between the line connecting the centers of the outer maxima and the line connecting the centers of the inner maxima [5,6]. The best photograph published to date is that obtained by a group of Russian scientists at Ivanovo [7,8]. This same group has shown, however, through studies on optical models, that although these x-ray photographs are strong evidence for the perpendicularity of the average direction of the molecules, they do not preclude fairly large local deviations from this perpendicular arrangement [8,9].

On the basis of a comparison of measured and calculated intensities of x-ray reflections from the smectic planes, Krigbaum, et al. [10] have argued that some smectic A phases have "tilted" structures, i.e., structures in which the long molecular axis makes a considerable angle with the normal to the smectic planes. This proposed tilted arrangement, however, has not yet been confirmed by more direct studies of the kind discussed in the previous paragraph.

All smectic A phases studied so far have been found to be optically uniaxial, and frequently they show the so-called pseudoisotropic or homeotropic texture, in which the smectic layers are parallel to the supporting surface and the optic axis is perpendicular to it.

3. Smectic C

Smectic C phases are very similar to smectic A phases: the molecules within a smectic layer are arranged in a disordered manner and are fairly free to move around; the layers are flexible and slide easily with respect to each other. The main difference between the two phase types is that in smectic C the molecules are tilted with respect to the layers whereas in smectic A they are perpendicular.

The most direct evidence of the tilted arrangement comes again from x-ray studies on monodomains [6,11,12], but strong supporting evidence comes from measurements of tilt angles by other techniques [12], from observations that the layer thickness in C phases is considerably less than the molecular length whereas in A phases the two are nearly equal [13], and from the fact that C phases are optically biaxial whereas A phases are optically uniaxial (an exception is formed by a group of uniaxial C phases discussed in Sec. II.C, "Cholesteric Phases," but the structure of these phases actually supports rather than disproves the concept of a tilted arrangement).

Some authors [14,15] have noted that there appear to be two kinds of C phases: those in which the tilt angle is <u>independent</u> of temperature and those in which the tilt angle <u>varies</u> with temperature (the tilt angle is the angle between the mean direction of the long molecular axis and the direction normal to the smectic plane). We have modified and expanded this distinction elsewhere [12,16] and will also discuss it in some detail in Sec. III.E. Sackmann and Demus [2] suggest that the tilt angle may also depend on the experimental conditions, and Luckhurst [17] has reported evidence for this. We feel that the data are still rather inconclusive (see also our discussion of C_1 phases in Ref. 12), and we expect that the tilt angle will be significantly influenced only by conditions like pressure and magnetic field strength, and only in those C phases that have a strongly temperature-dependent tilt angle.

4. Smectic B and H

Smectic B and H phases differ from A and C phases in that they have an ordered arrangement of the molecules in the layers rather than a disordered one. An ordered arrangement was first proposed by Herrmann and Krummacher [18,19], who suggested that the molecules rotated around their long axis, thus effectively creating cylindrical units, and that these cylinders were close packed. Such a packing would yield a hexagonal arrangement of the molecules in planes perpendicular to the long axis, and x-ray photographs of monodomains of B phases (Ref. 20 and Fig. 2) do indeed suggest hexagonal symmetry.

The ordered arrangement of the molecules appears to make the layers of B (and H) phases much more rigid than those of A and C phases [21], leading to the frequent appearance of mosaic textures for these phases. This paucity of curved layers in B phases is confirmed by the "broken-up" appearance of the smectic diffraction ring of B phases (Fig. 3a) as compared with the "continuous" appearance of these maxima in A and C phases (Fig. 3b).

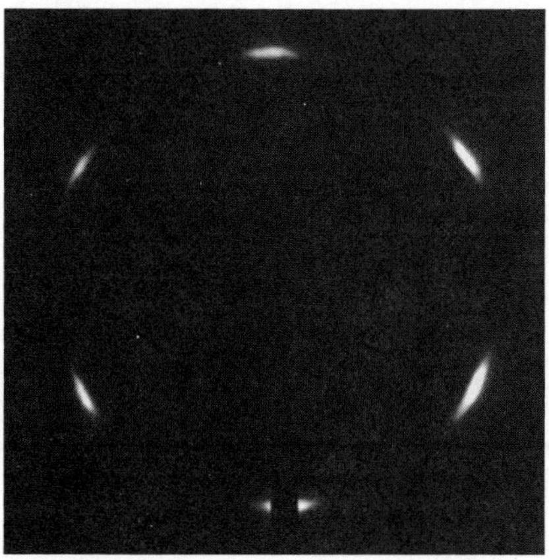

FIG. 2. X-ray diffraction photograph of a monodomain of the B_n phase of BBOA, taken with the incident beam approximately perpendicular to the smectic planes. With this arrangement there is no inner ring. The outer ring is broken up into six short streaks in an approximately hexagonal pattern.

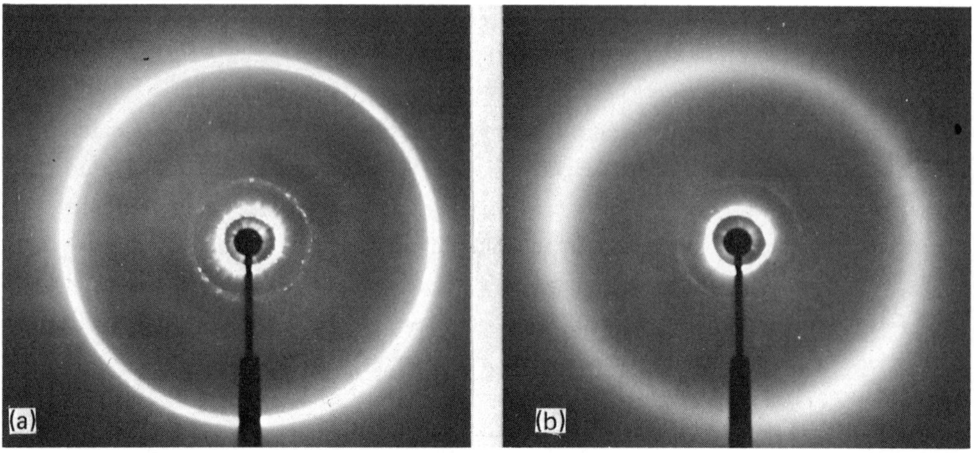

FIG. 3. X-ray diffraction photographs of the B phase (Fig. 3a) and the A phase (Fig. 3b) of 4-n-octyloxybenzal-4'-ethylaniline. The outer ring, originating from the interactions between molecules within a smectic layer, is sharp in the B phase, diffuse in the A phase. The inner rings, diffraction from the smectic layers, are quite spotty in the B phase, much more continuous in the A phase. (Reprinted from Ref. 69, by courtesy of the North-Holland Publishing Company.)

Originally it was assumed that the molecules in B phases were always normal to the layers, as in A phases (i.e., all B phases were assumed to be B_n phases, where the subscript n stands for normal), but later it became apparent that they could also be tilted, as in C phases: B_t phases, with t for tilted, or H phases. Tilt, by its very nature, introduces an element of asymmetry and thus eliminates the possibility of true hexagonal symmetry for the molecular packing in the layers of B_t and H phases. Thus, it is not surprising that the models proposed and/or the lattice parameters measured for B_t or H phases [3,22-24] indicate to a greater or lesser extent deviations from hexagonal packing in the plane perpendicular to the long axis of the molecules and even greater deviations in the plane of the smectic layer.

When we assigned the name "smectic H" [3] to the tilted, ordered smectic phase of BBEA (for the meaning of abbreviations of names of compounds, as used throughout this chapter, see Table 2), we did this on the basis of its x-ray diffraction pattern and the structure derived from it. It

TABLE 2. Abbreviations, names, and formulas of the compounds

No.	Abbreviation	Name	Formula
1.	AOAOB	4,4'-Di-n-alkoxyazoxybenzene	C_nH_{2n+1}-O-C_6H_4-N=NO-C_6H_4-O-C_nH_{2n+1}
2.	AOBA	4-n-Alkoxybenzoic acid	C_nH_{2n+1}-O-C_6H_4-COOH
3.	AOBCP	Bis-(4'-n-alkoxybenzal)-2-chloro-1,4-phenylenediamine	C_nH_{2n+1}-O-C_6H_4-CH=N-C_6H_3Cl-N=CH-C_6H_4-O-C_nH_{2n+1}
4.	BBEA	4-n-Butyloxybenzal-4'-ethylaniline	C_4H_9-O-C_6H_4-CH=N-C_6H_4-C_2H_5
5.	BBOA	4-n-Butyloxybenzal-4'-n-octylaniline	C_4H_9-O-C_6H_4-CH=N-C_6H_4-C_8H_{17}
6.	DeOBCP	Bis-(4'-n-decyloxybenzal)-2-chloro-1, 4-phenylenediamine	No. 3 with n = 10
7.	DoOBA	4-n-Dodecyloxybenzoic acid	No. 2 with n = 12
8.	DPTC	Di-n-propyl-p-terphenyl-4, 4''-carboxylate	C_3H_7-COO-C_6H_4-C_6H_4-C_6H_4-COO-C_3H_7
9.	EOAOB	4,4'-Diethoxyazoxybenzene	No. 1 with n = 2
10.	HpBPeA	4-n-Heptyloxybenzal-4'-n-pentylaniline	C_7H_{15}-O-C_6H_4-CH=N-C_6H_4-C_5H_{11}
11.	HpOAOB	4,4'-Di-n-heptyloxyazoxybenzene	No. 1 with n = 7
12.	NOBA	4-n-Nonyloxybenzoic acic	No. 2 with n = 9
13.	NOBCP	Bis-(4'-n-nonyloxybenzal)-2-chloro-1,4-phenylenediamine	No. 3 with n = 9

TABLE 2. Continued

No.	Abbreviation	Name	Formula
14.	OOBA	4-n-Octyloxybenzoic acid	No. 2 with n = 8
15.	OOBCP	Bis-(4'-n-octyloxybenzal)-2-chloro-1,4-phenylenediamine	No. 3 with n = 8
16.	PAP	p-Azoxyphenetole	Same as no. 9
17.	PPPOPP	2-(4-n-Pentylphenyl)-5-(4-n-pentyloxyphenyl)-pyrimidine	C_5H_{11}-C_6H_4-$C_4N_2H_2$-C_6H_4-O-C_5H_{11}
18.	TBBA	Terephthal-bis-(4-n-butylaniline)	C_4H_9-C_6H_4-N=CH-C_6H_4-CH=N-C_6H_4-C_4H_9
19.	TlOl	Thalium oleate	C_8H_{17}-CH=CH-C_7H_{14}-COOTℓ
20.	TlSt	Thalium stearate	$C_{17}H_{35}$-COOTℓ
21.	UOAOB	4,4'-Di-n-undecyloxyazoxy-benzene	No. 1 with n = 11

was reported later, however, that the H phase of BBEA was miscible in all proportions with a B phase [25,26] and that phase V of TBBA [22], a tilted, ordered smectic phase which gives x-ray diffraction patterns very similar to those of the H phase of BBEA [3,22,27], also is miscible in all proportions with a B phase [28]. Thus, on the basis of the miscibility rule, all these phases would have to be called B phases [29]. A very recent study [30], however, finds the earlier reports on the miscibility of BBEA [25,26] to be in error. This new study shows that the H phase of BBEA is not miscible in all proportions with the B phase used in the earlier work, but rather with a G phase. It confirms the miscibility of phase V of TBBA with a B phase, and it also reports that phase VI of TBBA (an ordered, tilted smectic phase found directly below phase V, on cooling) is miscible in all proportions with a G phase and with the H phase of BBEA. Thus, according to these new data, phase V of TBBA would be a B phase, but phase VI of TBBA and the H phase of BBEA would be G phases. The use of the name smectic H for the tilted, ordered smectic phases of this type is, however, quite widespread in the literature, and the name smectic H is also closely identified with the detailed structure described for the smectic H phase of BBEA [3]. To retain this identification with the current literature and the detailed structure on the one hand, and recognize on the other hand the new findings of Richter, et al. [30], we shall in the remaining part of this chapter use the symbol G(H) for the type of smectic phase found in BBEA, which up to now we identified as smectic H. For phase V of TBBA we will use the symbol B_t.

5. Smectic D

Only a few compounds have the smectic D phase, and only little is known about it. On the basis of x-ray diffraction patterns and the fact that the D phases were found to be optically isotropic, Diele et al. [31] proposed as a structure model a cubic close packing of spherical units, each unit consisting of several molecules (this would be a structure similar to that of the cubic phase in lyotropic liquid crystals). Winsor (Ref. 32, p. 57) proposes a similar structure, with "rotating units... composed of globular assemblies of parallel... molecules." Such a structure would not have the layers generally thought of as characteristic of smectic phases, and it has been questioned whether it should be called smectic [31,33].

The fact that in at least one compound [31,34] the D phase appears between an A and a C phase (between refers here to the temperature scale), raises the question of what, given the structure outlined above for the D phase, could be responsible for this change of a layered structure (smectic C) into a packing of spheres (smectic D) and then back into a layered structure (smectic A).

Another important question is raised by the fact that the D phase obtained by cooling the A phase is first obtained in a mosaic texture, which later changes spontaneously to an isotropic texture [2]. This suggests three possibilities:

1. The A-D transition exhibits supercooling, and the mosaic texture is really a texture of the supercooled A phase. This seems very unlikely, since smectic A phases have never been found to exhibit mosaic textures, and it would also be rather extraordinary that the texture of the A phase would change just at the A-D transition point.
2. There are actually three mesophases involved, and the mosaic texture signifies this third phase. The spontaneous transition to the isotropic texture would then indicate that this third phase is metastable with respect to the D phase.
3. The mosaic texture is indeed a texture of the D phase. Mosaic textures, however, are generally considered as an indication of a structure with rather rigid, planar layers [2,21], and moreover they indicate optical anisotropy (Ref. 2 indeed lists this smectic D texture as optically anisotropic). Neither layering nor anisotropy goes very well with the proposed cubic close packing of spheres. Anisotropy could possibly be explained by deformations of the packing, but it is not clear what could be causing such deformations nor how they could result in a rigid layered structure. Thus, since the mosaic texture cannot well be explained by a spherical packing, nor the isotropic texture by a layer structure [31], the conclusion seems to be that the two textures signify different structures.

Whether these two structures should both have the symbol D, i.e., whether we have case 3 or case 2, would have to be determined by miscibility studies. For the time being, assuming that both are D phases, we can differentiate between the two structures by using, e.g., D_ℓ for the layered structure and D_c for the cubic.

A third peculiar aspect of the D phase shows up in a paper by Pelzl and Sackmann [35] on birefringence in liquid crystals. According to their Fig. 5 and their Table 35, which give the results of refractive index measurements on a compound with a D and an A phase, the temperature range of the (isotropic) D phase overlaps with almost the entire temperature range of the (anisotropic) A phase! The authors appear not to have drawn attention to this most extraordinary phenomenon which would appear to support the idea of supercooling of the A phase mentioned above (point 1).

6. Smectic E

The first x-ray diffraction patterns reported for a smectic E phase [31] indicated (1) a high degree of order within the smectic layer, with a distinctly nonhexagonal lattice [33], (2) molecules perpendicular to the layers [2,33], and (3) a three-dimensional lattice [31,33]. This E phase was reported to be optically uniaxial [31]. Optically biaxial phases, miscible with an E phase, and therefore also E phases themselves, according to the miscibility rule, were reported by Coates et al. [36]. These authors suggest that these phases have tilted structures, and thus there would be E_t as well as E_n phases [6], just as there are B_t and B_n phases (for another explanation for the difference between uniaxial and biaxial E phases, see Sec. III.H.3).

The stable texture of the E_n phases is the mosaic texture [31], again indicating (as in the case of B phases) that the order within the layer causes the layers to become more rigid. The existence of a three-dimensional lattice in the E_n phase has recently been confirmed by the determination of the unit cell dimensions of the lattice in DPTC [37-39].

7. Smectic F and G

F and/or G phases have been reported in the phase sequences Is → A → C → G → Xℓ (Is = isotropic liquid, Xℓ = crystalline phase) and Is → A → C → F → G → Xℓ [2]. Only the last sequence has been described in detail so far (for PPPOPP) [40]. The data reported indicate a strong similarity between F and C [2,33,40]. In fact, there is at present no indication of what the main difference is between F and C. Assuming that both F and C have tilted structures, the only difference between them might be that the tilt in F is somewhat greater than in the corresponding C [6,40]. If this is true, the difference between the F and C phases of the compound reported in Ref. 40 would be no greater, and possibly less, than known differences among C phases [12,16] (see also Sec. III.E).

Based on the data in Ref. 40, the G phase appears to have an ordered arrangement of the molecules within the smectic layer (this is indicated by the sharpness of the outer rings), the molecules appear to be tilted (the layer thickness is much less than the molecular length obtained from models), and a three-dimensional lattice probably exists [2,6,33]. Elsewhere we suggested that the G phase might actually be very similar to the E_t phase, and we also raised the possibility of a similarity between the structures of these two phases and the structure of B phases [6]. New work on the structure of an E_n phase [37] has made us feel even more strongly that the structures of B_t, E_t, G, and H phases are probably very closely related and that the same is true for the structures of B_n and E_n phases.

Billard [41] has recently shown that the F and G phases of PPPOPP are completely miscible with the lower temperature smectic phases of some Schiff's bases. This opens up some interesting possibilities for the classification of these phases, since for one of those Schiff's bases (HpBPeA) the phase that is miscible with the F phase has been tentatively identified at our Institute [42] as a B_n phase and the phase miscible with the G phase as a B_t phase. These results would appear to confirm our suggestion above about the similarity of G and B_t, but they raise questions about the real identity of the F phase. As noted in Sec. II.A.4, in still more recent work [30] the G phase has now been shown to be completely miscible with what was up to now referred to as the H phase of BBEA, and we will use for it the symbol G(H).

B. Nematic Phases

In this section we shall not discuss twisted nematics (i.e., cholesterics), but even with this limitation on the meaning of the word nematic there is ample evidence that the usual definition of nematic phases, as phases in which the only element of order is the alignment of the long axes of neighboring molecules parallel or nearly parallel to each other, does not apply to all nematic phases in the same way.

The reason for this is that in many nematic phases there also exists short-range smecticlike order. This order is particularly evident in nematic phases in which the smecticlike order involves not only layered arrangements of the molecule but also a significant tilt of the molecules with respect to the layer normals [12,43-45]. We have called these phases "skewed cybotactic nematics" (N_{sc}) [6,46]. One would expect there to be also normal cybotactic nematics (N_{nc}) [6,46], i.e., nematics with the molecules parallel to the normals of the smecticlike groups, and this kind of order has indeed been found. Current evidence [47-49] suggests, however, that the smecticlike order in these cases persists only over a limited temperature range, directly above the S_A-N point, whereas for N_{sc} phases the smecticlike order can persist over the full stability range of the nematic phase [6,12,43,50].

To identify nematic phases with no significant degree of smecticlike order, we have proposed for those the name "ordinary nematics" (N_o) [6]. We noted later that Gray and Winsor (Ref. 32, p. 8) also use the term "ordinary nematic phase," but they appear to include under this name all nontwisted nematic phases, including N_{sc} and N_{nc} phases.

C. Cholesteric Phases

If an optically active material forms a nematic phase, the preferred direction of the long molecular axis in a uniformly oriented sample (a monodomain) is not constant over the whole sample, as it would be in a "regular" nematic, but rather it displays a continuous twist as one travels through the sample along the optic axis. Hence, the name "twisted nematic" or "spontaneously twisted nematic" was given to this kind of phase. Since the first compounds found to exhibit this phase were derivatives of cholesterol, the phase is also often called "cholesteric."

It was long believed that all phases showing the typical cholesteric properties (high optical rotation, iridescent colors) were all twisted nematics. Later data have shown, however, that twisted tilted smectics also behave optically as cholesterics [2,51-55]. Thus, it is quite possible that some phases that have been classified as cholesteric are actually twisted smectic [53,55]. To distinguish these two kinds of "optically cholesteric" phases, while emphasizing their cholesteric properties, we have proposed [6] for them the symbols Ch_N (nematic cholesteric, or chiral nematic) and Ch_S (smectic cholesteric, or chiral smectic). If classified on the basis of the miscibility criterion, Ch_N phases would be N phases and most Ch_S phases would be C phases, although recently some twisted G(H) phases have also been found [56,57].

A number of Ch phases have been reported to exhibit some rather interesting and puzzling phenomena. Friedel [1] observed that Ch phases obtained by cooling the isotropic phase "frequently appear in the form of a cloud which is not birefringent and which is colored dull purple to the naked eye" [58]. Gray [58] considered this cloud to be a "'homeotropic' condition" and noted that "the temperature at which this 'homeotropic' condition appears on cooling is in fact the same as the temperature at which the focal-conic texture becomes isotropic on heating." Saupe [59] refers to this homeotropic condition as "homogeneous, non-birefringent layers." Lehmann [60] contended that the homeotropic and focal conic conditions represented different mesophase types, but Gray [58] argued that this was not possible because "there is no reversible, discontinuous transition between them at a particular temperature" (as may be seen from the description above, the homeotropic condition was only observed on cooling, not on heating), and concluded that "at most they represent different textures of the same mesophase." More recently, however, Gray and Winsor (Ref. 32, p. 12) acknowledged that Gray's original argument appears to have been invalidated by the work of Price and Wendorff [61,62] who observed in cholesteryl myristate and in cholesteryl nonanoate a reversible change between the "turbid white modification" (the focal conic texture) and the higher temperature "blue transparent modification" (according to context and references the authors consider this to be the same as the homeotropic modification discussed above; Saupe [63] also considers the two to be the same). Observations by us [64] on cholesteryl nonanoate also indicated a reversible transition between the turbid white and the blue transparent modifications. Saupe [65] has described the blue transparent modification of Ref. 61 as "an optically isotropic modification" and considers it quite possible that this modification, in fact, represents a different mesophase [63]. Elser et al. [66] report a reversible transition in cholesteryl butyl carbonate, between the plane texture and a higher temperature texture which they describe as "platelets." Gray [58], Saupe [65], and Elser et al. [66] all note that the structure of these high-temperature modifications still remains unsolved. Winsor (Ref. 32, pp. 57, 58) suggests that these phases contain units consisting of "small molecular assemblies each with a twist as in the cholesteric plane texture" but "orientated at random."

Although it now appears evident that the high-temperature cholesteric modifications can be stable, this should not be interpreted to mean that they always are stable. A metastable high-temperature modification has been obtained by Fergason in large quantities [55]. Saupe [55] has reported it to be optically isotropic and noted that it was only faintly turbid and showed a yellow color; he proposed for it a structure based on a body-centered cubic lattice of singular points. Saupe [63] does not consider this model to be the general solution for the isotropic modification of the cholesteric phase, however, and also notes that the yellow color of Fergason's sample may have been due to slight decomposition.

III. A STRUCTURAL BASIS FOR CLASSIFICATION

A. Introduction

As mentioned in Sec. II.A.1, the classification system of the Halle group is based mainly on the miscibility rule, but use was also made of other characteristics: microscopic textures, x-ray diffraction patterns, and the order in which different phases appear in compounds with more than one smectic phase.

At first there were no conflicts between the results obtained from these different criteria, and it appeared that it would be possible to arrive at one generally valid classification scheme. A difficulty exists now, however, with regard to the B_t phase of TBBA. Its diffraction pattern and its structure are very similar to those of the G(H) phase of BBEA [3,24], much more so than to the diffraction pattern and structure of the B_n phase. The miscibility criterion, however, identifies the B_t phase as of the same type as the B_n phase and as different from the G(H) phase. Thus, a conflict has arisen between classification on the basis of x-ray diffraction patterns (i.e., on the basis of structure) and classification on the basis of miscibility.

The miscibility of B_t and B_n phases also leads to another conflict in the relationships between structure and miscibility. Since the B_t phase in question is tilted [3,27] and the B_n phase normal [28], it follows that in this case the miscibility rule does not differentiate between tilted and normal phases. This is quite surprising, since with other smectic phases the miscibility rule apparently does differentiate on the basis of tilt: A and C phases are never completely miscible, and the main difference between the structures of these phases is generally taken to be that A phases are normal and C phases are tilted. Thus, tilt appears to prevent mixing in one case (A and C phases) but not in the other (B_n and B_t phases) [33,67]. The puzzling part about this is that B_n and B_t phases have a higher degree of order than A and C phases. Usually, the more disordered phases are more miscible than the more ordered ones, but here the opposite seems to occur.

A possible solution of the first conflict mentioned above, miscibility of phases with different structures, might lie in the following. In the presentation of a recent paper [68], Sackmann showed a graph of smectic layer thickness versus composition for the B phase of mixtures of TBBA and a compound with a B_n phase. This graph consisted of two straight lines of different slope and a sharp discontinuity in slope at the point where the lines met. The slopes of the lines were such as to indicate that upon increasing the TBBA content the phase initially remains B_n, but that after the discontinuity the molecules adopt a tilted arrangement, with the tilt angle increasing from 0° until it reaches its maximum value for pure TBBA. This suggests to us the existence of a kind of second-order phase transition in this system, which would go unnoticed in the miscibility studies (a similar suggestion has been made elsewhere [30]). If this interpretation is correct, and if, because of this phase transition, one should regard the two compounds as having different phases rather than the same, this would mean that the results of miscibility studies would have to be treated with a great deal more caution than has been the case up to now.

The second conflict mentioned above, differentiation on the basis of tilt in one case but not in the other, might be resolved by our suggestion [67] that the principal difference between A and C phases is not tilt but rather the interaction between neighboring layers.

It is also possible, though, that the abovementioned conflicts are not so easily resolved, and also there is the ever present danger that new findings may introduce new conflicts (e.g., our suggestions about yet another type of B phase [69] and about various types of A and C phases [12,16]). For these reasons, it might be prudent to keep a classification on the basis of structure separate from the classification on the basis of miscibility. A similar kind of feeling has been expressed by Sackmann and Demus, who state in a recent paper [2] that in their classification system, based mainly on the miscibility rule, they "have avoided a nomenclature with structural connotations" (p. 269). This has been reemphasized by Sackmann [29], who strongly argued for separating considerations based on structure from those based on miscibility.

This implies that if one wants a classification based on structure one will have to build an independent framework. An early attempt in this direction was made by Hermann [70] and extended by Mabis [71]. Their classification system identifies 18 intermediate (or mesomorphic) phase types, but it appears that many of these are not realized in nature (as far as we can tell); also, many phase types which have been found and which one would like to distinguish from each other fall in Hermann's system in the same category. Thus, the system appears to make, on the one hand, too many distinctions, and on the other, too few. More recent work in this area comes from our Russian colleagues (e.g., Refs. 72-75), who have assigned detailed symmetry symbols to a number of liquid crystal phases. We feel, however, that these symbols do not take into account some of the "building principles" which now appear to be of importance in describing the structures of the various phases. We made a first step in the direction of a structural classification some time ago [33], dividing smectic phases into α, β, and γ phases, according to the degree of order in these phases, but this, obviously, falls far short of a complete structural classification scheme.

In the following sections we shall approach the subject of structural classification not by attempting to present a complete classification scheme but rather by discussing a number of the building principles which appear to influence liquid crystal structures. The relationships between the various liquid crystal phase types and these building principles have been summarized in Table 8 in Sec. IV.

B. Parallelism

Parallel will always refer here to the long axes of elongated or rodlike molecules or to the average planes of two-dimensionally extended or platelike molecules. Moreover, with parallel we will always mean approximately parallel.

Parallelism between neighboring molecules is the *conditio sine qua non* of liquid crystallinity. It does not play a major role in distinguishing between different types of liquid crystals, but there appears to be a lack of definition involved which warrants a few words of discussion here.

The degree of parallelism is usually expressed in terms of the orientational order parameter S, also called the "nematic order parameter." The formula for S can be written in several different but equivalent ways, e.g., as $S = \frac{1}{2}\langle 3 \cos^2 \theta - 1\rangle$. The lack of definition referred to in the preceding paragraph concerns the angle θ.

The angle θ and the function $\frac{1}{2}\langle 3 \cos^2 \theta - 1\rangle$ were already used in 1918 by Born [76] (p. 183 and p. 196, respectively). Born defines θ as the angle between the main axis of the molecule and "the preferred direction" in the sample. Zwetkoff [77] introduces the symbol S for the function $\frac{1}{2}\langle 3 \cos^2 \theta - 1\rangle$ and calls it the degree of long-range order and defines θ as the angle between the long axis of the molecule and "the symmetry axis of the macroscopic homogeneously oriented sample." Chatelain [78] defines θ with respect to "the isotropic axis of a nematic liquid crystal with only one isotropic axis." Maier and Englert [79] define θ relative to "the optical axis of a homogeneously oriented sample," and Maier and Saupe [80] use the same definition.

> It is of some historical interest to note here that Born [76] uses in his paper the formula $\frac{1}{2}\langle 3 \cos^2 \theta - 1\rangle$, that Zwetkoff [77] defines S with the same formula but also mentions the equivalent formulation $\langle 1 - (3/2) \sin^2 \theta\rangle$, that Chatelain [81] and Maier et al. [79,80,82,83] take over this second formula, but that most recent papers appear to have returned again to the original formulation of Born.

The point to which we would like to draw attention here is that in all the abovementioned papers [76-80] the definition of θ requires that the sample be a monodomain, whereas in more recent papers this requirement appears to have been dropped. The general position now seems to be that θ should be taken as defined with respect to the <u>local</u> preferred direction of the long molecular axis; this preferred direction usually varies throughout the sample [59]. Although this new definition of θ looks like a logical extension of the old one, a rigorous proof does not appear to have been given that this extension really can be made. Indeed, it seems to us that such a rigorous proof might be rather difficult, since it would appear rather hard to give a precise definition of "the preferred direction" in a situation where this direction varies with position and probably also with time. In a monodomain the local preferred direction might also vary slightly with position and time, and even in an isotropic liquid of rodlike molecules one could define a local preferred direction and end up with a nonzero S [84].

This lack of definition of θ for a random sample does not have any serious practical consequences, however, since for measurements of S one always uses uniformly oriented samples. Indeed, the preparation of a uniformly oriented sample has been noted by Saupe and Maier [83] as a prerequisite for the measurement of S.

C. Layer Formation

In smectic phases the molecules are not only parallel, but also arranged in such a way that the centers of adjacent molecules lie in a plane (just as parallelism, planarity is never absolute, i.e., there will always be deviations; also, the "planes" can be curved). Thus, the molecules in a smectic phase are arranged in layers, and neighboring layers are generally parallel to each other.

In the nematic phase the positions of the molecular centers have generally been considered to be without order (the planes of molecules that one usually talks about in discussing twisted nematic phases are, in our opinion, merely mathematical abstractions and have no physical reality whatsoever). This was later found not to be generally true, however. In 1970, we reported [43] a set of x-ray diffraction patterns of the nematic phase of OOBCP, which we could explain only in terms of "cybotactic groups," i.e., a short-range, but very definite, smecticlike ordering of the molecular centers in planes. In 1973, Chistyakov and Chaikowsky [44] adopted a similar explanation for the diffraction patterns of a series of AOAOBs, which they had reported earlier [85] and which were very similar to ours, and currently this smecticlike order (layer formation) in certain nematic phases appears to be generally accepted as an established fact [12].

Investigations of nematic-to-smectic phase transitions have shown that layer formation in nematic phases frequently occurs as a pretransition effect. In these cases the smecticlike order is near zero over most of the nematic range but increases rapidly with decreasing temperature on approaching the S-N point [47-49]. Nematic phases like those discussed in the preceding paragraph (those of OOBCP and the AOAOBs), however, show a different behavior of the smecticlike order [12]. A detailed x-ray diffraction study of HpOAOB [45] shows that the order decreases only relatively slowly with increasing temperature, and a conductivity study of this same compound [50] shows a "smectic" conductance behavior over the whole nematic range. Thus, if layer formation in nematic phases is always a pretransition effect of the S-N transition, it would in this case be extended over a rather long temperature range and have a relatively small temperature dependence near the transition point.

An even more extreme case of this is OOBCP. This compound has a monotropic S-N point 3 °C below the Xℓ-N point [86]. Yet, the smecticlike order is not only very strong just above the Xℓ-N point, but it also decreases only very slowly with temperature for the first 20-30° C [43,87], and, although it decreases faster at higher temperatures, it remains in existence over the full 120° C range of the nematic phase (and we have indications that the cybotactic groups persist even into the isotropic phase).

Another interesting fact regarding this kind of nematic phase is that a homolog of HpOAOB, EOAOB, has been reported [85] to have, above the Xℓ-N point, the x-ray diffraction pattern indicative of considerable smecticlike order, notwithstanding the fact that the extrapolated monotropic S-N point for this compound lies about 100° C below the Xℓ-N point.

Much more work is definitely needed in this area, but the present evidence seems to point to the following conclusions:

1. The nematic phases discussed in the last few paragraphs are sufficiently different in structure from other nematic phases to classify them as a separate subgroup. We have named them skewed cybotactic nematics (N_{sc} phases) to indicate both the smecticlike order and the fact that the molecules are not perpendicular to the layers [43] (see also Sec. II.B).
2. There may be two types of N_{sc} phases. Type I: phases in which the rate of change of the degree of smecticlike order with temperature increases continuously with decreasing temperature up to the transition to the lower temperature phase; this behavior seems to occur in compounds that also have a stable smectic phase (e.g., HpOAOB). Type II: phases in which the rate of change first increases

(upon cooling down from the N-Is point), but then decreases upon approaching the transition to the lower temperature phase; this behavior seems to occur in compounds with a direct Xℓ-N transition (e.g., OOBCP).

As a last remark on the subject of N_{sc} phases, it is interesting to note that in these phases the orientational order parameter S, which measures the degree of parallelism, is significantly greater than in other nematic phases. This is apparent from the much narrower outer x-ray diffraction maxima in aligned samples (see, e.g., Fig. 2 in Ref. 6) and from plots of S versus temperature (e.g., Fig. 1 in Ref. 79). This increase of parallelism with layer formation is in line with the generally larger S values for smectic phases as compared to nematic phases. The relationship between layer formation and phase types is summarized in Table 3.

D. Molecular Translation

Molecular translation, which in this section will have the restricted meaning of translation along the long axis of the molecule, and layer formation, as discussed in the previous section, may be considered as opposite sides of the same coin. And just as layer formation in nematic phases can be considered a smectic characteristic, so can molecular translation in smectic phases be considered a nematic characteristic.

TABLE 3. Relationship Between Layer Formation and Phase Type

Layers	Absent[a]	Short-range	Long-range
Phase type[b]	N_o	N_c	S

[a]Absent over most of the temperature range.

[b]N_o = ordinary nematic, N_C = cybotactic nematic, S = smectic (with the probable exception of S_D).

Molecular translations along the long axis are generally considered to be free in ordinary nematic phases but severely restricted in smectic phases. For the G(H) phases of two compounds, however, the observation of diffuse streaks on x-ray diffraction patterns of monodomains led to the conclusion that in these phases there was a considerable mobility of the molecules in the direction of their long axes and that these motions involved not just single molecules but whole columns or strings of molecules [3,27]. Although this type of motion has not been reported for any other smectic phases, we believe that translations along the long axis are fairly common in smectic phases. Evidence for this comes from the measurement of considerable self-diffusion in smectic phases in the direction of the long molecular axis by nuclear magnetic resonance (NMR) [88,89] and from x-ray data. Figures 1 and 4 are x-ray diffraction

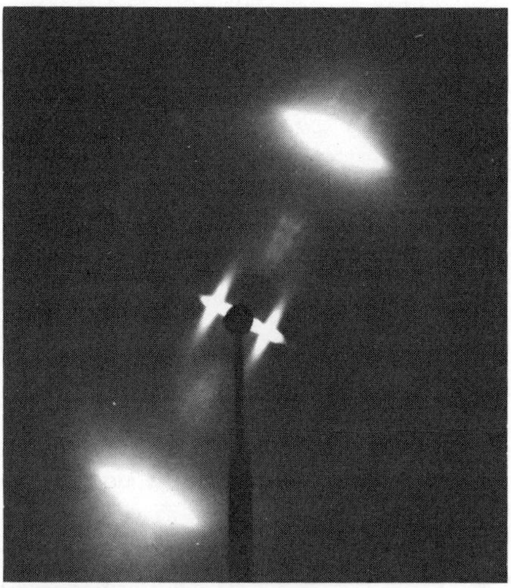

FIG. 4. X-ray diffraction photograph of a pseudomonodomain of the B_n phase of BBOA, taken with the incident beam perpendicular to the mean direction of the long molecular axis and parallel to the mean orientation of the smectic planes. The maxima in inner and outer rings are strongly overexposed to show better the streaks through the inner ring maxima, parallel to the imaginary line connecting the outer maxima.

photographs of a smectic A_n and a smectic B_n phase, respectively (we have used here the symbol A_n to emphasize that in this A phase the molecules are normal to the smectic planes; later, in Sec. III.H.1, we will present evidence indicating that there are also smectic A phases in which the molecules may be tilted with respect to the planes). On each of these two photographs one notices short, strong, diffuse streaks through the Bragg reflections from the smectic layers, in a direction perpendicular to the line one could draw between these Bragg reflections. These diffuse streaks are, in our opinion, indications of translational movement across the boundary planes of the smectic layers, and the fact that they are so short [as compared to the streaks observed for G(H) phases] suggests that these movements do not involve single molecules or single columns of molecules but rather groups of adjacent molecules moving in concert.

With different smectic A phases, we have observed these diffuse streaks in all gradations of intensity: strong (as in Fig. 2), medium, and apparently absent. Associating the streaks with molecular translations, as we have done above, one would expect a strong correlation between the intensities of the streaks and the diffusion or the electrical conductivity perpendicular to the smectic layers, but data concerning this do not appear to be available at present. It is also not yet clear whether it would be possible (or helpful) to make a subdivision of smectic phase types into groups with different levels of molecular translational mobility.

The existence of this type of motion, though, together with the layer formation in nematic phases discussed in the preceding section, suggest that the distinction between smectic and nematic phases might not be as clear-cut as is usually assumed.

E. Tilt

1. Introduction

Although Herrmann [90] already in 1933 argued that in the smectic phase of TlSt and TlOl the molecules were tilted, this idea apparently has not found much acceptance, and for a long time one generally continued to assume that in all smectic phases the molecules were perpendicular to the smectic planes. The now widely accepted view that in C phases the molecules are tilted whereas in A phases they are perpendicular has been an only rather recent development. In 1968, for instance, Saupe [59] could still write that "type C has a more highly ordered structure than type A, though it is not yet known in what respect," and in 1969 Sackmann and Demus [91] wrote that x-ray studies show no difference between C and A phases.

The turning point appears to have been around 1969-1970. In this period, Chistyakov et al. [20] proposed, on the basis of the x-ray diffraction pattern of an aligned sample and the optical diffraction pattern from a structural model, that in the smectic phase of NOBA the "molecules in planes have a concerted incline." They also suggested that this smectic phase was a C phase (that this phase is indeed a C phase seems virtually assured by the fact that the smectic phase of two of its homologs, OOBA and DoOBA, had been classified as such; see Ref. 92). Arora et al. [86] proposed that the lower temperature liquid crystal phase of a series of AOBCPs "has a layered structure in which the molecules are inclined to the layers, and that it corresponds to a ... smectic C" (the C assignment was based on textures).

Both of these papers gave only tentative conclusions and no numerical data. Almost simultaneously, however, two other papers were published that were more conclusive. Taylor et al. [93] presented optical measurements on the smectic phase of HpOAOB (a C phase, see Ref. 92), DeOBCP (designated C on the basis of texture), and NOBA (almost certainly C, see above) and concluded that in all three cases the tilt angle was 45°. De Vries [11] deduced from x-ray studies of the smectic phase of AOBCPs (designated C on the basis of textures; see Ref. 86) and of HpOAOB (a C phase; see Ref. 92) that in both cases the tilt angle was "about 45°."

The tilt angles reported in these last two papers were all about 45°. This apparently caused people to assume that <u>all</u> C phases would have tilt angles of 45°, for it is interesting to note that in most subsequent papers the authors argue that, although the actual tilt angles obtained by them might be different from 45°, they were not <u>significantly</u> different.

It now appears to us, however, that there exist genuine differences between the tilt angles of different C phases and that it is possible, on the basis of this tilt angle, to distinguish between a number of distinct subgroups. We shall discuss here the C_1, C_2, and C_3 types [12,16].

2. C_1 Phases

C_1 phases appear to have the following characteristics: (1) the tilt angle has little or no temperature dependence, (2) the tilt angle is large, roughly between 40° and 60°, and (3) on heating one gets an N_{sc} phase or an Is phase [12].

Presently known C_1 phases include all the phases originally reported to have a tilt angle of 45° (see previous section). From Ref. 12 we have taken what we believe to be the most accurate tilt angles for the various phases and listed them in Table 4. These data suggest that the tilt angle is fairly constant within a homologous series but that different series have significantly different tilt angles: around 42° for AOAOBs, around 50° for AOBCPs, and around 55° for AOBAs (an angle of 23° was recently reported for a member of the AOAOB series [96], but this was for a compound with

TABLE 4. Tilt Angles in C_1 Phases

Compound	Tilt (°)
HpOAOB	42[a], 43[b]
UOAOB	42[c]
NOBCP	49[b]
DeOBCP	50[d]
NOBA	55[e]

[a]Calculated from data in Ref. 45.

[b]From Ref. 12.

[c]Calculated from data in Ref. 94.

[d]From Ref. 95.

[e]Measured on Fig. 4a in Ref. 20.

very long alkoxy chains, and also it is not known whether this compound has an N_{sc} phase). The data presented in Table 4 are not conclusive, of course, but at least they are suggestive.

3. C_2 Phases

A completely different kind of C phase is the C_2 phase [12,16]. The characteristics are as follows. (1) The tilt angle is strongly temperature dependent, going to zero at the upper end of the stability range. (2) The maximum tilt angle is not very large (about 26°). (3) On heating one obtains an A phase (an A_2 phase; see Sec. III.F.2b).

The C phase of TBBA was for a long time the only C_2 phase on which data were available in the literature. It was first reported by Taylor et al. [14]. These authors noted that they had also some other C phases with a similar tilt-temperature relationship, and that these phases also gave A phases on heating, but more data on these compounds have not been forthcoming. Another C phase with C_2 characteristics has been reported recently by the Orsay group [97], and several more have been found by Mayer [98] and Neubert [99].

4. C_3 Phases

There are ample data in the literature on C_1 and C_2 phases, and the distinction between C_1 and C_2 phases may be regarded as well founded. There are few data on C_3 phases, however, and, as a consequence, our designation of certain C phases as C_3 phases will have to be somewhat tentative.

Before going into the actual discussion of C_3 phases, we would like to point out that the basis for this discussion will be data on the smectic layer thickness d. From calculations on other phases [A_n, C_1, B_t, and G(H)], for which the tilt angle and the d had been determined independently, we have found [100] that one can obtain quite accurate values for the tilt angle simply from d and the molecular length ℓ, using ℓ values calculated on the basis of a theoretical molecular model [5]. Thus, measurement of d appears to be a suitable basis for a discussion of tilt.

In a paper dealing with x-ray data on A, B, and C phases, Diele et al. [94] list a number of compounds which, on heating, first give a B, then a C, and then an A phase (compounds No. 2-5 in their Table 2). They report that these compounds have about the same d in all three phases. From this it seems reasonable to conclude that d is probably fairly constant over the full stability range of the C phase. If we further assume that the molecular configuration also stays more or less the same over the full C range (again a quite reasonable assumption), ℓ will also be more or less constant, and from this it follows that the tilt angle, which can be calculated from d and ℓ (see preceding paragraph), also will vary little. Thus, these C phases are different from the C_2 phases, which showed a large variation of tilt with temperature.

For one of their compounds (their No. 4), Diele et al. list d values for a series of temperatures including one in the C range. For this compound we have calculated ℓ, and, based on the d value given for the C phase, we find a tilt angle of 21°; the tilt angles for the other compounds appear to be similar. These tilt angles, then, are much smaller than those found for C_1 phases (Table 4). Moreover, the compounds discussed here have A phases whereas one of the characteristics of C_1 phases appears to be that they are found only in compounds without A phases [12,16]. Thus, the C phases discussed here also appear to be different from C_1 phases.

Finding these C phases different from C_1 as well as from C_2 phases, we have assigned them the symbol C_3 [16]. These C_3 phases appear to have the following characteristics: (1) the tilt angle has little or no temperature dependence, (2) the tilt angle is relatively small (around 20°), and (3) on heating one obtains an A phase (an A_3 phase, see Sec. III.H.1).

5. Summary

From the above discussion it appears that tilt is not only useful for distinguishing between tilted and nontilted phases, but that it also furnishes criteria for making distinctions between different types of tilted phases.

It is possible, of course, that more extensive studies will reveal that there are C phases with tilt angle characteristics at all kinds of intermediate levels in between those of the C_1, C_2, and C_3 phases outlined above, and in that case the distinction would probably have to be abandoned. But for the present it appears that it is possible to divide C phases into a limited number of subgroups with clearly distinguishable tilt angle characteristics (Table 5). The reasons for this, from the point of

TABLE 5. Different Kinds of C Phases

Phase type	Tilt angle (°)	Temperature Dependence[a]	Next Phase[b]
C_1	40-60	Small	N_{sc} or Is
C_2	0-26	Large	A_2
C_3	≈ 20	Small	A_3

[a]Of the tilt angle.

[b]With increase in temperature.

view of molecular packing, are not clear, and one of the challenges ahead will be to try to determine <u>why</u> molecules arrange themselves in these various different ways.

Another point we would like to make is that all three C categories discussed here (and also the C_4 type mentioned in Ref. 16) conform to the general criteria that C phases are tilted and optically biaxial.

F. Molecular Rotation

1. Types of Rotation

There is considerable disagreement in the literature as to whether or to what extent there is molecular rotation in liquid crystals (molecular rotation will in this paper always be considered to be rotation around the long axis). Part of this disagreement is real, part of it is due to misunderstandings caused by rather different meanings assigned to the word "rotation." It seems proper, therefore, to list here first some of the different kinds of motion that could be associated with the word "rotation" (limiting ourselves throughout to molecular motions about the long molecular axis).

1. <u>Free rotations</u>. All orientations of the molecule are equally probable, and each molecule goes through all of them in a relatively short period of time.

2. Oscillations. These could be small, like oscillations in solid crystals, or large, approaching full 360° rotations.
3. Hindered rotations. Here there is no limit on the total angular range of the motion, but some orientations are less probable than others. Or, in other words, the rotation of the molecule is influenced by potential barriers of limited height. As under point 2, there are an infinite number of gradations, depending on the number, the heights, and the positions of the potential barriers. With very low barriers the molecular motion approaches free rotation (as under point 1); with very high barriers the molecule will essentially oscillate (as under point 2) between these barriers. If the barriers are far apart the oscillations will have a large angular range; if the barriers are close together the oscillations will be small. In addition to the oscillatory motions, the molecules will occasionally "jump" across a barrier, with the frequency of jumping determined by the temperature and by the height of the barrier.

Of these three kinds of motion, the first, free rotation, can probably be excluded from consideration for liquid crystals, since recent data indicate that even in nematic and isotropic phases of mesogenic materials the rotational barriers are considerable [12]. The second kind of motion, oscillation only, can, in our opinion, also be ruled out because the more or less fluid character of liquid crystals makes it highly unlikely that jumps across the rotational potential barriers can be completely excluded, even in the rather rigid G(H) phases. This leaves only hindered rotations as a viable option, but actually this limitation of options is more apparent than real, because, as noted above, hindered rotations can have an infinite number of gradations, ranging all the way from an almost stationary, solid crystallike situation to almost pure oscillation and to nearly free rotation. Since change from one extreme to the other can be very gradual, it is very difficult to give precise definitions of what one means by this or that description of molecular rotation in liquid crystals (an additional complication will be discussed in Sec. III.G.5).

2. Relation to Structure

From the above it will be clear that in our opinion it would be hazardous, if not impossible, to distinguish between different types of liquid crystals merely on the basis of the kind of rotational motion of the molecules. Yet, there appear to be some important correlations between molecular motion and liquid crystal type, and some of these appear well worth discussing here.

a. Smectic B_t and G(H) Phases. Elsewhere [24] we have argued that the packing of the molecules in the B_t phase of TBBA is much looser than the packing in the G(H) phase of BBEA. These different molecular arrangements lead to lower and more widely separated rotational potential barriers in the TBBA phase, as compared to the BBEA phase. This would allow larger oscillations and more frequent jumps in the TBBA phase, which seems to be in essential agreement with recent experimental data [101].

b. Smectic A_1 and A_2 Phases. Just as we have made distinctions between different kinds of C phases (Sec. III.E), we have also made distinctions between different kinds of A phases [16]. We have defined A_1 phases by the following criteria: (1) the tilt angle is zero and (2) the material has no C phase. The characteristics of A_2 phases are: (1) the tilt angle is zero and (2) on cooling one obtains a C_2 phase. The sole difference between the two phase types, then, would appear to be the presence or absence of a C phase in the same material, which might be considered a rather arbitrary reason for distinguishing phase types. It seems, however, that there is also a significant difference in rotational freedom between A_1 and A_2 phases, and for this reason we would like to maintain, at least for the time being, the distinction between A_1 and A_2. The reasons for expecting a larger degree of rotational freedom in A_2 phases are twofold.

1. A_2 phases are formed by heating C_2 phases, and C_2 phases have a tilt angle which is strongly temperature dependent, approaching zero at the C_2-A_2 transition point. For this kind of C phase, McMillan [102] has developed a theory which couples the tilt to an orientational order parameter β. Within the C phase β decreases with temperature, indicating an increase in freedom of rotation as one approaches the A_2 phase. In the A_2 phase, therefore, a large rotational freedom is expected.

2. For the one A_2 phase for which data are available, the average intermolecular distance (between neighboring parallel molecules) is significantly greater than for A_1 phases [6,24].

Thus, for A_2 phases there appears to be both a reason to expect rotation (argument 1) and room to allow it (argument 2). For A_1 phases, on the other hand, there is no particular reason to expect considerable rotation, nor does there seem to be much room for it.

c. Rotation and Tilt. Relationships between rotation and tilt are discussed in the first two sections of the Appendix.

d. Rotation and Order. Some comments on the relationship between molecular rotation and the type of order within the smectic layer will be made in Sec. III.G.5.

G. Order Within Layers

1. The Hexagonal Model

The idea that the arrangement within a smectic layer can be either random or ordered is about as old as the idea that there are different kinds of smectic phases. When Herrmann and Krummacher [18] in 1932 found that x-ray diffraction photographs of some smectic phases had a diffuse outer ring whereas those of others had sharp ones, they concluded that the diffuse ring corresponded with the commonly assumed random molecular arrangement and that the sharp rings indicated an ordered arrangement. As a model for the ordered arrangement, Herrmann [19] suggested a close packing of cylinders. Such a packing has hexagonal symmetry, and the observation of x-ray diffraction patterns with hexagonal symmetry (see Ref. 20 and Fig. 2) has lent further support to Herrmann's model, which is now quite generally accepted.

2. Arguments Against Hexagonal Symmetry

Before going into a discussion of the hexagonal model, we have to note first that at present there are several different types of ordered smectic phases known, whereas in Herrmann's days there was only one. The current types are: B_n, B_t, E_n, possibly E_t, and G(H). About the structure of E_t [6,33,36] only very little is known, and it will not be discussed here, except in relation to classification (see Sec. III.G.6). Of E_n enough is now known [37-39,103] to be able to say that the structure is quite definitely not a hexagonal packing of cylinders (see also Sec. III.G.4).

This leaves only B_n, B_t, and G(H), and they, indeed, are the only ones for which a hexagonal packing has ever been more or less seriously considered. With respect to B_t and G(H), let us first point out that the mere fact that the molecules are tilted with respect to the smectic planes means automatically that the symmetry of the lattice can never be true hexagonal symmetry, because true hexagonal symmetry requires the symmetry axis to be perpendicular to the plane of the other axes (see also the first two sections of the Appendix). Second, precise lattice parameters have been determined for three B_t and G(H) phases [3,22], and in all three cases the arrangement of the long axes, in the plane perpendicular to these axes, has been found to deviate from a hexagonal lattice. Third, for two of these phases, x-ray diffraction patterns of random (i.e., polydomain) samples have been published [22] which indicate clearly [24] that the molecular packing has no hexagonal symmetry, not even in the projection along the direction of the long axes. Thus, theory and experiment agree in assigning a nonhexagonal lattice to B_t and G(H).

The only really serious candidate for the hexagonal model, then, is B_n. And, indeed, the x-ray diffraction patterns with hexagonal symmetry referred to in Sec. III.G.1 are almost certainly all from B_n phases. The hexagonal model does seem to offer a very logical explanation for the diffraction patterns of B_n phases, but yet we have always found it difficult to accept this model, because the hexagonal symmetry presupposes the existence of cylinders. Since liquid crystal molecules do not have cylindrical symmetry, one would have to assume a large degree of rotational motion of the molecules to "simulate" cylinders, and for such large rotational motions, we feel, there just is not enough room in the lattice. For the B_n phase used for Fig. 2, for instance, we have determined that the corresponding Bragg planes are 4.29Å apart, and with a hexagonal arrangement of the molecular axes this leads (Fig. 5) to a distance between the axes of 4.96Å. For most liquid crystal molecules the benzene rings are the most "noncylindrical" parts of the molecule, and thus an arrangement

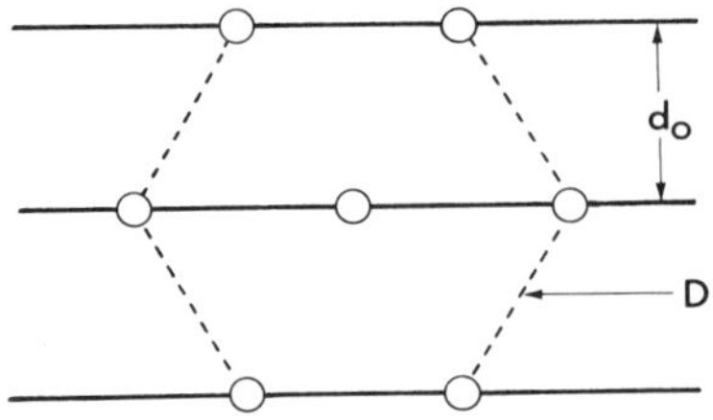

FIG. 5 . Schematic drawing showing the relationship between the intermolecular distance (D) and the interplanar spacing (d_o) for a hexagonal arrangement of molecules. (Reprinted from Ref. 6, by courtesy of the Indian Academy of Sciences.)

of benzene rings on a hexagonal lattice with a spacing of 4.96Å will give a good indication of the kind of packing requirements this 4.96Å distance entails. It appears that the only reasonable packing possible [6] is a herringbone packing of the kind illustrated in Fig. 6. A packing similar to this was first proposed by Gulrich and Brown [104] for a nematic phase and later used by De Vries and Fishel [3] as model for the packing in the G(H) phase of BBEA. We regard it as the most likely model for the short-range molecular arrangement in most liquid crystals [5,6]. It will be clear that with this kind of packing (Fig. 6) the rotational motions of individual molecules are strongly hindered, and that even cooperative rotational motions will be difficult (each molecule has <u>six</u> nearest neighbors!) and will have to be accompanied by considerable lateral motions of the molecules. These lateral motions would significantly disturb the long-range two-dimensional lattice within the smectic layer. The sharpness (Fig. 2) of the corresponding x-ray diffraction maxima, however, indicates that the long-range order is quite well preserved. A further indication of the long-range character of the order in the two-dimensional lattice is that the axes of this lattice apparently can extend throughout a monodomain of macroscopic size, as indicated by the limited extent of the outer ring maxima in x-ray diffraction patterns like Fig. 2. This, in our opinion, argues against the possibility of a large amount of molecular rotation in the B_n phase [6] and thus against "simulated cylindrical symmetry" of the molecules.

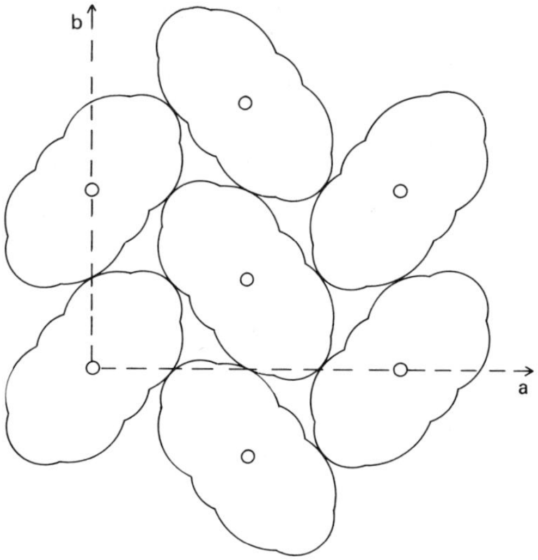

FIG. 6. Projection of a herringbone packing of space-filling models of benzene rings. The rings have one of their C-H bonds perpendicular to the plane of the paper (in this configuration they occupy the least amount of space in the projection); their centers (the small circles) are all in one plane parallel to the plane of the paper and arranged on a hexagonal lattice with a 4.96A distance between the centers of adjacent molecules.

3. The Pseudohexagonal Model

Having argued that the apparent hexagonal symmetry of B_n diffraction patterns cannot be due to cylindrical molecules simulated by molecular rotation, we would now like to give evidence that it is indeed possible to have a nearly perfect hexagonal arrangement of the molecular axes in a situation in which it is clear that the molecules do not simulate cylindrical symmetry. In short, above we have argued that cylindrical molecules are not possible, here we will show that they are not necessary.

Elsewhere [24] we have argued that the presence of tilt in a smectic layer considerably restricts the rotational motion of the molecule. To be specific, when there is a tilt, it is impossible that the "average symmetry of a molecule" (i.e., the symmetry obtained when the positions of all atoms in the molecule are averaged over all possible molecular orientations around the long axis, taking into account the probabilities of these orientations) along its long axis is a two-fold, or higher, symmetry axis (see Ref. 12 and the second section of the Appendix). The magnitude of this deviation from two-fold symmetry can be expected to be coupled with the magnitude of the tilt: for zero tilt the deviation will be zero, for large tilt the deviation will be large (see also the second section of the Appendix). In the G(H) phase of BBEA, the tilt is rather large (33.8°) [3], and thus the deviation from two-fold symmetry will also be rather large, i.e., the average symmetry deviates strongly from cylindrical symmetry. Yet, the arrangement of the molecular axes is nearly perfectly hexagonal: the nearest-neighbor distances, identical in a true hexagonal lattice, are 4.96 and 4.99Å [3], and the angles between the hexagonal axes, 120° in a true hexagonal lattice, are 120.16 and 119.68°. This proves, in our opinion, that cylindrical molecules are not a prerequisite for the occurrence of apparent hexagonal symmetry.

Of course, the above example would only explain an apparent hexagonal symmetry of the arrangement of the long axes, not a true hexagonal symmetry of the arrangement of the molecules. But we would like to argue the case that in a B_n layer the lattice is probably never truly hexagonal. The reasons for this are the following: On most published "hexagonal" x-ray diffraction photographs of B_n phases, the reflections have the form of streaks (e.g., Fig. 2), not the spots expected from a truly hexagonal lattice (a possible explanation for the streaks is given in Sec. III.H.2c). On the only published photograph that does show well-defined spots [33], the angles between the corresponding directions are not 60° (as required by a hexagonal lattice) but 61, 62, and 57° [105]. Moreover, the spots are not equivalent in appearance either. Another photograph with spot-shaped reflections, which was recently obtained in our laboratory [106] (Fig. 7), shows similar features: the angles are not 60°, and only four of the six reflections are equivalent.

We would like to suggest, then, that the molecular packing in a B_n layer is similar to that in Fig. 6: an approximately hexagonal lattice for the long axes, with a herringbone arrangement of the molecular planes. We realize that this packing has only orthorhombic symmetry and will be

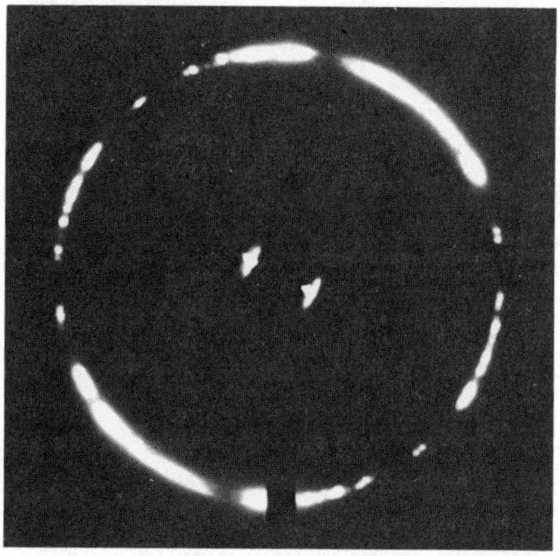

FIG. 7. X-ray diffraction photograph of the B_n phase of 4-n-heptyloxy-benzal-4'-n-pentylaniline. The outer ring contains several well-defined sharp diffraction maxima. (Courtesy of Dr. D.P. Lesser.)

optically biaxial, whereas B_n phases are known to be uniaxial. How an overall uniaxial structure can be obtained from these biaxial layers will be explained in Sec. III.H.2c.

4. Influence of Molecular Structure

It will be clear that in the B_n model, proposed above, the approximately hexagonal arrangement of the long axes is not a natural consequence of the packing mode but rather appears to be "accidental." We would like to suggest that this approximately hexagonal arrangement is the result of the shape and the packing characteristics of the molecules involved. The unit illustrated in Fig. 8a is frequently a major element in these molecules, and we believe that it is this unit which is mainly responsible for the pseudohexagonal characteristics of B and G(H) phases (both BBEA and TBBA, the only two compounds on which detailed studies of such phases have been made [3,22], contain this unit).

The influence of molecular structure on the molecular packing is illustrated by the recently proposed structure for the E_n phase of DPTC [37]. This structure assumes a herringbone packing very similar to that of Fig. 6, but, presumably due to the quite different molecular structure (Fig.

(a)

(b)

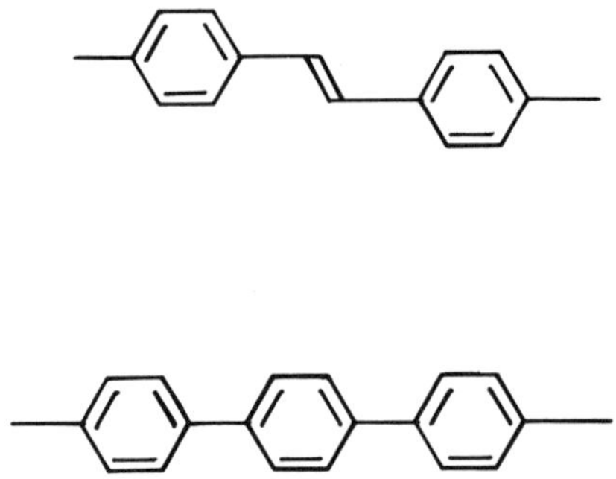

FIG. 8. Central part of the molecule (a) of many of the most common liquid crystals (the atoms in between the phenylene rings can be C, N, and O) and (b) of DPTC.

8b), this packing now has to be fitted to quite different unit-cell dimensions [37-39,103], and this leads to an arrangement of the molecular axes which deviate strongly from hexagonal symmetry.

A comparison of the relevant parameters is given in Table 6. It is clear that the differences in distances and in angles which measure the deviation from hexagonal symmetry are much larger in the E_n phase than in the B_t and G(H) phases. We believe these larger differences in the E_n phase of DPTC are caused by the particular structure of the molecules.

5. Relevance to Molecular Rotation

Above, we have come across two conditions which put special limitations on the rotational freedom of the molecules in certain smectic phases: the asymmetry caused by tilt, in phase types like C, B_t, and G(H) (Sec. III.G.3 and the second section of the Appendix), and the spatial limitations imposed by a two-dimensional lattice, in phase types like B_n (Section III.G.2).

In B_t and G(H) phases, both conditions operate simultaneously, but this does not always mean that they reinforce each other. The tilt, in general, actually relaxes the constraints imposed by the lattice, by moving the corresponding benzene rings of adjacent molecules away from each other. This effect appears to be very pronounced in the B_t phase of TBBA [24]. In other phases (e.g., in the G(H) phase of BBEA), however, other

TABLE 6. Data on the Lattice in the Smectic Layers of Four Ordered Phases

Phase	D(Å)[a]	ΔD (Å)[b]	ϕ (°)[c]	$\Delta\phi$ (°)[d]
E_n[e]	5.56 4.94	0.62	124.2 111.6	12.6
G(H)[f]	5.24 5.06	0.18	121.2 117.6	3.6
B_t[g]	5.18 5.13	0.05	120.3 119.4	0.9
G(H)[h]	4.99 4.96	0.03	120.2 119.7	0.5

[a]D = Nearest neighbor distances between the long axes.

[b]ΔD = Difference between the two D values of one lattice.

[c]ϕ = Angles between the pseudohexagonal axes of the lattice.

[d]$\Delta\phi$ = Difference between the two ϕ values of one lattice.

[e]E_n phase of DPTC [37].

[f]Phase VI of TBBA (values calculated from data in Ref. 22).

[g]Phase V of TBBA (values calculated from data in Ref. 22).

[h]G(H) phase of BBEA [3].

factors may counteract the effect of the tilt and restore the restraining effect of the lattice [24]. For this reason, we expect the B_t phase of TBBA to have more rotational motion than the G(H) phase of BBEA.

In E_n phases, the rotational motions are almost certainly restricted by the two-dimensional lattice, just as in B_n phases, but in addition there is a restriction of a more fundamental nature, like the asymmetry caused by the tilt in tilted phases. The pronounced deviation from hexagonal symmetry (see Sec. III.G.4) makes it quite clear that in the E_n phase the molecules do not pack like cylinders at all. This, in turn, indicates that the molecular rotation must be strongly restricted.

As a final comment on the effect of order on molecular rotation we would like to note the following. At the beginning of this section (Sec. III.G.5), we pointed out that molecular rotation may be expected to be limited by the existence of a two-dimensional lattice in the layer. The molecular packing mode of the lattice (Fig. 6), however, in a way contributes to molecular rotation. This seemingly paradoxical statement is explained as follows. It has been observed that even in B_n phases the molecules have a certain freedom of movement within the layer [89], and from Fig. 6 it will be clear that the translation of a molecule from one lattice position to an adjacent one quite often necessitates a concurrent rotation of the molecule. Such "translation-coupled rotation" could very well make a significant contribution to the total rotational motion of the molecule in B_n phases and also in all other liquid crystal phases (as noted in Sec. III.G.2 above, we consider Fig. 6 a good model for the short-range molecular packing in all liquid crystal phases).

6. Consequences for Classification

a. Tilted Phases. TBBA has [107,108], below its well-known B_t phase (called phase V in Ref. 22), another phase (phase VI) which has now been shown to be a G(H) phase [30]. If our models for the structures of these two phases and the G(H) phase of BBEA [3,24] are correct, the difference between the three phases would be that in phase V the molecules are on the average parallel, in BBEA they form a pronounced herringbone pattern, and in phase VI they form a not-so-pronounced herringbone pattern. Thus, the structure of phase VI would be in between those of phase V and BBEA, and it might well be possible that other intermediate structures will be found in other compounds, and that eventually there will be no clear distinction between the structures of B_t and G(H) phases. It should be kept in mind, though, that the B_t and G(H) phases of TBBA (phases V and VI, respectively) differ not only in the orientation of the molecules (parallel versus herringbone), but also in the degree of rotational freedom of the molecules about their long axes (see, e.g., Ref. 101): in the G(H) phase the molecules appear to be more or less stationary, as in our model of the G(H) phase of BBEA [3], whereas in the B_t phase the molecules probably undergo considerable oscillations about their mean positions, with occasional rotational jumps of 180 or 360° [109].

Another distinction which might eventually disappear (from the structural point of view) is that between E_t and G(H) phases. If E_t phases are simply tilted E_n phases, the main difference between E_t and G(H) would appear to be in the degree of deviation from a hexagonal arrangement of the molecular axes. At present this distinction seems still rather clear-cut (compare Table 6), but it is quite possible that further studies will make the distinction disappear through the discovery of intermediate structures.

b. Normal Phases. As mentioned in Sec. III.G.4, the structure proposed for the E_n phase [37] is very similar to the herringbone packing (Fig. 6) we have suggested for the B_n phase (see Sec. III.G.3 and Ref. 6). If both models are correct, a structure-based classification system would consider B_n and E_n as very closely related and possibly put them under the same heading.

7. Summary

The main points of our discussion in the section (Sec. III.G) can be summarized as follows:

1. A truly hexagonal molecular packing is very unlikely (Sec. III.G.2).
2. A herringbone packing is compatible with an approximately hexagonal lattice (Sec. III.G.3).
3. The extent to which the actual lattice (in a plane perpendicular to the long axes of the molecules) approaches a hexagonal lattice is strongly influenced by the molecular structure (Sec. III.G.4).
4. Rotation of the molecules about their long axes is restricted by space considerations and can be further limited by reduced symmetry due to tilt and/or reduced symmetry due to the two-dimensional lattice (Secs. III.G.2 and III.G.5).

From this we can formulate some propositions with regard to the structure of the various types of ordered smectic phases, and these are summarized in Table 7.

TABLE 7. Proposed Structures for the Ordered Smectic Phases

Phase type	Molecular direction[a]	Molecular packing	Deviation from hexagonal lattice[b]	Dimensionality of order
B_n	Normal	Herringbone	Very small	2 D or 3 D
E_n	Normal	Herringbone	Large	3 D
B_t	Tilted	Parallel	Small	3 D
G(H)	Tilted	Herringbone	Very small to medium	3 D
E_t	Tilted	Herringbone	Large	3 D

[a]Direction of the long molecular axis with respect to the smectic plane.

[b]For the lattice formed by the axes of the molecules in a single smectic layer.

H. Rotational Coupling

To simplify the discussion, we shall restrict ourselves in this section to smectic layers that have a uniform arrangement of the molecules. Such layers will always be flat, and we shall assume that neighboring layers are parallel to each other.

1. Tilted Layers

The existence of tilted smectic layers (Sec. III.E) brings with it the question of how these layers are stacked on top of each other. For, if there is a uniform tilt in a layer, there is a unique direction in this layer, namely the direction in which the long axes of the molecules are tilted away from the normal to the smectic planes. And if there is a unique direction in each layer, it is relevant to inquire about possible relationships between the unique directions in adjacent layers. The two extremes are: (1) complete correlation, i.e., the unique direction in each layer is completely determined by those of its neighbors and (2) no correlation at all. Since this correlation, or coupling, relates to directions which can only rotate with respect to each other about a common axis (the normal to the smectic layers), and to distinguish it from another type of coupling between smectic layers which we shall discuss later on, we shall use for this correlation the term "rotational coupling" [33]. In analogy with the nematic order parameter S (Sec. III.B), a rotational order parameter R can be introduced as R $\langle 2 \cos^2 \theta - 1 \rangle$, θ now being defined as the angle between the unique direction in a layer and a suitably defined average unique direction in a stack of layers. The R would run from 1 for complete correlation to 0 for no correlation.

Experimental evidence strongly suggests that C phases tend to have large values of R [67]: (1) C phases of optically inactive materials have always been found to be optically biaxial [2]; (2) all C phases investigated give a twisted structure when a small amount of optically active material is added [53,54]; and (3) it is possible to obtain smectic C monodomains with a uniform tilt direction [93]. G(H) phases have possibly (and, in our opinion, probably) even higher R values, for the sharp x-ray reflections obtained for a monodomain of the G(H) phase of BBEA [3] indicates an R of very close to 1; the more extended reflections from the B_t phase of TBBA [27] suggest a lower R value, but more data would be needed to be able to say whether this value is characteristic for that particular phase or only for that particular sample (see also below, in our discussion of rotational coupling between ordered layers).

Thus, there seems to be no doubt that extensive rotational coupling exists between tilted layers, presumably because of energy factors [67,110] involved in the packing of the layers. Entropy, however, would tend to favor "uncoupling" (i.e., loss of rotational coupling), and so it seems reasonable to expect that, as the temperature is raised, at a certain point uncoupling could occur while the tilt in the individual layers would be retained. We propose that this is precisely what happens at the C-to-A transition point of a number of compounds reported on by Diele et al. [94]

and which were mentioned already in Sec. III.E.4. These compounds have an A phase as well as a C phase (and also a B phase, but this will not be discussed here) and were reported to have essentially the same layer thickness d in both phases, with d being considerably less than the molecular length ℓ. Now, in all cases known to us in which d is considerably less than ℓ and which have been studied in some detail, the molecules are always tilted with respect to the layers, and the tilt calculated on the basis of the d and ℓ values corresponds always very well with independently determined tilt angles [100]. Therefore, the fact that for the abovementioned A phases d is considerably less than ℓ, is in our opinion a very strong indication that in these A phases the molecules are tilted (for one of the compounds we calculated from the d values in Ref. 94 is a tilt angle of about 19°), and from the fact that the d values in the A phases are about the same as the d values in the C phases we conclude that the tilt angles in both phases are also about the same. Since the A phases are optically uniaxial [94], we suggest that in these phases the tilt directions of the layers are uncoupled [16,67], whereas they were coupled in the C phases. When the tilt directions are uncoupled, they will be randomly distributed in all orientations perpendicular to the layer normal, and the sample will be uniaxial even though the individual layers are biaxial (Ref. 32, p. 6).

To distinguish these A phases, which according to the above argument would have a tilted arrangement of the molecules in the layers (A_t phases), from the A phases which have a perpendicular arrangement of the molecules (A_n phases) and which had been designated as A_1 and A_2 phases (Sec. III.F.2b), we have designated these A_t phases as A_3 phases [16].

If our description of A_3 phases as uncoupled, tilted phases is correct, the general distinction between A and C phases would not be that C phases are tilted and A phases are normal (as is commonly assumed), but rather that C phases are "coupled" and A phases "uncoupled." This has already been discussed above for C phases and for A_3 phases; A_1 and A_2 phases have no "unique direction" within the layer, so these phases are "automatically" uncoupled.

This change in structural distinction would have no influence on the optical distinction: C phases would still be optically biaxial (see Sec. III.E.5) and A phases optically uniaxial and able to give homeotropic (pseudoisotropic) textures.

2. Ordered Layers

a. Various Stacking Modes. Most of what has been said above for tilted layers is also applicable to the stacking of ordered layers, but in addition there is a very interesting effect peculiar to layers with a hexagonal type of lattice.

Let us take first a perfectly hexagonal two-dimensional lattice (Fig. 9) as a model for the molecular arrangement within an ordered smectic layer. Instead of only one unique direction (as in the case of tilt), there are now six equivalent "unique" directions: the directions indicated by the numbers 1, 2, and 3 in Fig. 9 and the directions opposite to these. Consequently, even with complete rotational correlation between the layers (R = 1) there are still six possible ways of stacking each layer on top of the previous layer.

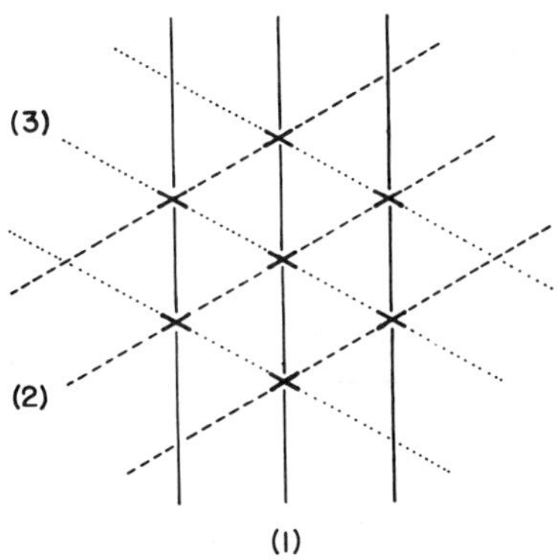

FIG. 9. Hexagonal array (crosses) with three sets of equivalent equidistant planes (perpendicular to the plane of the paper) indicated.

If we now go to the pseudohexagonal herringbone packing of Fig. 6, which is our preferred model for B and G(H) phases, we find that one of the three sets of lines from Fig. 9 has become unique, namely the lines parallel to the b axis in Fig. 6, but that the positive and negative directions along these lines are still equivalent. Thus, there are now only two possible ways to stack adjacent layers while maintaining complete correlation. It should be pointed out, however, that although one of the three sets of lines from Fig. 9 is now unique, all three sets of lines still have a lot in common, e.g., they are all going through molecular centers, all have the same repeat distance between these centers, etc. Thus, it may be surmised that the six ways of stacking, which were exactly equal for the hexagonal lattice, are still roughly equivalent for the pseudohexagonal herringbone packing.

b. Strength of Forces. Another factor which might play an important role in the stacking of ordered layers is that the forces responsible for the rotational coupling between ordered layers are apparently not very strong. A first indication for this comes from a comparison of Figs. 10 and 2. Both are from the same sample of a compound with the phase sequence $B_n \longrightarrow A_n \longrightarrow N$. This sample, contained in a glass capillary, was heated up into the N phase, put in a magnetic field perpendicular to the capillary axis, cooled down to a temperature in the B_n range, removed from the magnetic field, and put in an x-ray camera where it was kept at approximately the same temperature in the B_n range. X-ray diffraction photographs showed the sample to have an almost perfect alignment of the smectic planes parallel to each other and parallel to the capillary axis. A photograph obtained with these smectic planes perpendicular to the incident x-ray beam showed an outer diffraction ring with virtually constant intensity over its entire circumference (Fig. 10). This revealed that, notwithstanding the almost perfect order in the parallel alignment of the smectic planes, there was practically no order at all in the arrangement of the two-dimensional lattices of the individual layers with respect to each other (i.e., the R was ≈ 0 for the sample as a whole). After letting the sample crystallize at room temperature and reheating it to approximately the same temperature in the B_n range as used before (without applying any magnetic field), x-ray photographs showed that the smectic planes had maintained essentially the same degree of parallelism with respect to each other and to the capillary axis, but that the two-dimensional lattices of the layers had rearranged themselves into a much more ordered packing (large R), as evidenced by the well-defined reflections in the diffraction pattern obtained with the smectic layers perpendicular to the incident beam (Fig. 2). This behavior leads us to the conclusion that for this compound the rotational coupling forces are apparently sufficiently strong to maintain a high degree of rotational correlation once this correlation is induced by some other process (in this case probably crystallization; Fig. 2), but not strong enough to create rotational order in a sample with random orientation (Fig. 10).

FIG. 10. X-ray diffraction photograph of a pseudomonodomain of the B_n phase of BBOA (same sample as used in Fig. 4), taken with the incident beam perpendicular to the smectic planes (same arrangement as in Fig. 2). There are no intensity variations in the outer ring, although the sample was stationary.

A second indication about the relative weakness of the rotational coupling forces comes from a comparison of two x-ray diffraction photographs of aligned E_n phases, both obtained from the same compound, but from different samples and in different laboratories; for both photographs we shall use here schematic diagrams. The first diagram (Fig. 11) represents a situation similar to that of Fig. 2: the smectic planes are mainly perpendicular to the incident beam, and the photograph is indicative of the overall rotational correlation between the two-dimensional lattices of the individual smectic layers. The clear separation of the diffraction maxima (although they are not as well defined as in Fig. 2) and the very pronounced deviation from hexagonal symmetry indicate a fair degree of rotational correlation throughout the whole sample. Since the sample used was obtained by melting a single crystal of the corresponding solid phase [37,103], Fig. 11 shows, as did Fig. 2, that the rotational coupling forces are sufficient to maintain rotational order once it is there. Figure 12 is from a sample (oriented in a magnetic field by cooling from the isotropic phase) [31] in which the smectic planes were essentially parallel to the incident beam and to the horizontal direction. The very well-defined outer diffraction maxima, not only on the equator but also above and below it, indicate a high degree of order. In fact, in our opinion, they require local three-dimensional order. On the other hand, the fact that all these maxima are present on the same photograph, without rotation of the sample [31], indicates that there must be large overall rotational disorder. The most probable assignment of the indices of some of the reflections, assuming an average orientation of the sample similar to that corresponding to Fig. 11 (but rotated 90° around the horizontal), has been indicated in Fig. 12. But even with this most probable assignment, the simultaneous appearance of the $\bar{2}00$ and the 200 reflections requires that in different regions of the sample there are lattices rotated around the vertical by about 22° with respect to each other. The presence of the other reflections provides similar additional requirements for lattices rotated over various angles around the vertical. The most likely explanation for the simultaneous presence of all these reflections is that in the sample used for this

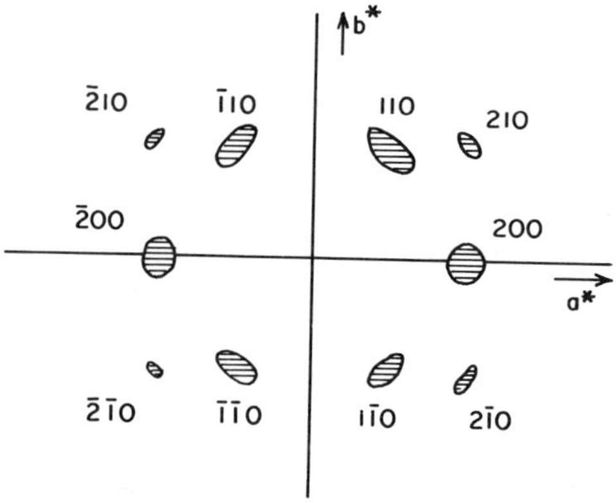

FIG. 11. Schematic representation of an x-ray diffraction photograph (from Ref. 37) of a monodomain of the E_n phase of DPTC, taken with the incident beam approximately perpendicular to the smectic planes (parallel to the c* axis). Probable reflection indices are indicated.

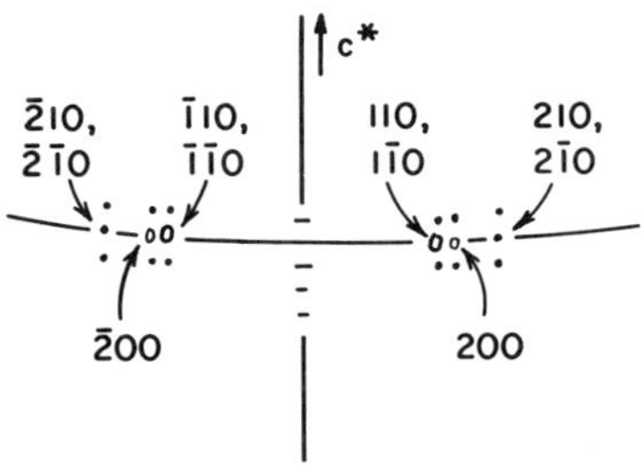

FIG. 12. Schematic representation of an x-ray diffraction photograph (from Ref. 31) of a pseudomonodomain of the E_n phase of DPTC, taken with the incident beam approximately parallel to the mean orientation of the smectic planes. Probable indices of some of the reflections are indicated.

photograph all orientations around the vertical (i.e., around the direction normal to the smectic planes) are present and, consequently, that the overall R was about zero. Our conclusion is then, again, that the rotational coupling forces were sufficient to maintain order (Fig. 11) but not strong enough to establish order (Fig. 12).

A comparison of available information of B_n and E_n phases shows an additional interesting feature, apparently common to both phase types: well-defined x-ray diffraction spots (Figs. 7 and 12), which are evidence of strict order, have been found only in samples with an overall grossly random arrangement of the individual domains; for samples with a fairly uniform overall alignment (monodomains), the reflections are extended streaks (Figs. 2 and 11). Simultaneous occurrence of good overall alignment (formation of a mondomain) and strict order on a local scale (sharp x-ray reflections) has so far only been observed for a G(H) phase [3].

c. Model for B_n Phases. For the structure of a single layer of a B_n we propose, as discussed in Sec. III.G.3, a packing based on the pseudohexagonal herringbone model of Fig. 6. Since this kind of structure would be expected to be optically biaxial, whereas B_n phases are known to be optically uniaxial, something will have to be done to make this model acceptable for B_n phases. Doucet et al. [22] have proposed that the order represented in Fig. 6 is only a short-range order, not maintained over long distances. This could indeed explain the uniaxial nature of B_n phases, but we submit that it is rather unlikely that with only short-range order in the molecular packing it would be possible to have a long-range order in the positions of the molecular axes. The long-range order of the axes appears to be well established by the following facts: (1) the outer x-ray diffraction ring is sharp (Fig. 10) and (2) the axes of the pseudohexagonal lattice can extend throughout the whole sample (Fig. 2). Moreover, there are many similarities between B_n phases and G(H) and E_n phases, and G(H) as well as E_n phases have a unique direction in the smectic layer (tilt direction and orientation of lattice, respectively) which can extend throughout the layer (see Ref. 3 and Fig. 11). This unique direction is probably strongly coupled to the orientation of the a and b axes (Fig. 6) of the herringbone-type packing [24,37], and thus this orientation is probably well defined and constant throughout the layers in G(H) and E_n phases and by inference also in B_n phases.

For these reasons we feel that it is far more likely that the herringbone packing in the B_n phase is long range, not short range, and we propose the following model for a monodomain of a B_n phase: (1) Each smectic layer is flat and has a molecular packing similar to Fig. 6, extending over a long range, and therefore with a well-defined direction for the a axis and for the b axis. (2) All layers are parallel. (3) There is long-range rotational coupling with respect to the pseudohexagonal axes but not with respect to the orthogonal axes (the a and b axes in Fig. 6). With this last statement we mean that the two-dimensional pseudohexagonal lattices, determined by the positions of the centers of the molecules in each layer (Fig. 9), are stacked on top of each other in such a way that the 1 direction of each layer is approximately parallel to one of the hexagonally equivalent directions of the preceding layer (i.e., the direction of 1, 2, 3, $\bar{1}$, $\bar{2}$, or $\bar{3}$, with $\bar{1}$ indicating the direction opposite to that of 1, etc.), with little or no preference for any one of these directions. The resulting structure would have overall hexagonal symmetry and thus be optically uniaxial. Also, since the angles between the pseudohexagonal axes (i.e., between the directions of 1 and 2, 2 and 3, and 3 and $\bar{1}$) would in general not be exactly 60°, the proposed stacking would lead to reflections in the form of streaks (Fig. 2) rather than well-defined spots.

The sometimes observed spotlike x-ray diffraction maxima (Fig. 7, and Fig. 1b of Ref. 33) suggest that in these instances there are domains in which there is a high degree of rotational correlation between the orthogonal axes of the layers, rather than just between the pseudohexagonal axes as proposed above under point 3. So far, such monodomains have always been very small and were found only in polydomain samples. If it would be possible to obtain larger monodomains of this kind, we would expect them to be optically biaxial and also to show other indications of deviations from hexagonal symmetry. Some of these indications have been already intimated above (Sec. III.G.3), where we noted that the angles between the directions in which the spotlike x-ray diffraction maxima were found differed significantly from the 60° required by hexagonal symmetry, and where we also pointed out that there were quite noticeable differences in the appearances of the maxima.

3. Monodomains and Pseudomonodomains

In the last part of Sec. III.H.2c we came across two kinds of monodomains for the B_n phase: those characterized by coupling between the orthogonal axes and those characterized by coupling between the pseudohexagonal axes. This poses the question of whether the name monodomain indeed applies to both cases or only to one.

An answer to this question requires a more precise definition of the term "monodomain." The difficulty with respect to this definition, caused by the introduction of the concept of rotational coupling or rather by the implicit possibility of absence of rotational coupling, is best illustrated with the case of molecules tilted with respect to the smectic layers. If, in a stack of planar, parallel smectic layers, there is no correlation between the tilt directions in adjacent layers, can this stack still be regarded as a monodomain? One could even ask whether this stack is still a liquid crystal, for even the basic assumption, that neighboring molecules are parallel to each other in liquid crystals, does not hold any more in this case, at least not when one considers neighboring molecules from different layers.

In Sec. II.A.2 we defined a monodomain as a region of uniform orientation. We would like to redefine this now as follows: a monodomain of a liquid crystal phase is a region over which the full potential order of that phase is realized. A region which is optically uniform and thus would have the appearance of a monodomain, but which does not satisfy the above definition of a monodomain, we would like to call a pseudomonodomain.

With these definitions, and with our earlier proposed models for A_3 (Sec. III.H.1) and C_3 (Sec. III.E.4) phases, the above described stack of uncoupled tilted layers could be a monodomain if the phase was A_3 but a pseudomonodomain if the phase was C_3 (or C, in general). Similarly, the sample that gave the x-ray diffraction pattern of Fig. 10 would be a pseudomonodomain of a B_n phase, but the sample that gave Fig. 2 could be a monodomain.

In the preceding sentence we have used the word "could" on purpose, because at this point we do not consider it clear what the "full potential order" of the B_n phase is. Our model for the B_n phase, developed in Sec. III.H.2c, considers the "streaky" x-ray reflections as inherent to the structure of the B_n phase. Levelut and Lambert [27], although using a different model, came to the same conclusion. If this is correct, Fig. 2 would represent a genuine monodomain. But the spotlike reflections observed in photographs like Fig. 7 pose a problem: is the spotlike shape caused by a higher degree of order in the domains, or is it merely the result of the small size of the domains and the inherent smaller amount of disorder? This problem will have to be solved before an answer can be given to the question raised in the first paragraph of this section.

The importance of making a proper distinction between monodomains and pseudomonodomains is borne out by some conflicting data on E_n phases. Diele et al. [31] and Sackmann and Demus [2] report that the E_n phase in general and the E_n phase of DPTC in particular are optically uniaxial. The monodomain x-ray diffraction pattern of DPTC (Fig. 11) obtained by Chung [37,103], however, shows that this monodomain has an orthorhombic lattice and does not show the apparent hexagonal symmetry observed with B_n monodomains (Fig. 2); consequently, the sample should be optically biaxial. We would like to suggest here that E_n phases indeed are biaxial, but that, as a consequence of the rather weak rotational coupling forces (see Sec. III.H.2b), samples will often have long-range rotational disorder and thus appear uniaxial (in our discussion of Fig. 12 we concluded that the sample used indeed had this kind of long-range rotational disorder; Biering et al. [111] also propose rotational disorder in the stacking of the layers as the reason for the observed uniaxial optical behavior). In other words, the optically studied samples that were found to be uniaxial were pseudomonodomains, and genuine monodomains will be optically biaxial. In this context it seems significant that Demus et al. [112] report that, in two compounds with an E_n phase directly below an A phase, they could not obtain pseudoisotropic textures of the E_n phase, not even by cooling pseudoisotropic textures of the A phase.

I. Translational Coupling and Three-Dimensional Order

It is generally stated in the literature that smectic layers slide easily over each other. For certain smectic phases, G(H) [3] and E_n [112], however, it has been reported that they are very viscous, implying that in these phases it is very difficult to make the layers slide. This indicates a strong coupling between the positions of adjacent layers. Since this coupling limits translation, we will call it translational coupling.

Since B_t, E_n, and G(H) phases have well-developed two-dimensional lattices in their layers and also rotational coupling between those lattices, adjacent layers in monodomains of these phases have identical, parallel lattices. The apparent strong translational coupling (see preceding paragraph) indicates a strong preference for a particular translational relationship between the lattices of neighboring layers, with each layer-lattice shifted over a fixed distance in a given direction with respect to its neighbors [33]. Most probably, these shifts between neighboring layers will all be equal, and the two-dimensional lattices of the layers will be stacked in a regular manner, resulting in a three-dimensional lattice for the monodomain. Such three-dimensional lattices indeed appear to exist in B_t [22,27], E_n [6,31,33], and G(H) [3,12] phases.

The existence of a three-dimensional lattice suggests that these smectic phases could also be classified as solid crystalline phases. We prefer to retain their classification as smectic phases, however, for the following reasons [2,3,105]:

1. These phases can be made to flow.
2. On heating, these phases go over in other liquid crystal phases, and on cooling these phase transitions are reproduced with no noticeable supercooling; phase transitions involving solid phases almost always show considerable supercooling.
3. In a plot of transition temperatures versus chain length in a homologous series, it has been shown that the temperatures of transitions involving these phases fall on a smooth curve with the temperatures of transitions between other liquid crystal phases.

4. The mosaic textures of these phases strongly resemble those of smectic B_n phases.
5. B_t phases are miscible in all proportions with B_n phases [28,30] (most B_n phases appear to have only two-dimensional lattices and are definitely not crystalline).
6. A Mössbauer study [113] showed the solid phase to be "decidedly more rigid" than the supercooled G(H) phase at the same temperature.

If these three-dimensional smectic phases are indeed not solid crystalline phases, it would be of considerable interest to know what then constitutes the difference between the two.

It has been suggested that the transition from the solid phase to the B_t phase of TBBA is connected with a correlated motion of the molecules around their long axes [22], but in this paper and elsewhere [12] we have already argued that rotational motions in B_t phases can at the most be only limited.

It has also been suggested [22,114] that the order in the molecular packing (e.g., the herringbone-type packing of Fig. 6) extends only over very short distances. We should like to counter this suggestion by a two-point argument. First, the sharpness and the number of the x-ray reflections [in particular in the case of the G(H) phase of BBEA] shows conclusively, in our opinion, that the three-dimensional lattice, formed by the centers of the molecules, is well established and long range. The second point concerns the orientations of the molecules in this lattice. B_t and G(H) phases have tilted molecular arrangements, and this tilt has to be the result of the particular way in which the molecules pack. In other words, the tilt, and the direction in which the tilt occurs, must have a very direct relationship with the molecular packing: a unique direction of tilt requires a unique direction in the packing. In our models (see Fig. 6 and Ref. 24) we have taken this unique direction to be the b axis and assumed that the direction of tilt is perpendicular to this axis, but other choices and assumptions could be made. The important point is that in B_t or G(H) monodomains the direction of tilt has been found to be constant throughout the monodomain [3,27], and as a consequence the unique direction in the packing also has to be constant throughout the monodomain. This in turn means that the particular packing mode which produced this unique direction also has to remain the same throughout the entire monodomain, not just over a short distance; in Fig. 6 this means that the direction of the b axis, which is characterized by the fact that adjacent molecules in this direction are all parallel, maintains this unique character throughout the whole monodomain. Direct evidence for the maintenance of the local molecular packing throughout the three-dimensional lattice is provided by the x-ray diffraction pattern of the B_t phase of TBBA [24].

For the E_n phase, an argument for a unique direction can be made similar to the one above for the B_t and G(H) phases, only now the unique direction in the packing is not required by tilt, because there is none, but by the distinctly nonhexagonal character of the lattice. Direct evidence for the maintenance of a unique packing mode of the molecules throughout the entire monodomain is given by Fig. 11.

Thus, in our opinion, the question of the difference between three-dimensional smectic phases and solid crystalline phases remains unanswered. It may well be that the main difference will turn out to be simply a much higher degree of thermal motion in the smectic phases, as suggested by the strong diffuse maxima on the x-ray diffraction photographs of B_t [27], E_n [37,103], and G(H) [3] phases. Or it may be, as has been suggested, e.g., in Ref. 96, that the difference is a "melting" of the alkyl chains. It is also quite possible that a single answer does not exist, for present data on B_t and G(H) phases [3,22,24,115] indicate that these phases can have quite different degrees of order.

J. Twist

Spontaneous twist can occur in nematic phases as well as in tilted smectic phases C [52-55] and G(H) [56,57]. These twisted nematic (cholesteric; Ch or Ch_N) and twisted smectic (Ch_S) [6] phases are always the result of the presence of optically active molecules. They are found as mesophases of pure optically active compounds, but can also be obtained from the corresponding nontwisted phases through the addition of an optically active material [53,54].

Since twisted nematic phases are completely miscible with other nematic phases [2], it is to be expected that the pitch (i.e., the distance over which the direction of the long molecular axis rotates a full 360°) can be varied in these mixtures in a continuous manner from infinity (in the pure nematic material) to the pitch of the pure cholesteric material. This appears to be a classic example of a case in which there is a continuous change in property but where it is nevertheless useful to make a distinction between, and use different names for, the two extremes: N, with no optical rotation, and Ch_N, with large optical rotation.

So far, in smectics, twist has only been reported for phases in which the molecules in the layers were tilted. It seems to us, however, that the real requirement for the possibility of twist in smectic phases is not tilt but rather the presence of a unique direction in the smectic layer and rotational coupling between the unique directions in adjacent layers. With optically active materials this unique direction can then be twisted in a continuous manner around the layer normal, resulting in a twisted smectic phase. C, B_t, and G(H) satisfy these requirements, but they are not the only ones. Since the E_n phase also has a unique direction and rotational coupling (see Sec. III.I) it should also be possible to make twisted E_n phases. And if there are B_n phases with sufficient rotational coupling between the unique directions of the layers (see Sec. III.H.3), it might even be possible to obtain twisted B_n phases.

K. Antiparallel Order

The asymmetry of the molecules of most mesogenic materials is such that, even with complete parallelism of the long axes and complete freedom of rotation around these axes, each molecule still has two distinctly different orientations: pointing "up" and pointing "down" (in the case of vertically aligned molecules). All experimental evidence indicates that in the nematic phase these molecules are arranged in an antiparallel fashion, i.e., that half of them point up and half down. For smectic phases, models with parallel arrangements (i.e., all molecules in one layer pointing in the same direction) have been suggested as theoretically possible, but experimentally they have not been observed [55]. It is true, of course, that molecules of compounds like benzoic acids probably line up parallel, but through hydrogen bonding the effective building units of the smectic layer in these cases are double molecules which have a center of symmetry which eliminates any distinction between up and down.

It thus seems fair to assume that in smectic layers asymmetric molecules are also arranged in an antiparallel manner [116]. At first sight it might seem that in layers with a regular two-dimensional lattice there would be a number of significantly different antiparallel arrangements possible. As mentioned already in Sec. III.G.5, however, even in ordered layers there is probably still a significant amount of diffusion within the smectic layers, and this diffusion would eradicate the differences between these arrangements.

IV. SUMMARY AND CONCLUSIONS

Although we do not want to present a structure-based classification scheme at this point, we would like to summarize here the results from Sec. III and to draw some tentative conclusions. Table 8 gives the relationships between the structural aspects discussed in Sec. III and all known thermotropic phase types. Our tentative conclusions are the following:

1. There are no significant differences between the molecular packing within a single layer of B_n and E_n phases.
2. There are no significant differences between the molecular packing of E_t and G(H) phases.
3. There is a definite distinction between normal and tilted B phases (B_n and B_t, respectively).
4. There are significant differences among A phases, notably between A_1 and A_2 on the one hand (normal phases) and A_3 and A_4 on the other hand (tilted phases).
5. There are significant differences among C phases: C_1 has a large, temperature-insensitive tilt; C_2 has a strongly temperature-dependent tilt; C_3 and C_4 have a moderate, temperature-insensitive tilt.
6. For several optically uniaxial smectic phases the individual smectic layers are biaxial (A_3, A_4, E_n, and possibly B_n).
7. There are significant differences among N phases (N_o, N, N_{nc}).

TABLE 8. Structural Aspects of the Various Phase Types[a]

LF	RC	O	Ti	TC	Tw	MR	Phase type
-	0	0	0	0	-	+	N_o
-	0	0	0	0	+	+	Ch_N[b]
SR	0	-	-	0	-	+	N_{nc}
SR	0	-	+	0	-	(+)	N_{sc}
+	0	-	-	0	0	+	A_2
+	0	-	-	0	0	(+)	A_1
+	-	-	+	0	0	(+)	A_3, A_4
+	-	+	+	-	-	(-)	B_3, (F)
+	+	-	$\pm$	0	-	(+)	C_2
+	+	-	+	0	-	(+)	C_3, C_4, (F)
+	+	-	++	0	-	(+)	C_1
+	+	-	+	0	+	(+)	Ch_{SC}[b]
+	+	+	-	-	-	(-)	B_1
+	+	+	-	+	-	(-)	E_n
+	+	+	+	-	-	(-)	B_2
+	+	+	+	+	-	(-)	E_t, G(H)
+	+	+	+	+	+	(-)	Ch_{SH}

[a]The structural aspects are abbreviated in the column headings as follows: LF = layer formation; RC = rotational coupling; O = order within the layers; Ti = tilt; TC = translational coupling; Tw = twist; MR = molecular rotation. The meaning of the symbols in the table is as follows: 0 = irrelevant; - = absent; SR = short range only; + = present; ++ = strong; ( ) = only to a certain extent; $\pm$ = strongly dependent on temperature.

[b]For Ch we have only listed the simplest variant, e.g., for Ch_N no reference has been made to the possibility of cybotactic groups, for Ch_{SC} no reference has been made to the different types of tilt.

V. APPENDIX: RELATIONSHIPS BETWEEN TILT, MOLECULAR ROTATION, AND ORDER

Rotation in Normal Smectic Phases

Considering that in the molecules of most liquid crystals the long axis is not an axis of rotational symmetry, we feel that it is rather unlikely that these molecules would arrange themselves perpendicular to smectic planes if there is not a special reason for doing so. Saupe [55] has expressed a similar opinion and concluded that there would be many tilted phases. We would like to go in the other direction with our conclusion and deduce from the fact that there are several smectic phases in which the molecules stand perpendicular to the smectic planes that there must be a special reason for doing so. This reason, we would like to suggest, is that in these phases there is sufficient hindered rotation, rotational disorder about the long axis, or other rotational disorder, to preclude any specific direction of tilt in the smectic layer. To be more specific, the hindered rotation and/or the rotational disorder has to provide a certain kind of symmetry when the structure of the layer is averaged over time and/or place: the symmetry necessary to have the optic axis perpendicular to the smectic plane. And the kinds of symmetry necessary and sufficient for having an axis perpendicular to a plane are known (compare, e.g., crystallographic space groups): the axis has to be an n-fold symmetry axis (with $n > 1$), or the plane has to be a mirror plane (a screw axis or a glide mirror plane would also give a perpendicular axis, but these symmetries seem inappropriate for liquid crystals). If this kind of symmetry is present, the arrangement will be perpendicular. Conversely, a perpendicular arrangement suggests that such symmetry is present, and thus it is indicative of a corresponding molecular rotation and/or rotational disorder [24].

Tilt as a Rotational Constraint

Above we saw that molecules will be perpendicular to the smectic planes if molecular rotation and/or rotational disorder create for the average molecular arrangement an n-fold symmetry axis (perpendicular to the smectic plane) or a mirror plane (parallel to the smectic plane). Consequently, this kind of rotation has to be absent in tilted phases (smectic as well as cybotactic nematic).

One can also turn this argument around. In a tilted phase the surroundings of a molecule cannot have an n-fold symmetry axis parallel to the long axis or perpendicular to the smectic plane, or a mirror plane parallel to the smectic plane, because the uniform tilt gives rise to a single unique direction. Consequently, the potential energy field at the molecular position also cannot have these symmetry elements, and as a result the rotational motion and/or disorder of the molecule cannot have these symmetry elements either. Thus, the presence of tilt restricts the rotational freedom of motion of the molecule [24].

Since the degree of asymmetry of the surroundings of a molecule is coupled directly to the magnitude of the tilt, the restriction of the rotational freedom will also be coupled directly to the magnitude of the tilt: the larger the tilt, the greater the restriction. An example of this is McMillan's dipole model for the smectic C phase [102].

Tilt in a Two-Dimensional Lattice

In a monodomain, the direction of tilt will be the same throughout the sample. The direction of tilt, therefore, will be a unique direction. The only symmetry elements compatible with uniform tilt are a two-fold axis perpendicular to the direction of tilt and parallel to the smectic plane, and a mirror or glide mirror plane through the direction of tilt and perpendicular to the smectic plane. Consequently, if the arrangement of the molecules in the smectic layer has a two-dimensional lattice, the only symmetry element allowed in the lattice is this same mirror or glide mirror plane. Assuming the mirror or glide mirror plane to be present, the lattice will be orthogonal, with one axis parallel to the mirror plane. Since the mirror plane was parallel to the direction of tilt, this means that one axis will be parallel to the direction of tilt and the other perpendicular to it.

In our detailed study of the lattice of the G(H) phase of BBEA [3] we did indeed find that the lattice in the smectic plane was orthogonal and that the direction of tilt was parallel to one of its axes. For the B_t phase of TBBA, Doucet et al. [22] report similar results. For the only other case in which the two-dimensional lattice in the smectic layer has been determined accurately, the E_n phase of DPTC [37-39,103], there is no tilt, but the lattice in the layer is again orthogonal.

These experimental data support our assumption, made above, that the two-dimensional lattice in the layers of an ordered smectic phase will be orthogonal and that the tilt, if present, will be along one of the axes of this lattice.

The models we have proposed for the molecular packing in ordered smectic phases are also in accord with these assumptions: Fig. 6 has orthogonal axes, and both of them are parallel to glide mirror planes, and the models for B_t and G(H) phases [3,24] have orthogonal axes of which one is parallel to a glide mirror plane and to the direction of tilt.

Rotation in a Two-Dimensional Lattice

The various aspects of this have already been dealt with in Sec. III.G.2 (space limitation effect) and Sec. III.G.5 (limitation because of symmetry considerations).

RECENT DEVELOPMENTS

Our chapter on structure and classification was submitted to the editor in its final form in June, 1976. Science has not been standing still since then, fortunately, and so we appreciate the opportunity given us by the editor and the publisher to comment here on some of the recent developments.

Tilted Smectic A Phases (Sec. III.H.1)

Reports on compounds with the phase sequence A-C-B have now appeared in the literature [117-119], for which the smectic layer thickness d in the A phase is significantly less than in the B phase. This probably means that the conclusion of Diele et al [94], that the d spacings in the A and B phases were essentially equal for a number of such compounds, is not correct. Their other conclusion, however, that the d spacings of the A and C phases of these compounds were essentially equal and significantly less than the molecular length, has been confirmed by these new studies. We still feel that the most probable explanation for this is that the molecules in both the A and the C phase are tilted, but we would like to modify slightly our previous descriptions of this type of A phase, the A_3 phase. We now define the A_3 phase as a smectic phase in which the molecules are tilted but in which there is no long-range order of the direction of the tilt. Wulff [120] has developed a theoretical model for this kind of phase.

Bimolecular Smectic A Phases

A number of papers [121-123] have reported smectic A phases in which the layer thickness is significantly greater than the molecular length but less than twice this length. It has been proposed that the smectic layers in these phases are bilayers with partly interdigitated molecules.

Tilted Smectic B Phases (Sec. II.A.4 and III.A)

Gray has recently pointed out [124] that in some cases the region of nonmiscibility between two different phase types can be very narrow and might be easily overlooked. Because of this, and because of the possibility of undetected second-order-type phase transitions discussed in Sec. III.A, it now appears very likely that it will become generally accepted that tilted B and normal B are two distinctly different phase types.

Smectic C_3 Phases (Sec. III.E.4)

A recent study by us [118] confirmed by further data on the same compound [125], and the investigation [126] of another compound with the phase sequence A_3-C_3-B_n, show that the layer thickness in the C_3 phase is not constant but shows a moderate increase with increasing temperature, in an approximately linear fashion. The differences between C_2 and C_3 now appear to be as follows: (1) In C_2 the d value approaches the molecular length at the high-temperature end of the phase range, in C_3 it does not. (2) The temperature dependence of d in C_3 is small compared to that in C_2. (3) The temperature dependence of d is approximately linear in C_3, but strongly nonlinear in C_2.

Three-Dimensional Smectic Phases (Sec. III.I)

On rereading the chapter, it now appears to me that the question of the fundamental difference between crystals and three-dimensional smectic phases might well be answered by their different behavior toward incorporating optically active molecules in their structure. A crystal structure of an optically active compound has a single three-dimensional lattice, just like any other crystal structure, but the lattice of a three-dimensional smectic phase becomes twisted if the molecules are optically active (Sec. III.J).

It might also be noted here that crystal structures generally differentiate between the up and down orientations of a molecule, but that smectic structures probably do not do this (Sec. III.K.).

Note Added in Proof: Some very significant advancements have taken place since we submitted our section on recent developments in March 1978. We want to note here three of these developments.

First, phase V of TBBA now appears to be a G phase, and phase VI of TBBA appears to be a new kind of phase and has been assigned the symbol H (H. Sackmann, paper presented at the Seventh International Liquid Crystal Conference, Bordeaux, France, 1978).

Second, following a lead by Leadbetter (A.J. Leadbetter, paper presented at the NATO Advanced Study Institute on the Molecular Physics of Liquid Crystals, Cambridge, England, August 1977) and Leadbetter and Richardson (A.J. Leadbetter and R.M. Richardson, Mol. Phys. 35:1191 (1978)) we have proposed that in all smectic A phases the molecules have significant tilt angles, i.e., all smectic A phases appear to have the structure proposed above for A_3 phases (A. De Vries, A. Ekachai, and N. Spielberg, Mol. Cryst. Liq. Cryst. Lett., 49:143 (1979)). Data on four compounds studied in our Institute indicate the average molecular tilt angle to be about 20°. This diffuse-cone model for the smectic A phase has been expanded to cover the smectic C phase and the A-C phase transition (A. De Vries, Mol. Cryst. Liq. Cryst. Lett., 49:179 (1979). At present it is not clear whether there still is a significant difference between A_2-C_2 and A_3-C_3 transitions.

A third advancement is the development of a complete classification of all known layered smectic phases, based on symmetry and order (A. De Vries, J. Chem. Phys., in press). A notable difference between this classification system and the tentative conclusions reached in Section IV is that in the new classification system there are definite differences between the structures of B_n and E_n phases, and between those of E_t and G(H) phases.

Acknowledgment

Work supported in part by the National Science Foundation under Grant No. DMR-74-13173.

REFERENCES

1. G. Friedel, Ann. Phys. (Paris), 18:273 (1922).
2. H. Sackmann and D. Demus, Mol. Cryst. Liq. Cryst., 21:239 (1973).
3. A. de Vries and D.L. Fishel, Mol. Cryst. Liq. Cryst., 16:311 (1972).
4. W. Bragg, Nature (London), 133:445 (1934).
5. A. de Vries, Mol. Cryst. Liq. Cryst., 11:361 (1970); and in Liquid Crystals 3, Part I, (G.H. Brown and M.M. Labes, eds.), Gordon and Breach, London, 1972, p. 457.
6. A. de Vries, Pramana, Suppl. No. 1:93 (1975).
7. A.D. Inozemtseva, Kristallografiya, 17:656 (1972) (in Russian); Soviet Phys. Cryst. (English transl.), 17, 566 (1972).
8. B.K. Vainshtein, I.G. Chistyakov, and A.D. Inozemtseva, Kristallografiya, 17:484 (1972) (Russian); Sov. Phys. Cryst. (English transl.), 17:424 (1972).
9. E.A. Kosterin, Kristallografiya, 17:639 (1972) (Russian); Sov. Phys. Crsyst. (English transl.), 17:549 (1972).
10. W.R. Krigbaum, J.C. Poirier, and M.J. Costello, Mol. Cryst. Liq. Cryst., 20:133 (1973).
11. A. de Vries, Acta Crystallogr., A25:S135 (1969).
12. A. de Vries, J. Phys. (Paris), 36:Cl-1 (1975).
13. A. de Vries, Mol. Cryst. Liq. Cryst., 20:119 (1973).
14. T.R. Taylor, S.L. Arora, and J.L. Fergason, Phys. Rev. Lett., 25:722 (1970).
15. R.A. Wise, D.H. Smith, and J.W. Doane, Phys. Rev., A7:1366 (1973).
16. A. de Vries, Abstracts of the Vth International Liquid Crystal Conference, Stockholm, Sweden, June 1974, p. 150.
17. G.R. Luckhurst, Symp. Faraday Soc., 5:96 (1971).
18. K. Herrmann and A.H. Krummacher, Z. Kristallogr., A81:317 (1932).
19. K. Herrmann, Z. Kristallogr., A92:49 (1935).

20. I.G. Chistyakov, L.S. Schabischev, R.I. Jarenov, and L.A. Gusakova, Mol. Cryst. Liq. Cryst., 7:279 (1969); and in Liquid Crystals 2, Part II, (G.H. Brown, ed.), Gordon and Breach, London, 1969, p. 813.
21. D. Demus, Wiss Z. Univ. Halle, 21:41 (1972).
22. J. Doucet, A.M. Levelut, and M. Lambert, Phys. Rev. Lett., 32:301 (1974).
23. R.J. Meyer and W.L. McMillan, Phys. Rev., A9:899 (1974).
24. A. de Vries, J. Chem. Phys., 61:2367 (1974).
25. W.Z. Urbach and J. Billard, Paper No. 34, Fourth International Liquid Crystal Cónference, Kent, Ohio, August, 1972.
26. W. Urbach, Thesis, University Paris VI, 1973.
27. A.M. Levelut and M. Lambert, C.R. Acad. Sci., Paris, Ser. A.B., B272, 1018 (1971).
28. D. Demus, Ref. 54 in our Ref. 2.
29. H. Sackmann, Vth International Liquid Crystal Conference, Stockholm, Sweden, June, 1974.
30. L. Richter, D. Demus, and H. Sackmann, J. Phys. (Paris), 37:C3-41 (1976).
31. S. Diele, P. Brand, and H. Sackmann, Mol. Cryst. Liq. Cryst., 17:163 (1972).
32. Liquid Crystals and Plastic Crystals, Vol. 1, (G.W. Gray and P.A. Winsor, eds.), Ellis Horwood, Chichester, 1974.
33. A. de Vries, Mol. Cryst. Liq. Cryst., 24:337 (1973).
34. G.W. Gray, Symp. Faraday Soc., 5:94 (1971).
35. G. Pelzl and H. Sackmann, Symp. Faraday Soc., 5:68 (1971).
36. D. Coates, K.J. Harrison, and G.W. Gray, Mol. Cryst. Liq. Cryst., 22:99 (1973).
37. D.B. Chung, Ph.D. Thesis, Kent State University, Kent, Ohio, 1974.
38. S. Diele, Phys. Status Solidi, A25:K183 (1974).
39. J. Doucet, A.M. Levelut, M. Lambert, L. Liébert, and L. Strzelecki, J. Phys. (Paris), 36:C1-13 (1975).
40. D. Demus, S. Diele, M. Klapperstück, V. Link, and H. Zaschke, Mol. Cryst. Liq. Cryst., 15:161 (1971); and in Liquid Crystals 3, Part I, (G.H. Brown and M.M. Labes, eds.), Gordon and Breach, London, 1972, p. 493.

41. J. Billard, C.R. Acad. Sci., Paris, Ser. B., 280:573 (1975).
42. D.P. Lesser, M.E. Neubert, and A. de Vries, unpublished work, 1975.
43. A. de Vries, Mol. Cryst. Liq. Cryst., 10:219 (1970).
44. I.G. Chistyakov and W.M. Chaikowsky, Kristallografiya, 18:293 (1973) (Russian); Sov. Phys. Cryst. (English transl.), 18:181 (1973).
45. W.L. McMillan, Phys. Rev., A8:328 (1973).
46. A. de Vries, Mol. Cryst. Liq. Cryst., 10:31 (1970).
47. W.L. McMillan, Phys. Rev., A7:1673 (1973).
48. L. Cheung and R.B. Meyer, Phys. Lett., 43A:261 (1973).
49. L. Léger, Phys. Lett., 44A:535 (1973).
50. F. Rondelez, Solid State Commun., 11:1675 (1972).
51. M. Leclercq, J. Billard, and J. Jacques, Mol. Cryst. Liq. Cryst., 8:367 (1969); and in Liquid Crystals 2, Part II (G.H. Brown, ed.), Gordon and Breach, London, 1969, p. 579.
52. W. Helfrich and C.S. Oh, Mol. Cryst. Liq. Cryst., 14:289 (1971).
53. J. Billard, Bull. Soc. Fr. Mineral. Cristallogr., 95:206 (1972).
54. W.Z. Urbach and J. Billard, C.R. Acad. Sci., Paris, Ser. A.B, B274:1287 (1972).
55. A. Saupe, Mol. Cryst. Liq. Cryst., 7:59 (1969); and in Liquid Crystals 2, Part I (G.H. Brown, ed.), Gordon and Breach, London, 1969, p. 59.
56. R.B. Meyer, L. Liébert, L. Strzelecki, P. Keller, and J. Doucet, Abstracts of the Vth International Liquid Crystal Conference, Stockholm, Sweden, June, 1974, p. 71.
57. R.B. Meyer, L. Liébert, L. Strzelecki, and P. Keller, J. Phys. (Paris) Lett., 36:L-69 (1975).
58. G.W. Gray, Molecular Structure and the Properties of Liquid Crystals, Academic, London, 1962, p. 41.
59. A. Saupe, Angew. Chem., 80:99 (1968) (German); Angew. Chem. (Int. Ed), 7:97 (1968).
60. O. Lehmann, Z. Phys. Chem. (Leipzig), 56:750 (1906).
61. F.P. Price and J.H. Wendorff, J. Phys. Chem., 75:2839 (1971).
62. F.P. Price and J.H. Wendorff, J. Phys. Chem., 76:276 (1972).
63. A. Saupe, private communication, 1974.
64. A. de Vries, unpublished work, 1966.

65. A. Saupe, Ann. Rev. Phys. Chem., 24:441 (1973).

66. W. Elser, J.L.W. Pohlmann, and P.R. Boyd, Mol. Cryst. Liq. Cryst., 20:77 (1973).

67. A. de Vries, J. Phys. (Paris) Lett., 35:L-139 (1974).

68. D. Demus, S. Diele, and H. Sackmann, "Remarks on the relations of polymorphism, miscibility, and structure of liquid crystals," Vth International Liquid Crystal Conference, Stockholm, Sweden, June, 1974.

69. A. de Vries, Chem. Phys. Lett., 28:252 (1974).

70. C. Hermann, Z. Kristallogr., 79:186 (1931).

71. A.J. Mabis, Acta Crystallogr., 15:1152 (1962).

72. B.K. Vainshtein and I.G. Chistyakov, Dokl. Akad. Nauk SSSR, 153:326 (1963) (Russian); Sov. Phys. Doklady (English transl.), 8:1044 (1964).

73. I.G. Chistyakov, Kristallografiya, 8:859 (1963) (Russian); Sov. Phys. Cryst. (English transl.), 8:691 (1964).

74. E.A. Kosterin and I.G. Chistyakov, Kristallografiya, 14:321 (1969) (Russian); Sov. Phys. Cryst. (English transl.), 14:252 (1969).

75. I.G. Chistyakov, in Advances in Liquid Crystals, Vol. 1 (G.H. Brown, ed.), Academic, New York, 1975, pp. 143-168.

76. M. Born, Ann. Phys. (Leipzig), 55:177 (1918).

77. W. Zwetkoff, Acta Physicochim. USSR, 16:132 (1942).

78. P. Chatelain, Acta Crystallogr., 4:453 (1951).

79. W. Maier and G. Englert, Z. Phys. Chem. (Frankfurt am Main), 12:123 (1957).

80. W. Maier and A. Saupe, Z. Naturforsch. A, 15:287 (1960).

81. P. Chatelain, Bull. Soc. Fr. Mineral. Christallogr., 78:262 (1955).

82. W. Maier and A. Saupe, Z. Naturforsch. A, 13:564 (1958).

83. A. Saupe and W. Maier, Z. Naturforsch. A, 16:816 (1961).

84. W.G.F. Ford, J. Chem. Phys., 56:6270 (1972).

85. I.G. Chistyakov and W.M. Chaikowsky, Mol. Cryst. Liq. Cryst., 7:269 (1969); and in Liquid Crystals 2, Part II (G.H. Brown, ed.), Gordon and Breach, London, 1969, p. 803.

86. S.L. Arora, J.L. Fergason, and A. Saupe, Mol. Cryst. Liq. Cryst., 10:243 (1970); and in Liquid Crystals 2, Part II (G.H. Brown, ed.), Gordon and Breach, London, 1969, p. 563.

87. A. de Vries and S.B. Qadri, to be published.
88. R. Blinc, J. Pirš, and I. Zupančič, Phys. Rev. Lett., 30:546 (1973).
89. R. Blinc, International Conference on Liquid Crystals, Bangalore, India, December, 1973.
90. K. Herrmann, Trans. Faraday Soc., 29:972 (1933).
91. H. Sackmann and D. Demus, Fortschr. Chem. Forsch., 12:349 (1969).
92. D. Demus and H. Sackmann, Z. Phys. Chem. (Leipzig), 222:127 (1963).
93. T.R. Taylor, J.L. Fergason, and S.L. Arora, Phys. Rev. Lett., 24:359 (1970).
94. S. Diele, P. Brand, and H. Sackmann, Mol. Cryst. Liq. Cryst., 16:105 (1972); and in Liquid Crystals 3, Part I (G.H. Brown and M.M. Labes, eds.), Gordon and Breach, London, 1972, p. 481.
95. M. Lefèvre, J.L. Martinand, G. Durand, and M. Veyssié, C.R. Acad. Sci., Paris, Ser. A.B, B273:403 (1971).
96. F. Poldy, M. Dvolaitzky, and C. Taupin, J. Phys. (Paris), 36:C1-27 (1975).
97. I.W. Smith, Y. Galerne, and S.T. Lagerwall, Abstracts of the Vth International Liquid Crystal Conference, Stockholm, Sweden, June, 1974, p. 171.
98. W.J. Mayer, M.S. Thesis, Kent State University, Kent, Ohio, 1976.
99. M.E. Neubert, private communication (1975).
100. A. de Vries, unpublished work, 1974.
101. B. Deloche, J. Charvolin, L. Liebert, and L. Strzelecki, J. Phys. (Paris), 36:C1-21 (1975).
102. W.L. McMillan, Phys. Rev., A8:1921 (1973).
103. D.B. Chung, A. de Vries, J.W. Reed and G.H. Brown, to be published.
104. L.W. Gulrich and G.H. Brown, Mol. Cryst., 3:493 (1968).
105. A. de Vries, unpublished work, 1972.
106. D.P. Lesser and A. de Vries, to be published.
107. Z. Luz and S. Meiboom, J. Chem. Phys., 59:275 (1973).
108. G.W. Gray, Ref. 3 in our Ref. 22.
109. A. de Vries, Gordon Research Conference on Plastic Crystals and Molecular Freedom, Santa Barbara, California, January, 1976.
110. P.G. de Gennes and G. Sarma, Phys. Lett., 38A:219 (1972).

111. A. Biering, D. Demus, G.W. Gray, and H. Sackmann, Mol. Cryst. Liq. Cryst., 28:275 (1974).
112. D. Demus, K.H. Kölz, and H. Sackmann, Z. Phys. Chem. (Leipzig), 252:93 (1973).
113. R.E. Detjen, D.L. Uhrich, and C.F. Sheley, Phys. Lett., 42A:522 (1973).
114. A.M. Levelut, Vth International Liquid Crystal Conference, Stockholm, Sweden, June, 1974.
115. A.M. Levelut, J. Doucet, and M. Lambert, J. Phys. (Paris), 35:773 (1974).
116. J.W. Shelton and Y.R. Shen, Phys. Rev., A5:1867 (1972).
117. W.H. de Jeu and J.A. de Poorter, Phys. Lett., 61A:114 (1977).
118. A. de Vries, Mol. Cryst. Liq. Cryst. Lett., 41:27 (1977).
119. J. Doucet and A.M. Levelut, J. Phys. (Paris), 38:1163 (1977).
120. A. Wulf, Phys. Rev., A17: 2077 (1978).
121. G.W. Gray and J.E. Lydon, Nature, 252:221 (1974).
122. W.L. McMillan, Phys. Rev., A9:1720 (1974).
123. A.C. Griffin and J.F. Johnson, J. Amer. Chem. Soc., 99:4859 (1977).
124. G.W. Gray, NATO Advanced Study Institute on The Molecular Physics of Liquid Crystals, Cambridge, 1977.
125. A. de Vries, A. Ekachai, and N. Spielberg, to be published.
126. A. de Vries, A. Ekachai, and N. Spielberg, J. Phys. (Paris), Proceedings of the VII International Liquid Crystal Conference, Bordeaux, France, 1978, in press.

Chapter 2

NEMATIC MESOPHASES

Franklin D. Saeva*

Xerox Corporation
Webster Research Center
Webster, New York

I. INTRODUCTION

The term liquid crystal refers to the phase that lies between the rigidly ordered solid phase where the mobility of individual molecules is restricted and the isotropic phase where molecular mobility and a general lack of

*Present Affiliation: Research Laboratories, Eastman Kodak Company, Rochester, New York, 14650

molecular order exists. Liquid crystals may be of the nematic, cholesteric, or smectic type and either be thermally (thermotropic) or solvent induced (lyotropic). This chapter will deal exclusively with nematic mesophases of the thermotropic type and is intended to be a rather selective review reflecting the author's points of view and interests.

Nematic liquid crystals (NLCs) derive their name from the Greek word for thread which indicates the appearance of the mesomorphic phase produced by rapid cooling of the isotropic melt. This rather unique threadlike structure is due to optical discontinuities produced by variations in molecular organization. Nematics are perhaps the most common mesophase type formed by many rod-shaped organic molecules. The molecular structural requirements for nematics have not been quantitatively defined; however, the importance of relatively weak, i.e., $\Delta G° \ll 5$ kcal/mol, short-range intermolecular interactions of the van der Waals type, i.e., dipole-dipole, induced dipole-dipole, and induced dipole-induced dipole forces, cannot be overemphasized in the author's opinion. Intermolecular hydrogen bonding ($\Delta G° < 9$ kcal/mol) often leads to an interaction between molecules which prevents the type of order required for the nematic mesophase.

II. MESOPHASE STRUCTURE

The molecular alignment within a classic nematic mesophase is described schematically in Fig. 1. The long molecular axes of the individual molecules lie essentially parallel to one another. This spontaneous alignment or orientation process produces highly birefringent material, i.e., $n_{\parallel} \neq n_{\perp}$ where n is the refractive index of light polarized parallel or perpendicular to an external reference. The most common texture observed in NLCs, under a polarizing microscope, is that of the schlieren or threaded texture shown in Fig. 2. The threads result from a change in the preferred molecular alignment in the vicinity of the axis of the thread. NCLs exhibit <u>positive</u> birefringence ($\Delta n = n_{\parallel} - n_{\perp}$), indicating that the optic axis of the nematic phase coincides with the preferred direction of the long axis of the molecule as shown schematically below. The alignment of molecules within a nematic mesophase is not precisely parallel because

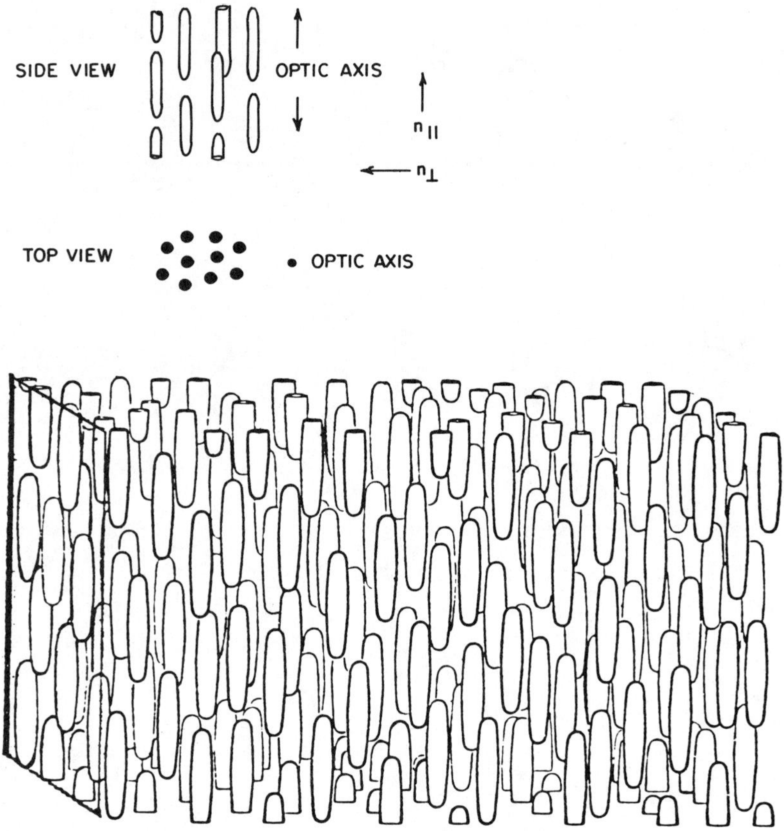

FIG. 1. Schematic diagram of a classic nematic mesophase.

of thermal motion. A measure of the degree of parallel alignment is obtained by a determination of the order parameter S by means of the following expression for rod-shaped molecules:

$$S = \frac{1}{2}(3 \cos \theta - 1)$$

Theta (θ) is the angle between the long molecular axis and some external reference, such as the direction of an applied electric field, for example [1]. For perfect alignment of molecules parallel to an external reference, S = 1; for a completely random isotropic phase, S = 0; and for perpendicular alignment, S = 1/2.

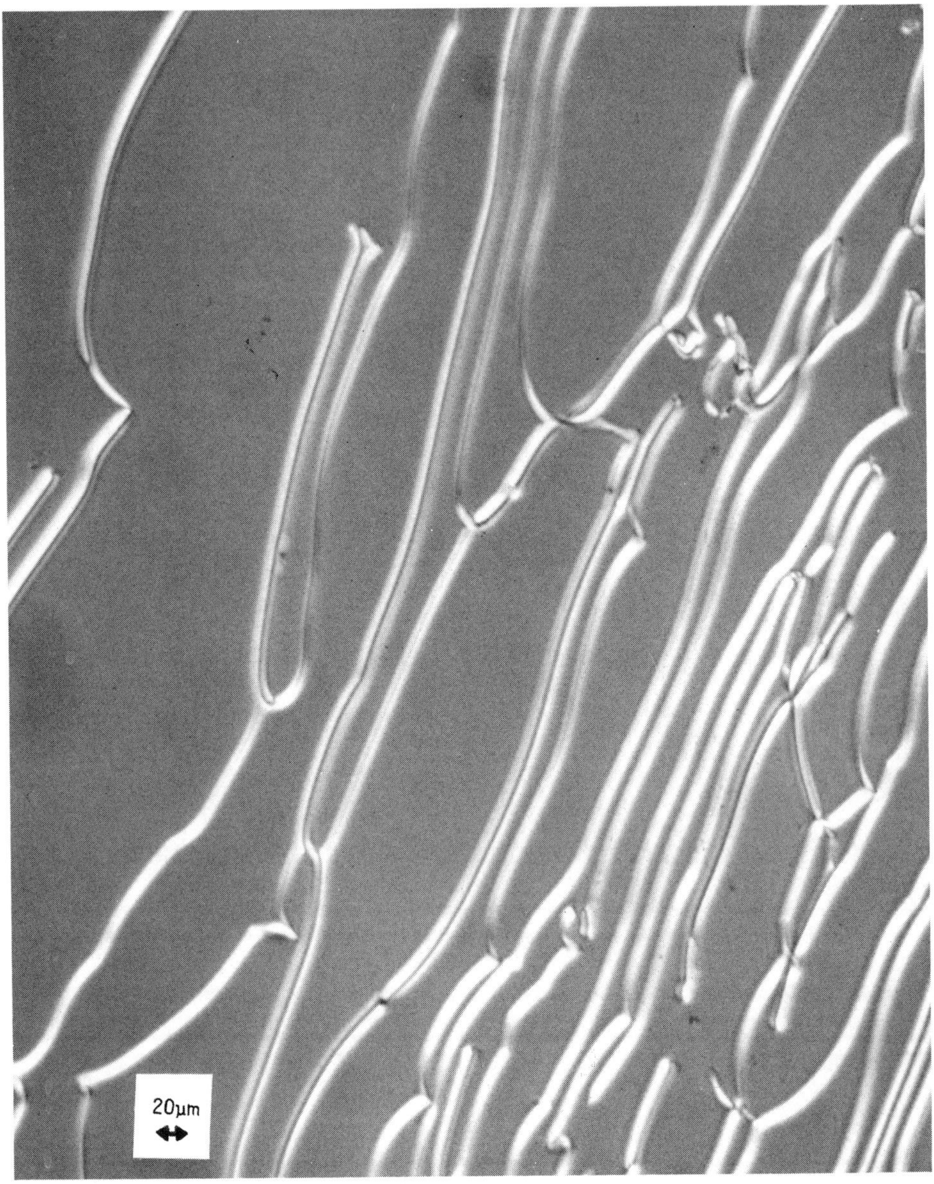

FIG. 2. Schlieren or threaded texture of a nematic mesophase (magnification X 200).

The temperature range of a thermotropic liquid crystal mesophase may be observed visually in a melting point capillary or on a hot stage attachment to a polarizing microscope. Crystalline p-azoxyanisole (I), for example, melts at 119.5° C to a nematic phase which exists until 135° C when the mesophase is transformed into an isotropic liquid [2].

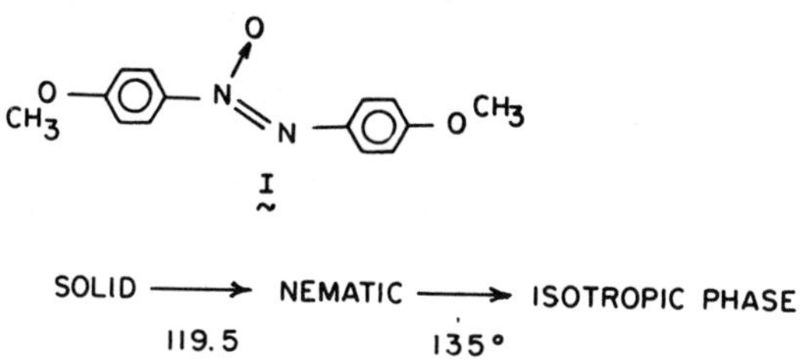

III. NEMATIC PHASES

Molecular ordering within a classic nematic phase is such that there is no regular arrangement between neighboring molecules, as shown below.

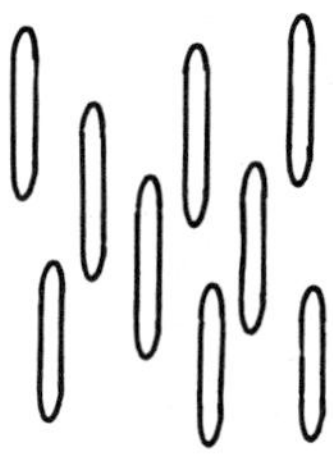

de Vries [3-5] has observed evidence from x-ray microscopy for more than one type of nematic phase and suggests that they are the following three types:

1. Skewed cybotactic phase in which molecules are arranged in groups such that centers of the molecules in each group lie in a plane which makes an angle α with the mean direction of molecules in the group. The angle α in this case is very different from 90°;

2. Normal cybotactic phase which is similar to the skewed but where α is close to 90°;

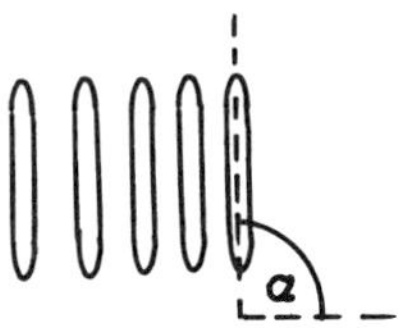

3. Classic nematic phase.

In the specific nematic phase of bis-(4'-n-octyloxybenzal)-2-chloro-1,4-phenylenediamine [5] at low temperature, a large majority of the molecules are arranged in groups and they appear fairly rigid. At higher temperatures, the groups become less prominent but do not disappear completely.

IV. OPTICAL PROPERTIES

The molecular alignment and consequent optical properties of nematic mesophases can be controlled by surface treatment of a substrate such as a glass slide. Rubbing of the surface by the Chatelain method [6,7] with either a piece of cotton or cloth produces molecular alignment along the direction of rubbing. The rubbing procedure produces grooves in the

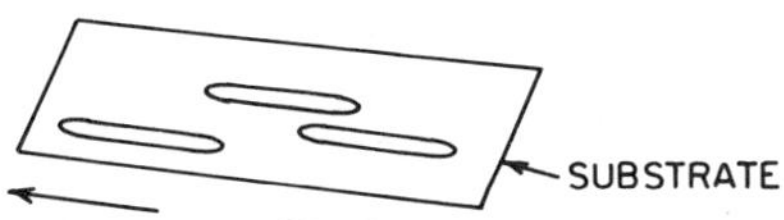

substrate [8,9], and the rodlike liquid crystal molecules preferentially orient their long molecular axes parallel to the grooves. Janning [10] has demonstrated that a silicon monoxide (SiO) film deposited at an angle provides an oriented surface for alignment of nematic liquid crystals which is considerably more uniform than produced by the rubbing technique. In the Janning technique a grooved surface coating of SiO produces molecular alignment by a physical mechanism similar to that of the rubbing process.

Guyon et al. [11] have found that molecular alignment depended on the angle θ between the direction of the evaporation and the direction normal to the film surface. When θ was between 0 and 45° no preferred orientation of the NLCs is obtained. Angles of θ between 45 and 80° produced alignment perpendicular to the projection of the evaporation beam. When angles between 80 and 90° were used, molecular alignment parallel to the plane of incidence of the beam was obtained. The molecular alignment of NLCs can also be controlled by surface-active agents [12,13] such as lecithin and siloxanes.

Alignment perpendicular to the substrate can also be achieved by simple treatment of a glass slide with, for example, sulfuric acid and rinsing with distilled water [14]. The nematic molecules lie preferentially perpendicular to the substrate, as schematically described below:

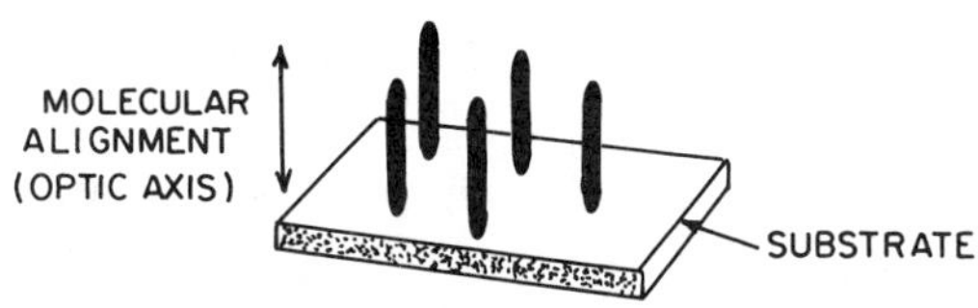

The optical properties of nematics are dependent on the type and orientation of the nematic phases relative to the polarization of incoming light and viewing direction. A nematic liquid crystal placed between rubbed surfaces placed at an angle of 0° relative to one another produces an aligned mesophase which is schematically described below:

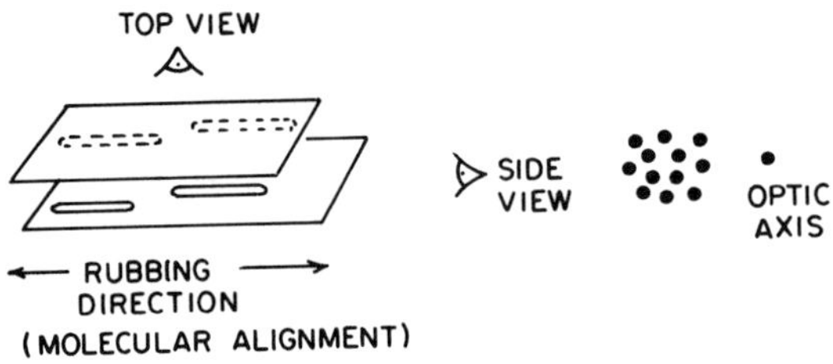

Viewing this mesophase structure from the top between crossed polarizers results in the observation of a highly linearly birefringent phase whose birefringence is dependent on the sample orientation with respect to the incoming linear polarized light. When the molecular alignment and polarization of the incoming light are parallel, the optical transmission between crossed polarizers is at a minimum. Viewed from the side between crossed polarizers, on the other hand, the mesomorphic phase is nonbirefringent since now the viewing direction is also the direction of the optic axis.

Placing the nematic liquid crystal between rubbed surfaces which are at some angle other than 0° results in a twisted nematic phase which has some rather unique optical properties [15-19]. Linear birefringence is observed for this phase between crossed polarizers, independent of viewing direction (side or top view) as well as sample orientation. Circular dichroism has also been observed in twisted nematic mesophases, indicating their cholesteric character [20].

Areas of reverse twist have been observed in twisted nematic mesophases where the angle between the rubbed surfaces is 90° [17,18,20]. The presence of areas of reverse chirality gives a patchy appearance to layers when viewed between crossed polarizers. The addition of chiral molecules to the twisted nematic mesophase provides an internal dissymmetric perturbation which directs the chirality of the twist in a single direction [18]. However, it is the author's opinion that these mesophases are now cholesteric in nature.

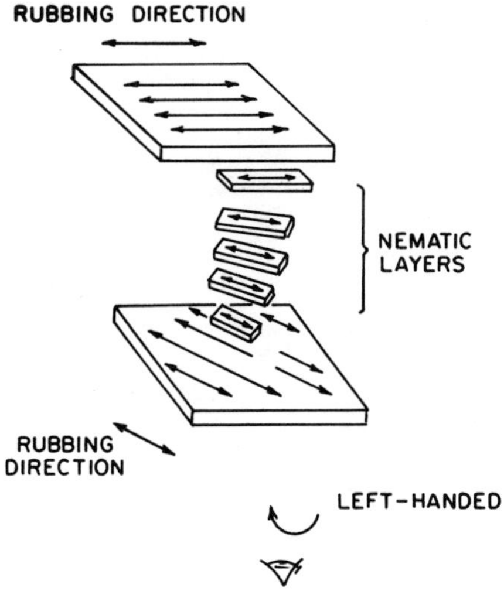

V. ELECTRIC PROPERTIES

Nematic liquid crystalline phases exhibit a variety of electro- and magneto-optic effects produced by either field-induced or ion flow molecular realignment through the mesophase structure.

A. Dielectric Field Effects

An anisotropic dielectric or diamagnetic component will tend to align such that the largest component of dielectric or diamagnetic constant lies in the direction of the applied field. Nematic molecules behave as dipolar species with $\varepsilon_{\|} \neq \varepsilon_{\perp}$ ($\mu_{\|} \neq \mu_{\perp}$) where ε and μ are the dielectric and diamagnetic susceptibilities parallel and perpendicular to the long molecular axis. When $\varepsilon_{\|} > \varepsilon_{\perp}$ ($\mu_{\|} > \mu_{\perp}$) the NLC long molecular axis will tend to align with the applied field. For example, p-methoxy-p'-cyanobenzylideneaniline (Ia) is a highly polarizable molecule with a positive dielectric anisotropy, i.e., $\varepsilon_{\|} > \varepsilon_{\perp}$, that will align its long molecular axis with an applied field. p-Methoxy-p'-n-butylbenzylideneaniline (IIa), on the other hand, with $\varepsilon_{\perp} > \varepsilon_{\|}$,

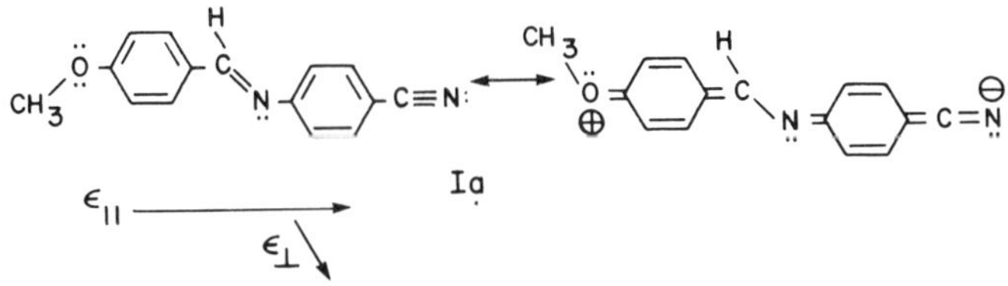

Ia

will tend to align its long molecular axis perpendicular to the applied field. A variety of electro-optic effects of liquid crystals are based on the electric field molecular realignment. Using the surface treatment to produce a molecular alignment which is orthogonal to that produced by the applied field one can interconvert readily between the two molecular reorientations with millisecond response times. The alignment of nematics by external electric fields has been recently reviewed by Gray [21], and was first observed by Freedericksz [22]. One can observe a variation in the linear birefringence or light absorption (linear dichroism) by reorienting nematic molecules with a positive dielectric anisotropy ($\varepsilon_{||} > \varepsilon_{\perp}$) as described schematically below. Nonliquid crystalline dye molecules may be aligned by the nematic molecules to produce useful display applications [23,24] which will be discussed in more detail in Chap. 12.

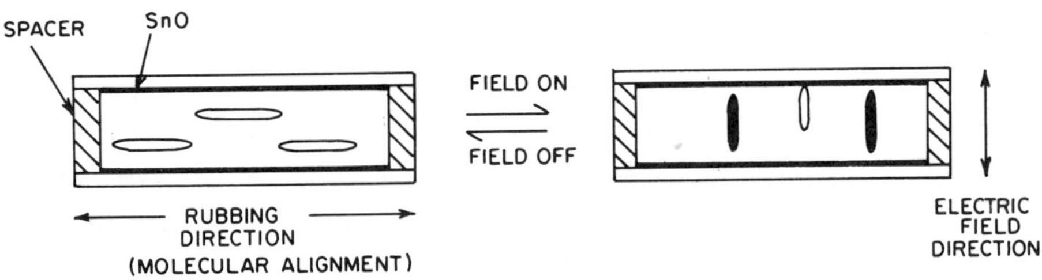

B. Hydrodynamic (Current) Effects

The application of an electric field to a layer ($\approx$ 10-100 μm) of a nematic mesophase between transparent conducting electrodes, e.g., SnO or InO, produces a uniform domain structure [25,26], providing the mesophase possesses a negative dielectric anisotropy. Under the proper conditions of layer thickness and material resistivity (> $10^{10}\ \Omega$/cm), an applied ac or dc electric field produces the domain structure by a conduction-induced alignment mechanism [27].

More recently, a new electrohydrodynamic effect in nematics called the variable grating mode [28-31] has been reported. The onset of the domain structure occurs at a characteristic threshold voltage of ≈ 5-10 V, independent of layer thickness, but the domain density is inversely proportional to film thickness near threshold and varies linearly with voltage above threshold [29]. In this dc effect, the domains consist of regions of vortex-type fluid motion and are visible as a result of spatially periodic variation of the anisotropic refractive index of the nematic phase. As the voltage is increased beyond that necessary for domain formation, the mesophase begins to stir due to the movement of ions, producing a highly turbid light-scattering medium which is called dynamic scattering [32].

VI. MOLECULAR STRUCTURAL EFFECTS

A. Benzylideneanilines

There has been a tremendous upsurge of interest in liquid crystals in the past decade or so due to the discovery of many technological applications of these intriguing compounds. Much of this research has been directed toward the synthesis of new compounds of a certain mesophase type with the desired liquid crystal range, physical, optical, and electric properties. The monograph [33] written by Gray in 1962 greatly assisted this renaissance of activity in the liquid crystal field.

The results of considerable synthetic effort have led to theories concerning the relative importance of specific molecular structural features to mesophase stability. Some of these features are molecular rigidity, linearity, polarizability, and the enhancement of polarizability by conjugation [34].

An important factor in the synthesis of new liquid crystalline compounds is the ease and simplicity of the actual chemistry involved. For this reason derivatives of aromatic anils (benzylideneanilines) of the general structure (II) have received considerable attention since the final synthetic step involves an acid-catalyzed condensation reaction. The effect of terminal substitution on the mesomorphic behavior of some benzylideneanilines has been reported by Castellano and coworkers [35]. They report that replacement of a terminal methoxy group with $-SCH_3$ (i.e., an exchange of S for O) results in a loss of mesomorphic behavior. Also, groups with dipoles acting along the long molecular axis, e.g., NO_2 and Cl, tend to disrupt mesomorphic behavior.

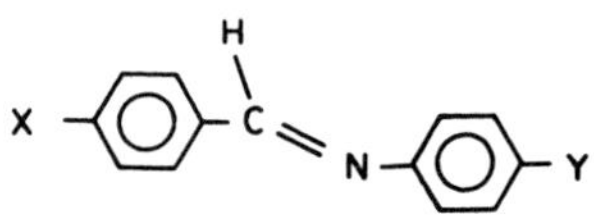

II

Initially, synthetic efforts emphasized the design of room temperature stable mesomorphic compounds. A series of alkoxybenzylideneaminoacetophenones was reported by Arora et al [36] to have melting points in the 90°C range. Young [37] synthesized a class of nitrones (benzylideneaniline N-oxides) with crystal → nematic transitions greater than 100°C, where the importance of molecular polarizability as well as conformational rigidity to mesophase stability was suggested. Prior to the discovery of room-temperature mesomorphic compounds one had to resort to the use of mixtures of NCLs that melted above room temperature (see above) [38]. In 1969 the first room-temperature mesomorphic compound was reported. Its synthesis was a result of judicious substitution on II [39]. The compound N-(4-methoxy benzylidene)-4-n-butylaniline (IIa) exhibited nematic liquid crystalline behavior between 22 and 48° C [39]. Azo and azoxy derivatives of I were subsequently reported to possess somewhat higher solid-to-nematic transition temperatures [40] than IIa. The discovery of IIa and the formation of extended range nematics from binary systems of N-(4-n-alkoxybenzylidene)-4-n-alkylanilines [41,42] greatly

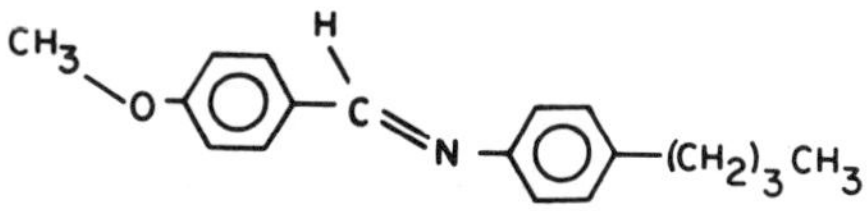

IIa

stimulated liquid crystal research activities since potential applications of liquid crystals appeared more realistic and could be achieved without elaborate temperature control. Researchers continued to synthesize new mesomorphic compounds which would offer advantages over available NLCs by exploring the effect of structure on NLC behavior. In 1971 Young and coworkers [43,44] synthesized a series of chlorostilbene derivatives which possess lower crystal-to-mesomorphic temperatures than do the planar stilbenes.

Knaak and Rosenberg [45] published a survey of reported nematic (N) to isotropic (Is) temperatures of NLCs with the following structure (III). They describe a procedure for estimating the N → Is transition temperature whereby the mesomorphic behavior of unknown compounds may be estimated.

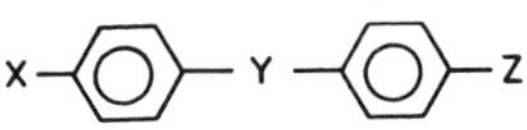

III

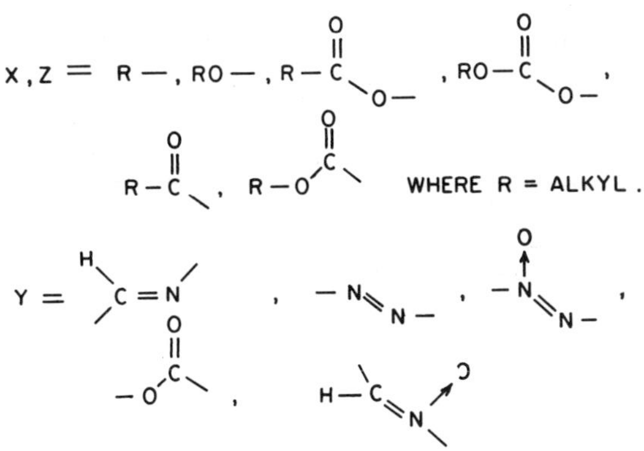

The benzylideneanilines, although readily synthesized from the substituted benzaldehyde and aniline derivatives, suffer from their chemical instability. They readily hydrolyze in the presence of trace acid and moisture. Labes and coworkers [46] have found that the presence of an ortho-hydroxy group in the starting benzaldehyde derivative stabilizes the resulting anil (IV) to hydrolysis relative to II. This increased chemical stability is due to intramolecular H bonding.

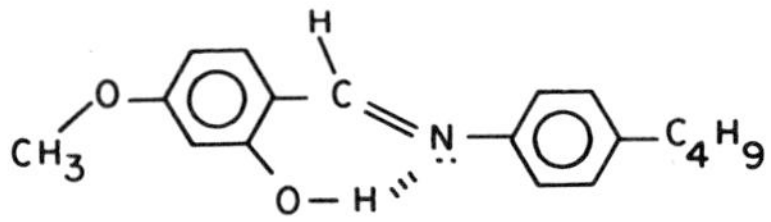

IV

B. Esters and Carbonates

NLCs of the ester and carbonate type offer increased chemical stability over the benzylideneaniline and chlorostilbene compounds; however, room-temperature mesomorphic compounds had to be designed. In 1972, Verbit and Tuggey [47] reported the synthesis of a series of ester derivatives with the general structure V. The nematic stability and phase transition temperatures for the two series have led to a number of conclusions concerning the effect of molecular structure on liquid crystalline behavior. The nematic-isotropic transitions for the acetylene compounds were more than 100°C lower than those of the corresponding benzenoid derivatives,

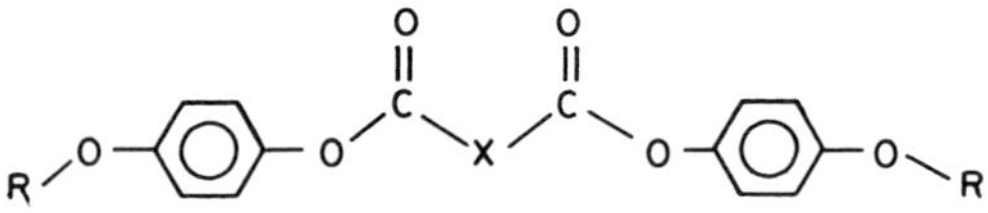

V

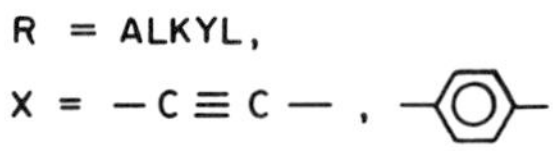

while the melting or crystal-nematic transitions of the acetylenic esters range from 60-100° C lower. Jones and Ratto [48] synthesized a series of alkyl and alkoxycinnamic acid esters with crystal to nematic transition temperatures less than 60° C.

Van Meter and Klanderman reported [49] the effect of terminal alkyl and alkoxy substitution on the mesomorphic properties of 60 phenyl benzoate esters with the following general structure (VI). They report

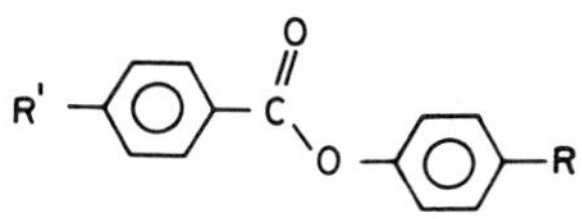

VI

R, R' = ALKYL, ALKOXY

crystal-nematic temperatures as low as 50° C and show an effect of the central linkage as well as O versus S in the terminal substituent. The relative thermal stability of the nematic mesophase as defined by the upper limit temperature, with changes in the central linkages and symmetrical substitution, decreases as follows:

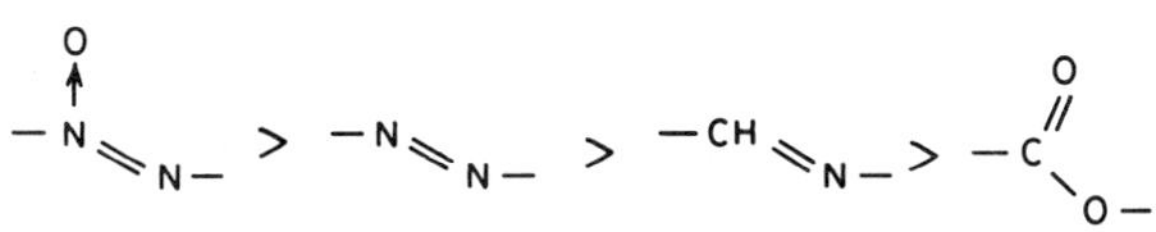

Also, as previously indicated for anils [4], replacement of oxygen by sulfur in a terminal substituent in VI, e.g., $-OCH_3 \rightarrow -SCH_3$, resulted in a destabilization of the nematic mesophase. Steinstrasser reported [50] the syntheses of 19 phenyl esters of the general structure VII with crystal to nematic transition temperatures above room temperature. Compounds in this series possess a positive dielectric anisotropy, i.e., $\varepsilon_{\parallel} > \varepsilon_{\perp s}$ where ε is the dielectric constant, indicating that the long molecular axis of VII tends to align parallel to an applied electric field. The electrical behavior of mesophases will be discussed in a later section of this chapter.

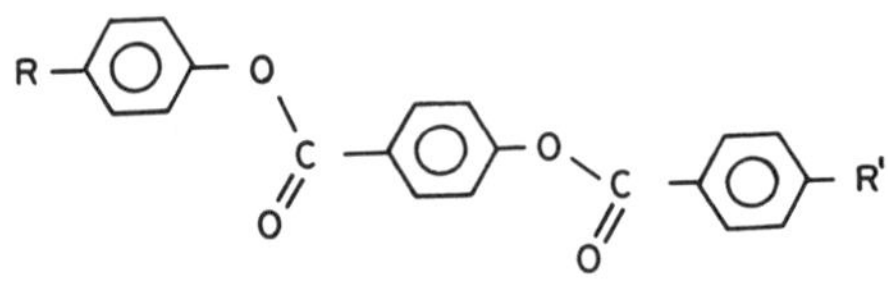

VII

R, R' = ALKYL, ALKOXY

Schroeder [51] describes two nematic phenyl esters which possess phenol and amino terminal groups. Normally the presence of these groups inhibits mesomorphic behavior by providing nonlinear molecular arrangement through H bonding. Apparently the linear carbonate structural effects dominate those due to H bonding.

Young [52] utilized lateral methyl group substitution in a series of phenyl esters of general structure VIII to lower the crystal-to-nematic transition temperatures. Crystal-to-nematic transition temperatures as low as 43 °C were observed with large (> 70 °C) nematic ranges. Van Meter and Klanderman [53,54] have reported the synthesis of seven low-melting nematic phenyl 4-benzoylbenzoates with melting points as low as 39 °C and nematic ranges between 36 and 101 °C. The lowering of the crystal-nematic temperatures was produced by lateral substitution, i.e., a replacement of hydrogen by chloro on an aromatic ring. Lateral substitution has been used effectively [53] as a means of increasing molecular dissymmetry, which presumably alters the molecular packing in the crystalline state and in turn reduces the crystal-to-liquid crystalline transition temperature.

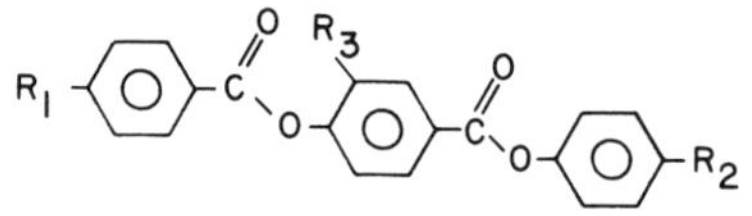

VIII

R_1, R_2 = H, ALKYL, OR ALKOXY
R_3 = CH_3

Verbit and coworkers [55] have reported the synthesis of 20 nematic esters of structure IX in an attempt to interrelate changes in molecular structure and mesomorphic behavior. The mesomorphic behavior of this series of compounds indicates the importance of intermolecular separation and stereochemistry on nematic stability and demonstrates that π-electron density in the central group is not a necessary condition for mesophase formation.

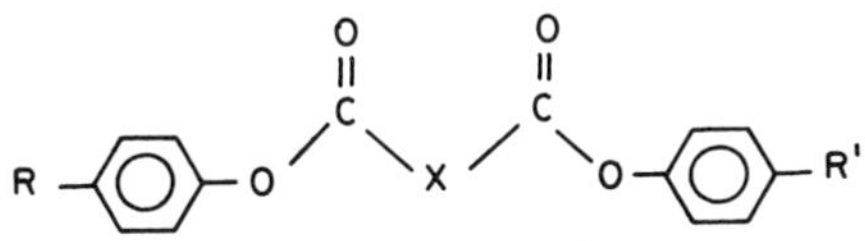

IX

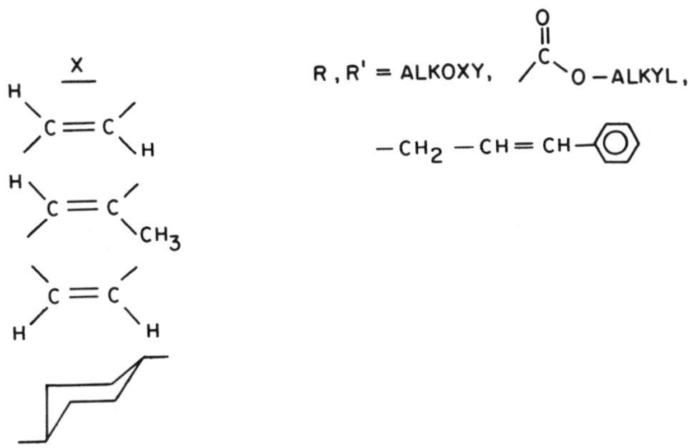

C. Cyanobiphenyl Derivatives

A variety of applications require the liquid crystals to be stable to numerous external stimuli. Liquid crystals of increased stability (chemically and photochemically) have recently been reported by Gray [56,57]. This class of mesomorphic compounds are derivatives of cyanobiphenyls (structure X). These compounds are colorless and relatively stable chemically and photochemically, and eutectic mixtures give wide-range nematics covering room temperature. They also possess some very desirable electrical properties.

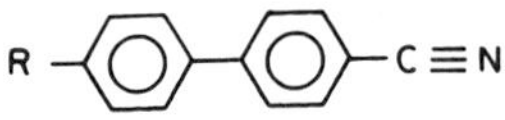

X

R = n-ALKYL, n-ALKYL 

More recently, Dewar and Riddle [34] have concluded from their investigations of molecular structural variations versus mesomorphic behavior that the predominant factor that influences nematic stability is geometric, with polarity having little or no beneficial effect.

VII. APPLICATIONS

The numerous applications of liquid crystals have been the impetus for the increased research activity in this area. Although the applications of liquid crystal phenomena will be covered in detail in Chap. 12, a brief account of the applications of nematics will be presented here. The use of nematics in a variety of modes employing the dielectric and hydrodynamic effects previously discussed have been outlined by Kmetz [58], Sussman [59], and Scheffer [60]. The anisotropic mesophase properties have been utilized by Kelker [61], Dewar and Schroeder [62], and Cook and Spangelo [63] as column substrates for gas-liquid chromatography (glc), and have been used for the separation of geometric isomers, which is a difficult task since they differ very little in polarity and boiling point. In the glc separation of the geometric isomers of xylene the ortho isomer is eluted first, the meta second, and the para third. The order of isomer elution is related to their molecular shape rather than their boiling points, which in the latter case would be para, meta, and then ortho if adequate resolution could be achieved. The para isomer more closely resembles the liquid crystal molecules in shape and therefore interacts most strongly with the mesophase and is eluted last.

The anisotropic ordering in nematics also has been extensively utilized for a variety of spectroscopic applications such as the determination of the polarization of optical transitions of solutes ordered by the mesophase from absorption [64-66] and emission [67] polarization experiments. A relatively new application of NLCs is their use in optical waveguides. Channin [68] has demonstrated the coupling of light in waveguide modes of a nematic liquid crystal.

Also, nuclear magnetic resonance (NMR) [65] and electron paramagnetic resonance (EPR) [69] experiments in anisotropic nematics has provided useful scientific information concerning molecular structure.

RECENT DEVELOPMENTS

The synthesis of novel nematogenic compounds continues to be an active area of research. Materials with broad nematic ranges, chemical stability, and desired dielectric anisotropy appear to be the ultimate goal of these research efforts. Karamysheva et al. [70] have recently reported the synthesis and mesomorphic properties of a series of cyanophenyl-4-alkylbiphenyl-4-carbonates and cyanophenyl-4-(4-alkylphenyl)cyclohexanecarboxylates. The biphenyl carbonates with an n-pentyl, n-hexyl, or n-heptyl terminal group possess nematic ranges greater than 100°C with crystal → nematic transitions slightly above 100° C. In comparison, the phenylcyclohexane carbonates were found to possess reduced nematic ranges and lower crystal-nematic transition temperatures than the biphenyl analogs. Compounds in both series were found to exhibit a positive dielectric anisotropy.

Pavluchenko et al. [71] have synthesized a series of weakly nematogenic 6-alkoxynicotinic acids with crystal-nematic temperatures below 100°C. The isomeric 5-alkoxypicolinic acids, however, were found not to be mesomorphic. They also report the nematic mesophase behavior of some 2,5- and 2,6-disubstituted benzoxazoles which possess in general small positive dielectric anisotropies.

Gray [72] and Cox [73] have reported the synthesis of a series of 4-cyano-4'-alkyltolanes which were found to be weakly nematogenic. In contrast, the alkyl- and alkoxy-cyanoaryl esters exhibit broad nematic mesophases [74]. Gray and Coates [74] reported the liquid crystal properties and dielectric behavior of 45 cyanoaryl esters where the aryl moiety of the ester was either 1,4-phenylene, 2,6-naphthylene, or 4,4'-biphenylene. All the compounds were found to be nematogenic with a large positive dielectric anisotropy.

Neubert and Maurer [75] have synthesized a homologous series of terephthal-bis-n-alkylanilines with C_1-C_8 and C_{10} alkyl groups. The compounds with C_1-C_7 terminal groups exhibited nematic mesophases with crystal-nematic transition temperatures well above 100° C. De Jeu and Van Der Veen [76] have recently attempted to understand the relationship between molecular structure and nematic-isotropic transition temperature (T_{NIs}). They found that the effect of terminal substituents on T_{NIs} for members of a homologous series could be understood in terms of the change in the anisotropic dispersion forces. The effect of the central group on T_{NIs} could not be explained in these terms, however.

Gray and Coates [77] have described the synthesis and mesomorphic properties of some 4-n-alkylphenyl-2'-chloro-4-(6-n-alkyl-2-naphthoyloxy)-benzoates which possess wide nematic ranges. In contrast, a series of p-n-alkoxybenzylidene-p-bromoanilines with alkoxy groups from C_1 to C_{18} were found not to be nematogenic [78].

THE FUTURE

It is the author's opinion that liquid crystal research activities will continue to increase in intensity. The future of liquid crystal research appears to be tied to technological applications and the realization of its relevance to biological processes. Applications are very much dependent on mesophase physicochemical properties, such as chemical, electrochemical, mesophase, and photochemical stability, as well as their dielectric and diamagnetic properties. The chemist will have to keep in mind the above when seeking new liquid crystalline compounds to synthesize.

An area which is generating considerable interest is the use of the organization within liquid crystal media to direct the course of chemical reactions other than those produced in isotropic solution. The implications of this concept for synthetic chemistry are vast. Polymerization in and on liquid crystal mesophases has provided some encouraging experimental observations [79-80]. The increase in percentage of isotactic component obtained by running the polymerization in a mesophase structure versus an isotropic solvent has been explained on the basis of steric organization of the reactive species within the mesophase [81]. Polymerization of liquid crystalline monomers while in the mesophase structure have been found to yield polymers which retain the morphological structure of the mesophase [82-86]. Barnett [87] and Brown and Bacon [88,89] have reported the effects of mesomorphic solvent on the product ratios and kinetics of unimolecular thermal reactions to be different from isotropic media. These results should be tempered with the report of Dewar and Nohlovsky 90 that the activation parameters for the Claisen rearrangement of cinnamyl phenyl ether show no discontinuity on passing from the nematic mesophase to the isotropic liquid.

Future activity in the liquid crystal field will most certainly continue to provide exciting scientific and technological applications of organized systems. Hopefully a greater insight to some important biological processes will also evolve.

ACKNOWLEDGMENT

Stimulating discussions with Dr. H.W. Gibson and J.M. Pollack are gratefully acknowledged.

REFERENCES

1. W. Maier and A. Saupe, Z. Naturf., A, 15:287 (1960).
2. M.J.S. Dewar, J.P. Schroeder, and D.C. Schroeder, J. Org. Chem., 32:1692 (1967).
3. A. de Vries, Acta Crystallogr., A25:S135 (1969).
4. A. de Vries, Mol. Cryst. Liq. Cryst., 10:31 (1970).
5. A. de Vries, Mol. Cryst. Liq. Cryst., 10:219 (1970).
6. P. Chatelain, Acta Crystallogr., 4:453 (1951).
7. P. Chatelain, Bull. Soc. Fr. Mineral. Cristallogr., 77:323 (1954).
8. G.D. Dixon, T.P. Brody, and W.A. Hester, Appl. Phys. Lett., 24:47 (1974).
9. D.W. Berreman, Mol. Cryst. Liq. Cryst., 23:215 (1973).
10. J.L. Janning, Appl. Phys. Lett., 4:173 (1972).
11. E. Guyon, P. Pieranski and M. Boix, Appl. Eng. Sci. Lett., 1:19 (1973).
12. L.T. Creagh and A.R. Kmetz, Mol. Cryst. Liq. Cryst., 24:59 (1973).
13. J.C. Dubois, M. Gazard, and A. Zann, Appl. Phys. Lett., 24:297 (1974).

14. I.G. Chistyakov, Rost. Kristallov., Akad. Nauk SSSR, 4:74 (1964).
15. M. Schadt and W. Helfrich, Appl. Phys. Lett., 18:127 (1971).
16. T.J. Scheffer, J. Appl. Phys., 44:4799 (1974).
17. V.A. Tsvetkov, N.A. Morozov, and L.K. Vistin, Sov. Phys. Cryst., 19:185 (1974).
18. E.P. Raynes, Electronic Lett., 9:101 (1973).
19. D.W. Berreman, J. Opt. Soc. Amer., 63:1374 (1973).
20. F.D. Saeva and G.R. Olin, J. Amer. Chem. Soc., 98:2709 (1976).
21. G.W. Gray, Molecular Structure and Properties of Liquid Crystals, Academic, London, 1962.
22. V. Freedericksz and V. Zolina, Trans. Faraday Soc., 29:919 (1933).
23. G.H. Heilmeier and L. Zanoni, Appl. Phys. Lett., 13:91 (1968).
24. G.H. Heilmeier, J.A. Castellano, and L.A. Zanoni, Mol. Cryst. Liq. Cryst., 8:293 (1969).
25. R. Williams, J. Chem. Phys., 39:384 (1963).
26. R. Williams, Nature (London), 199:273 (1963).
27. W. Helfrich, J. Chem. Phys., 51:4092 (1969).
28. L.K. Vistin, Kristallografiya, (Russian) 15:594 (1970) also Sov. Phys. Cryst., (English transl.), 15:514 (1970).
29. W. Greubel and U. Wolff, Appl. Phys. Lett., 19:213 (1971).
30. P.A. Penz and G.W. Ford, Phys. Rev., A6:414 (1972).
31. J.M. Pollack and J.B. Flannery, in Liquid Crystals and Ordered Fluids, Vol. II, (J.F. Johnson and R.S. Porter, eds.), Plenum, New York, 1974, p. 557.
32. G.H. Heilmeier, L. Zanoni, and L. Barton, Proc. IEEE, 56:1162 (1968).
33. G.W. Gray, Molecular Structure and Properties of Liquid Crystals, Academic, New York, 1962.
34. M.J.S. Dewar and R.M. Riddle, J. Amer. Chem. Soc., 97:6658 (1975).
35. J.A. Castellano, J.E. Goldmacher, L.A. Barton, and J.S. Kane, J. Org. Chem., 33:3501 (1968).
36. S.L. Arora, T.R. Taylor, and J.L. Fergason, J. Org. Chem., 35:1705 (1970).
37. W.R. Young, Mol. Cryst. Liq. Cryst., 10:237 (1970).
38. J.P. Schroeder and D.C. Schroeder, J. Org. Chem., 33:591 (1968).
39. H. Kelker and B. Scheurle, Angew. Chem., 81:903 (1969).

40. H. Kelker, B. Scheurle, R. Hatz, and W. Bartsch, Angew. Chem., 82:984 (1970).
41. J.B. Flannery and W. Haas, J. Phys. Chem., 74:3611 (1970).
42. D.L. Fishel and Y.Y. Hsu, Chem. Commun., 1557 (1971).
43. W.R. Young, A. Aviram, and R.J. Cox, J. Amer. Chem. Soc., 94:3976 (1971).
44. W.R. Young, A. Aviram, and R.J. Cox, Angew. Chem. (Int. Ed.), 10:410 (1971).
45. L.E. Knaak and H.M. Rosenberg, Mol. Cryst. Liq. Cryst., 17:171 (1972).
46. H. Hirata, S.N. Waxman, I. Teucher, and M.M. Labes, Mol. Cryst. Liq. Cryst., 20:343 (1973).
47. L. Verbit and R.L. Tuggey, Mol. Cryst. Liq. Cryst., 17:49 (1972).
48. F. Jones and J. Ratto, J. Chem. Soc. Chem. Commun., 841 (1973).
49. J.P. Van Meter and B.H. Klanderman, Mol. Cryst. Liq. Cryst., 32:271 (1973).
50. R. Steinstrasser, Angew. Chem., 84:636 (1972).
51. D.C. Schroeder and J.P. Schroeder, J. Amer. Chem. Soc., 96:4347 (1974).
52. W.R. Young and D.C. Green, Mol. Cryst. Liq. Cryst., 26:7 (1974).
53. J.P. Van Meter and B.H. Klanderman, Mol. Cryst. Liq. Cryst., 32:285 (1973).
54. J.P. Van Meter and B.H. Klanderman, J. Amer. Chem. Soc., 95:626 (1973).
55. L. Verbit, R.L. Tuggey, and A.R. Pinhas, Mol. Cryst. Liq. Cryst., 30:201 (1975).
56. G.W. Gray, K.J. Harrison, and J.A. Nash, J. Chem. Soc. Chem. Commun., 431 (1974).
57. G.W. Gray, J. Phys. (Paris), 1:337 (1975).
58. A.R. Kmetz, IEEE Trans., ED-20:954 (1973).
59. A. Sussman, IEEE Trans., PHP-8:24 (1972).
60. T.J. Scheffer, J. Appl. Phys., 44:4799 (1973).
61. H. Kelker, Ber. Bunsenges. Phys. Chem., 67:698 (1963); Z. Anal. Chem., 198:254 (1963).

62. M.J.S. Dewar and J.P. Schroeder, J. Amer. Chem. Soc., 86:5235 (1964).
63. L.E. Cook and R.C. Spangelo, Anal. Chem., 46:122 (1974).
64. G.P. Ceasar and H.B. Gray, J. Amer. Chem. Soc., 91:191 (1969).
65. A. Saupe, Mol. Cryst. Liq. Cryst., 1:3 (1972).
66. E.D. Cehelnik, R.B. Cundall, and C.J. Timmons, Proc. R. Soc., Ser. A., 335:387 (1973).
67. S. Sakagami, A. Takase, M. Nakamizo, and H. Kayiyama, Bull. Chem. Soc. Jap., 46:2062 (1973).
68. D.J. Channin, Appl. Phys. Lett., 22:365 (1973).
69. B.L. Bales, J.A. Swenson, and R.N. Schwartz, Mol. Cryst. Liq. Cryst., 28:143 (1975).
70. L.A. Karamysheva, E.I. Kovshev, and M.I. Barnik, Mol. Cryst. Liq. Cryst., 37:29 (1976).
71. A.I. Pavluchenko, N.I. Smirnova, V.V. Titov, E.I. Kovshev, and K.M. Djumaev, Mol. Cryst. Liq. Cryst., 37:35 (1976).
72. G.W. Gray and A. Mosley, Mol. Cryst. Liq Cryst., 37:213 (1976).
73. R.J. Cox and N.J. Clecak, Mol. Cryst. Liq. Cryst., 37:241 (1976).
74. D. Coates and G.W. Gray, Mol. Cryst. Liq. Cryst., 37:249 (1976).
75. M.E. Neubert and L.J. Maurer, Mol. Cryst. Liq. Cryst., 43:313 (1977).
76. W.H. De Jeu and J. Van Der Veen, Mol. Cryst. Liq. Cryst., 40:1 (1977).
77. D. Coates and G.W. Gray, Mol. Cryst. Liq. Cryst. Lett., 41:119 (1978).
78. R.A. Vora, Mol. Cryst. Liq. Cryst., 44:83 (1978).
79. Y.B. Amerik, I.I. Konstantinov, B.A. Krentzel, and E.M. Malachaev, Vysokomol. Soedin., A9(19):2591 (1967).
80. A. Blumstein, in Advances in Macromolecular Chemistry, Vol. 2, (W.M. Pasika, ed.), Academic, New York, 1970.
81. A. Blumstein, N. Kitagawa, and R. Blumstein, Mol. Cryst. Liq. Cryst., 12:215 (1971).
82. E. Perplies, H. Ringsdorf, and J.H. Wendorff, Makromol. Chem., 175:553 (1974).
83. E. Perplies, H. Ringsdorf, and J.H. Wendorff, Ber. Bunsenges. Phys. Chem., 59:861 (1955).

84. A. Blumstein, R.B. Blumstein, S.B. Clough, and E.C. Hsu, Macromolecules, 8:73 (1975).
85. L. Strzelecki and L. Liebert, Bull. Soc. Chim. Fra., 2:597, 603,605 (1973).
86. L. Liebert and L. Strzelecki, C.R. Acad. Sci., Paris, Ser. A.B, 647 (1973).
87. W.E. Barnett and W.H. Sohn, Chem. Commun., 1002 (1971).
88. W.E. Bacon and G.H. Brown, Mol. Cryst. Liq. Cryst., 6:155 (1969).
89. W.E. Bacon and G.H. Brown, Mol. Cryst. Liq. Cryst., 12:229 (1971).
90. M.J.S. Dewar and B.D. Nahlovsky, J. Amer. Chem. Soc., 96:460 (1974).

Chapter 3

CHOLESTERIC MESOPHASES

Harry W. Gibson

Xerox Corporation
Webster Research Center
Webster, New York

I. INTRODUCTION

The history of cholesteric liquid crystals coincides with that of the liquid crystal field inasmuch as this class was the first to be discovered. In 1888, Reinitzer reported the synthesis and thermal behavior of a series of cholesteryl esters (1) [1]. Among these are the first recognized examples of liquid crystals, specifically cholesteric liquid crystals. The term cholesteric arises because of this original association of cholesterol derivatives with this mesophase. As we shall see, however, the mesophase is by no means restricted to members of the cholesterol family. In this

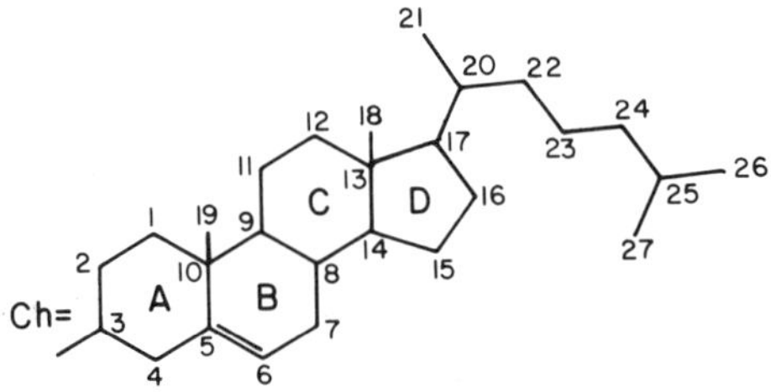

ChOOCR

I

chapter, an attempt will be made to summarize the major findings in this field, though by no means will this constitute an exhaustive coverage of this burgeoning area. Work done in the past 10-15 years, which coincides with the exponential growth of interest in liquid crystals, will be emphasized at the expense of earlier work, which has been adequately reviewed [2]. This is reflected by the fact that 104 of the 174 individual literature citations are for after 1969. The author's bias toward some areas and ignorance in others will no doubt be reflected, and apologies are made to those contributors whose work is not cited.

II. STRUCTURE

Cholesteric mesophases are described as being comprised of helical aggregates of molecules. The helical array arises from a chiral perturbation [3], usually the presence of a center of optical activity in the molecules that comprise the mesophase. This chiral perturbation minimizes relative to other arrangements the energy content of a stacked array in which each molecule is skewed at some average angle θ with respect to its neighbors above and below. Of course, two such helices, one right-handed and the other left-handed (Fig. 1), are possible, but one is of lower energy than the other because of the chiral nature of the perturbation.

To quantify the degree of twist, the pitch P of the helix is defined as the longitudinal distance in which the angle of twist has made one complete 360° revolution (Fig. 1). Thus, the pitch of a helix is analogous to the pitch of a screw. It is measured optically (see Sec. IV). The optical experiments also allow one to assign the handedness (chirality) of the helix.

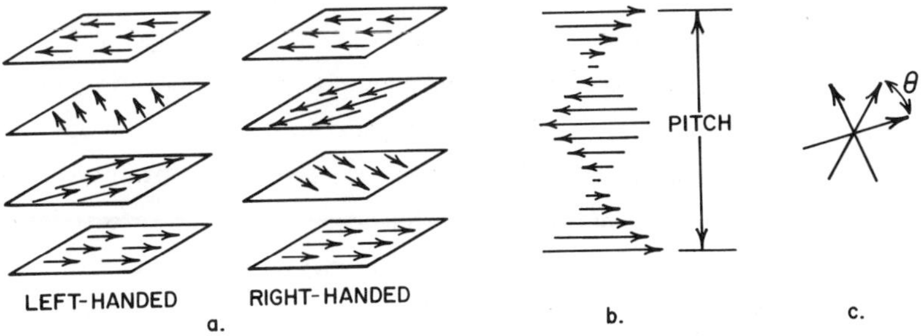

FIG. 1. Left- and right-handed helical structures showing pitch P and displacement angle θ.

The cholesteric mesophase has been described as a nematic planar structure in which there is a twist angle θ from one layer to the next. Because of this, the cholesteric mesophase is often called a twisted nematic mesophase [4-6].

It has been suggested that the cholesteric mesophase is a special case of the nematic structure, which also consists of helical aggregates of molecules, but since the two helical forms are of equal energy, there is no net chiral influence [3]. The net result is that both forms are present in equal amounts, and the nematic mesophase is structurally the same as the compensated cholesteric. Compensated cholesterics result when right- and left-handed cholesterics are combined to produce a mixture with no net handedness, i.e., infinite pitch, at a given temperature. This situation [7], then, is precisely analogous to that which exists with chiral single molecules, where a 1:1 mixture of enantiomers (nonsuperimposable mirror image stereoisomers) possesses no net optical activity, but the individual enantiomers possess equal but oppositely signed optical activity [8]. Thus, for example, it has long been known that addition of low concentrations of optically active substance to nematics results in conversion to cholesteric mesophases [3,7,9-18]. The inverse pitch is a linear function of the concentration at low concentrations of the optically active compound. In this situation the optically active compound is the chiral perturbation which lowers the energy of one of the two helical forms relative to the other.

Additionally, it has been demonstrated that incorporation of a resolved chiral center into the molecules comprising the nematic mesophase also gives rise to a cholesteric liquid crystal [7,15-30]. Perhaps the most vivid demonstration of this phenomenon is the result obtained with compound 2 [30]. This compound was prepared and resolved into its two enantiomers. While the racemic compound (the mixture of equal amounts of the two enantiomers) is nematic, the resolved enantiomers are cholesteric, as indicated by the optical properties [30]. This is a particularly striking example because the chirality derives only from replacement of a hydrogen atom by a deuterium atom; these two atoms differ only slightly in size, but, of course, deuterium has two neutrons in its core whereas hydrogen has only one. Yet this chiral perturbation is sufficient to generate a helical (cholesteric) structure in the mesophase. This has been used as support for the idea that force fields and not steric interactions are the controlling factor in the cholesteric mesophase [30].

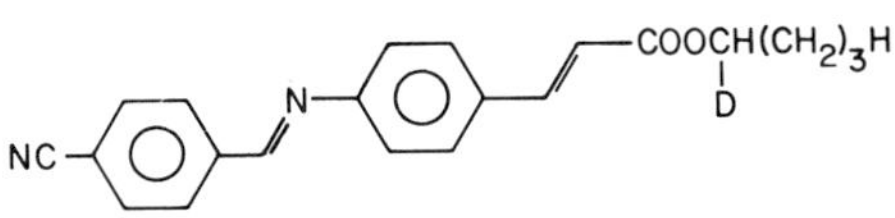

2

Yet another chiral influence that can swing the energy balance from the nematic mesophase in favor of the cholesteric is a purely mechanical perturbation arising from interfacial interactions. This is the situation in nematic mesophases that are placed between glass slides which have been rubbed unidirectionally and placed at an angle to one another [31,32]. This old phenomenon [31] has been rediscovered [32] and forms the basis of some new applications (see Chap. 12). The molecules at the glass surfaces are aligned in the direction of rubbing [33-36]. These layers align the adjacent layers, and so on, so that the given angle is transversed once in the mesophase. This results in a helical organization of the mesophase and confers upon it the cholesteric structure [31,32].

There are two distinct ways the experiment can be done. The first is to place the glass slides at the desired angle in the first place [32]. The second is to place the glass slides at 0° relative angle with respect to each other and mechanically twist them through a range of angles [31]. The second method encompasses the chiral mechanical perturbation as well as the surface interactions. Hence, it should lead to a greater initial metastable preference for one-handedness than the first method, which is based only on surface effects.

The close relationship between the nematic and cholesteric mesophases is also evidenced by the miscibility criterion, which states that when two liquid crystals are miscible in all proportions they belong to the same class (and subclass) of mesophase [37]. Nematics and cholesterics are miscible and on this basis are members of the same class of mesophase. Moreover, the two phases have never been found in coexistence in a single pure compound. Thus, the difference between them may be viewed as a matter of degree of chirality, the nematic being the special case of no net chirality. And furthermore, as we shall see, it is possible to convert the cholesteric mesophase to a nematic structure by means of electric and magnetic fields.

Modern theoretical treatments developed initially for nematic structures have been applied to the cholesteric mesophase. These modern treatments grew out of earlier efforts to apply continuum theory [38-40]. The underlying assumption is that for every point in the mesophase there is a preferred orientation of the long molecular axis and this preferred orientation varies continuously throughout the mesophase. Of the two somewhat different theories evolved, one is based upon micropolar continuum mechanics [41] and the other on anisotropic fluid theory [42,43]. The two treatments have been compared in detail, and the salient differences in fundamental assumptions have been delineated [44].

The cholesteric mesophase can exist in three different textures. These differ only in the degree and direction of ordering relative to the observer. If the helices are at a 90° angle to the light observation path, the focal conic, or undisturbed (since it forms upon cooling the isotropic liquid without perturbation) texture results. This texture is birefringent but optically inactive. Though the helical axes are normal to the light path, they are arranged within that plane in a focal conic manner, along the axes of elipses and hyperbolae [2,6]. In the Grandjean (planar or disturbed) texture, which forms from the focal conic by mechanical shear, the helices are more or less all aligned with their axes parallel to the light path [2,6]. This is the texture that most strikingly exhibits the unique optical effects of the mesophase, the iridescence, optical activity, circular dichroism, etc. (see below). The orientations are shown schematically in Fig. 2. A third texture is the homeotropic state, which often, but not always, intervenes prior to the focal conic texture upon cooling the isotropic liquid. This texture is transparent, has only very slight birefringence, and no optical activity. The orientation of this structure is not well understood [2].

Recently, through use of a very clever experimental technique, a fourth texture has been produced. Application of pressure to the cholesteric mesophase contained in a narrow capillary produces a cylindrically symmetric texture, in which the helical axes are radially oriented; there is also a sense of twist within each cylindrical surface, but this twist varies from one such surface to another [45]. Optical studies allow direct determination of the birefringence [45].

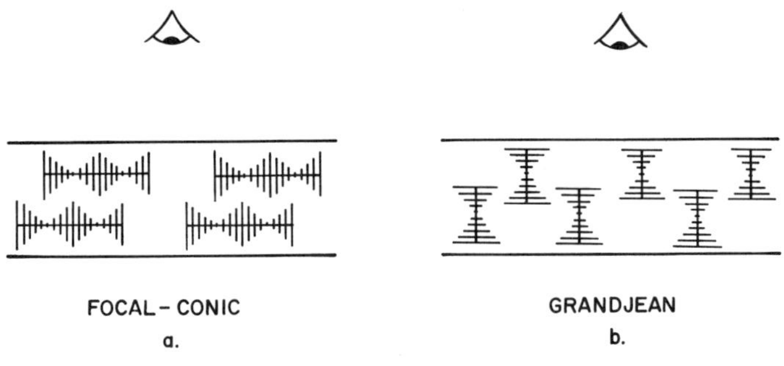

FIG. 2. Orientations of helical (long lines) and molecular axes (short lines) relative to observer for (a) focal conic texture and (b) Grandjean texture.

III. PHASE CHANGES

There are two types of cholesteric systems: thermotropic and lyotropic. As the name implies, thermotropic mesophases form when a solid is heated to the melt. If the mesophase forms directly it is called enantiotropic; monotropic mesophases are those that form only upon supercooling the isotropic liquid. On the other hand, lyotropic systems are the result of solvent interaction with the molecules. In the main, lyotropic cholesteric systems are polymeric and of biological origin. These are discussed later in Chap. 5 and 8 and will not be described here.

Cholesteric compounds may exhibit smectic mesophases as well as the cholesteric mesophases; in fact, this is quite common. In these cases the cholesteric mesophase exists at higher temperatures than the smectic mesophase. Since it has a higher energy content, the cholesteric mesophase possesses a lower degree of order than the smectic mesophase.

The conversions from the crystalline or isotropic liquid phases to the liquid crystalline state are true phase changes, as are the interconversions of the smectic and cholesteric mesophases. These transformations are all completely thermally reversible. In the case of the mesophase-to-crystal transition, supercooling is possible and is usually observed. The transitions between the other phases (smectic, cholesteric, isotropic liquid) cannot be supercooled.

The cholesteric and nematic mesophases cannot coexist in a single composition (without the intervention of external electric or magnetic fields, Sec. V). As we have seen, the structures of the two are qualitatively the same. Mixing of the two phases gives rise to a cholesteric mesophase due to the chiral nature of the cholesteric component.

Quantitative thermal measurements by differential scanning calorimetry have been made on a variety of cholesteric liquid crystals. It is usually possible to discern the identity of the mesophase from such measurements. Generally, it has been found that the enthalpy associated with the cholesteric-to-isotropic transition is 2-3% of the total enthalpy for the crystal-to-isotropic liquid [46,47]. In contrast, the enthalpy change (ΔH) involved in the conversion of a smectic mesophase to an isotropic liquid is 10-15% of the ΔH for the crystal-to-smectic transition [46,47]. This difference is a consequence of the higher heat content of the cholesteric mesophase relative to the smectic. This thermal technique can thus serve as a diagnostic when only a single mesophase is present.

Supercooling of the cholesteric mesophase below the melting point produces a mestastable system. Some liquid crystals form glasses [48-51], some of which are reportedly stable for years. This stability is due apparently to the high viscosity, which retards the reorganization necessary for crystallization [50,51].

Nucleation theory derived for condensation of liquids from the gas phase and crystallization from isotropic liquids has been successfully applied to crystal nucleation [52] and growth [53] from the cholesteric mesophase. Interfacial energies for the nucleation process are an order of magnitude less than such values for crystallization from supercooled isotropic phases. Direct correlation of nucleation rates with structure does not seem to exist due to offsetting contributions from the individual thermodynamic parameters [52]. However, the interfacial energies, those associated with the interface between mesophase and crystal, are correlated with structure and also with pitch [54].

The formation of the cholesteric mesophase from the isotropic liquid has been studied in detail [55,56]. Also, the conversion of the initially formed homeotropic texture to a focal conic state has been carefully examined. The formation of the homeotropic state from the isotropic liquid involves disklike nuclei that grow in a rodlike fashion [55]. The conversion to the focal conic state corresponds to an increase in the number of focal conic domains and not the number of helices in each domain [56].

IV. OPTICAL PROPERTIES

The longitudinal dimension of the helical aggregates of the cholesteric mesophase is on the order of the wavelength of visible light. This is the source of the unique optical properties of this phase, more specifically those associated with the Grandjean or planar texture.

The first consequence of this structure is the extremely large optical activity, leading to optical rotations in the range of 10^3-10^5 degrees per millimeter for visible light [4,7,58-65] (Fig. 3a). This is enormous in comparison to the optical rotations of 10^{-2}-10^2 degrees per millimeter of optically active isotropic liquids [8].

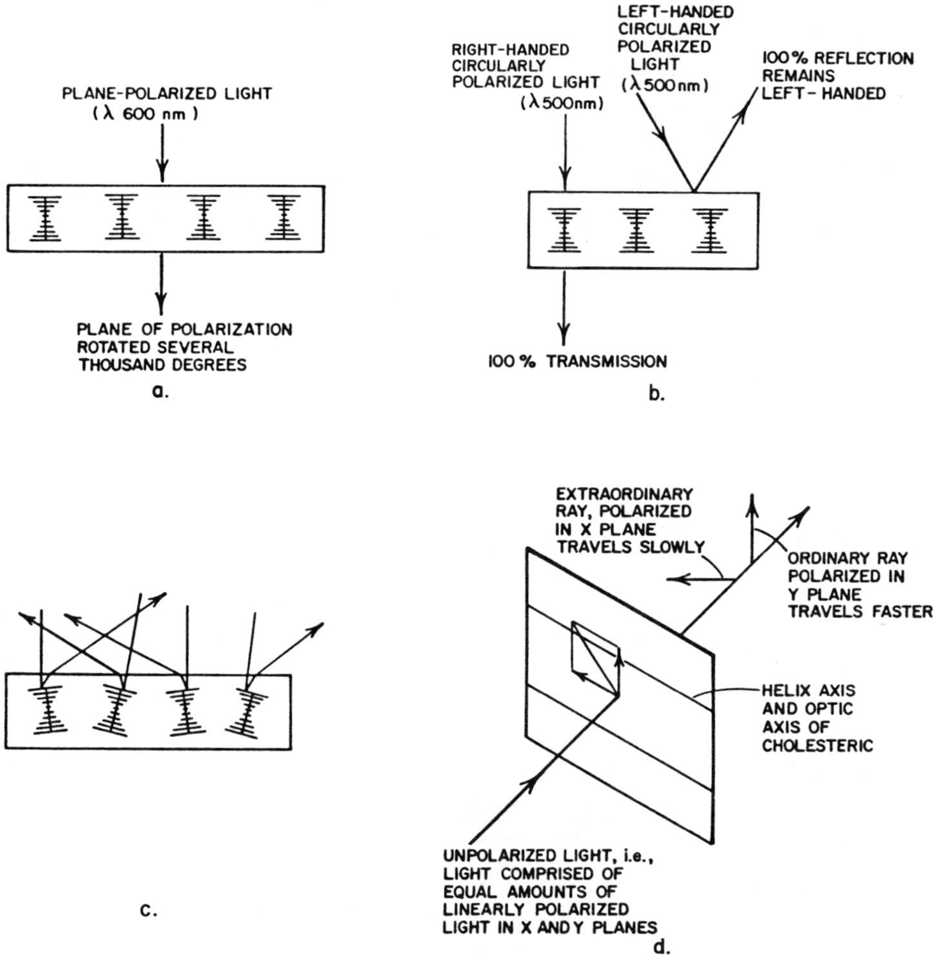

FIG. 3. Optical properties of a hypothetical cholesteric mesophase with right-handed helical structure and pitch corresponding to a reflective wavelength λ_o = 500 nm; (a) rotation of plane-polarized incident light, (b) circular dichroism, (c) scattering of nonnormal incident light, (d) definition of negative optic axis relative to helical and molecular axes.

The second optical property characteristic of the helical organization is the circular dichroism effect. Incident light is resolved into its two circularly polarized components, left and right, at a given wavelength [7,66] (Fig. 3b). The wavelength depends upon (1) the pitch of the material, which in turn is a function of the constitution and temperature of the mesophase, and (2) the angle of incidence [7,66]. At this wavelength, one handedness of circularly polarized light is completely reflected while its complement is completely transmitted [7,66]. A truly unique feature of this phenomenon is that the polarization of neither handedness is altered. With normal reflective surfaces the handedness of circularly polarized light is reversed upon reflection, whereas with cholesterics it retains its original chirality.

There has been some confusion in the literature as to the definition of handedness of helical arrays. This arises from the fact that two conventions are employed for the description, one by the chemists and one by the physicists. As the situation has evolved cholesteric liquid crystals have been labeled left- or right-handed on the basis of the sense of the reflected light. Thus, a liquid crystal whose helix sense is right-handed, such as the hypothetical case of Fig. 3, is defined as left-handed. An effort will be made in this chapter to refer to the handedness of the reflected light. By this definition most steroidal esters are right-handed. The cholesteryl halides and 5α-cholest-8-(14)-en-3β-ol derivatives are left-handed (see below). The helix sense (Fig. 1) is defined by examining the sense of the circle described by a given end of the molecule in a direction toward the observer; Fig. 1 then gives the chirality of the helices depicted and not the chirality of the reflected circularly polarized light.

The so-called reflective wavelength λ_o is related to the pitch of the cholesteric helix at normal incidence by the equation $\lambda_o = nP$, where n is the mean refractive index [66]. This affords a direct means of determination of the pitch P; λ_o can be measured spectrophotometrically since at that point only 50% of normal unpolarized light is transmitted. In many cases, the reflective wavelength or pitch band can be observed visually.

The third optical property that distinguishes the cholesteric mesophase is the Bragg-type scattering of incident white light by the layered structure of the mesophase to produce characteristic iridescent colors [7,66] (Fig. 3c). The colors vary with the nature of the sample, temperature, and angles of incidence and observation.

The fourth optical characteristic that differentiates cholesteric liquid crystals from those of smectics and nematics is that the structure is uniaxial and optically negative. This means that the refractive index is a minimum along the optic axis, which is normal to the local optic axes of the nematic layers and to the molecular axes (Fig. 3d).

The oft-quoted de Vries equation and derived theory [31,67-69] interrelate these phenomena and describe the experimental observations very well [15,63,68,69]. The de Vries equation is discussed in more detail in Chap. 1, along with the effect of the mesophase on the absorption properties of solutes. Spectroscopic studies of the mesophase itself using infrared dichroism are described in Chap. 10.

The relationship between pitch and composition of mixtures is a subject of both practical and theoretical interest. Early work on mixtures of two cholesteric compounds showed a linearly additive effect of the "helical twisting power" or inverse pitch of the two components on a weight percent basis [70]. This was later extended to more complex mixtures [71]. However, exceptions do occur [5,72-76]; for example, many systems comprised of cholesteric and nematic compounds do not obey such a linear equation [5,72-76]. To treat this situation, terms ascribed to interaction of the components have been added. This treatment has been found to describe binary and ternary systems satisfactorily [77,78]; again, this set of authors has favored the use of weight percentages [72,78]. It is also necessary to ascribe a helical twisting power to the achiral nematic [77-79].

Another recent treatment of mixtures of cholesteric and nematic compounds invokes the formation of aggregates through association; both the number of aggregates and the ratio of the two components that make up the aggregate depend upon the relative numbers of molecules [80].

This treatment, then, is based upon mole percentages rather than weight percentages. Inherent in this hypothesis is the variation of pitch as a function of the nature of the aggregates.

Thus, both treatments invoke the idea that interaction of molecules A and B need not lead to a pitch intermediate between A-A and B-B.

Examples of such cases have been reported for binary mixtures of cholesteryl esters (1) and aromatic anils [5,74,76]. In these systems the pitch goes through a minimum as a function of composition. A noteworthy example is provided by n-amyl p-(4-cyanobenzylideneamino)cinnamate (CBAC) (3) and cholesteryl 2-(ethoxyethoxy)ethyl carbonate (CEEC) (4). A plot of pitch versus composition for this system comprised of two right-handed components not only possesses a minimum but also contains two compensation points, due to the fact that from 25-90 mol% CEEC the reflected light is left-handed (denoted by negative pitches) [74,76]! These results are rationalized on a mole fraction basis in terms of an extension of Goosen's theory of dispersion energies [12] which treats twist in terms of dipole-dipole and dipole-quadrapole interactions [76]. Good agreement between theoretically predicted and empirical parameters is found, even for the CEEC-CBAC system [76]. The theory does not require that the achiral nematic systems possess an inherent handedness. This is an intuitively and theoretically desirable feature, since such compounds in the absence of chiral perturbation have no net rotatory power.

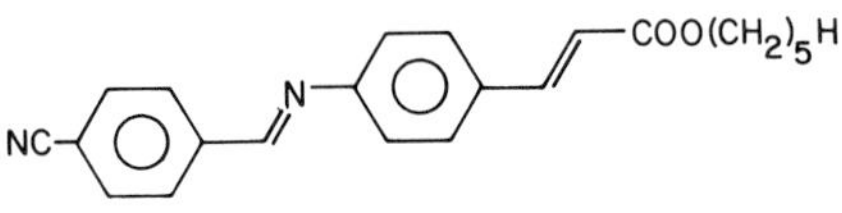

3

$ChOOC(OCH_2CH_2)_3H$

4

There is, however, one disturbing feature of this combination of experiment and theory. The theoretical equations implicitly assume constant temperature. Yet, the pitches were all measured at temperatures 10° C below the cholesteric isotropic temperature in each individual sample [74,76], presumably to avoid the intervention of the smectic phase (see below). Thus, the experimental results contain a temperature-dependent pitch contribution, while the theory is for constant temperature pitch variation with composition. An earlier attempt to rationalize the pitch as a function of composition, which seems to possess a minimum even though both components [cholesteryl propionate (5) and MBBA (6) are left-handed, on the basis of Goosen's theory [12] contains the same flaw [5]. In view of the rather large dependence of pitch on temperature often observed (see below), the data in these two publications is probably not solely due to composition variation. In fairness, it must be said that this problem is not easily surmounted since in most cases the entire composition range does not have a common temperature for the cholesteric phase. The application of Goosen's theory to composition dependence nonetheless has not yet been explicitly verified.

$ChOO(CH_2)_2H$

5

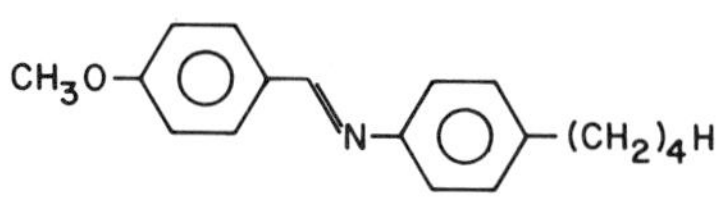

6

The mathematical formalism of Goosen's theory [12] and the concept of molecular aggregation [80] seem to be equivalent, the latter providing a physically visualizable representation of the former. Thus, interaction of two right-handed species to give a left-handed system is the result of new interactions, which lead to a new molecular association pattern.

The point should be made, then, that no molecule has an intrinsic twisting power; the molecule has certain conformational possibilities and electrostatic properties. The nature of the helical aggregate results from minimization of the total energy of the system, i.e., the molecule and its environment. In different environments, different pitches will result because of varying interactions. Of course, for similar molecules such changes will tend to be small. Thus, while the concept was useful in the past and is philosophically attractive, the idea of intrinsic twisting power is probably best discarded so that attention can be focused on the energy minimization process involved in the intermolecular interactions.

Within the temperature regime of the cholesteric mesophase, the structure of the mesophase is responsive to temperature variations. As the temperature is raised there is more thermal motion in the molecules comprising the mesophase. As a result, their molar volume increases. Each molecule has a larger volume requirement. This can affect the helical structure in two opposing ways: (1) the intermolecular distance along the helix axis increases, tending to increase the pitch; and (2) the displacement angle θ between adjacent molecules in the helical stack increases, tending to decrease the pitch (Fig. 4). Mechanism 2 has been treated theoretically on the basis of anharmonic intermolecular forces [81].

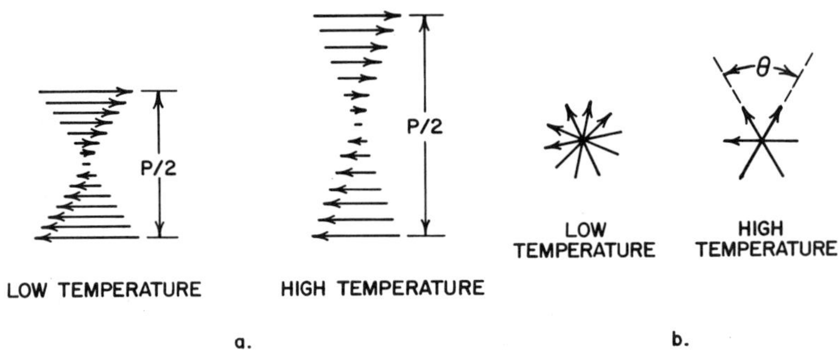

FIG. 4. Possible effects of temperature on pitch: (a) alteration of intermolecular spacing resulting in an increase of pitch with temperature, (b) alteration of displacement angle θ resulting in a decrease of pitch with temperature.

In pure single cholesteryl compounds, the pitch always decreases with increasing temperature [2,57,58,82]. This is also the case with the chiral nematics mixtures studied to date [29], though the effect is very small. Mechanism 2 governing the displacement angle θ therefore dominates here over the longitudinal helix expansion mode 1. It has been noted that only a 7-min change in the displacement angle θ causes the pitch band to scan the whole visible spectrum [66].

In all mixtures comprised of components of the same handedness, the pitch decreases with increasing temperature [66]; a report to the contrary [83] has been shown to be due to the presence of impurities of opposite handedness [58].

However, some systems containing cholesteric compounds of opposite handedness and mixtures of cholesterics and nematics show anomalous temperature effects in that the pitch increases with temperature. An example of such a case is provided by the cholesteryl chloride (7)-cholesteryl nonanoate (8) systems with more than 25% chloride [66,75]. For some compositions of cholesteryl propionate (5) in the nematic p-methoxybenzylidene-p-(n-butyl)aniline (MBBA) (6) the pitch increases with temperature, while with others it decreases; the handedness does not change [5].

ChCl

7

$ChOO(CH_2)_8H$

8

In cases involving mixtures of right- and left-handed cholesterics the pitch increases to infinity at the compensation or nematic temperature, denoted by T_N, and then decreases again with the opposite handedness [7], e.g., Fig. 5a. A 36 wt% mixture of cholesteryl myristate (9) in cholesteryl chloride (7) displays such an effect [84,61], as do ternary mixtures derived from it [10,60,64,85]. Thus, anomalous temperature dependence, i.e., pitch increase with temperature, is observed in only two distinct cases: (1) mixtures of a cholesteric compound with a nematic compound and (2) mixtures of right- and left-handed cholesterics.

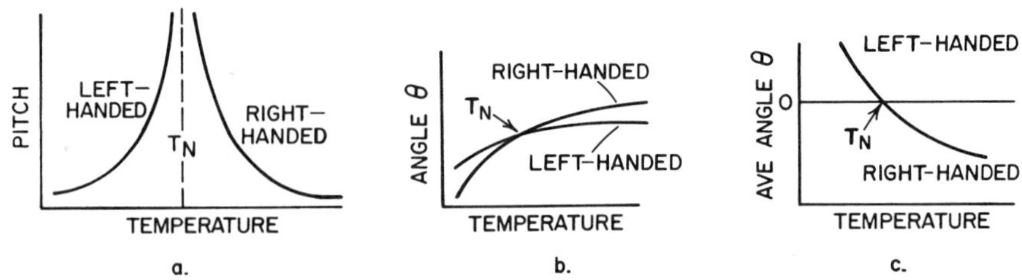

FIG. 5. Effect of temperature on the pitch of a hypothetical cholesteric mixture comprised of equimolar amounts of right- and left-handed components: (a) dependence of overall pitch on temperature, (b) temperature dependence of displacement angle θ for the right- and left-handed components, (c) temperature dependence of average overall displacement angle θ resulting from situation in (b).

$ChOOC(CH_2)_{13}H$

9

The explanation of the first case relates to the dependence of pitch on composition [80] discussed in the previous section. In that model the pitch is considered dependent upon the nature of the associated aggregate. The stoichiometry of the associated aggregate is dependent upon the composition of the mixture. In such a dynamic situation an equilibrium between various aggregates would presumably exist; therefore, the distribution of aggregates would be expected to be temperature dependent. In this general way the pitch would be expected to be temperature dependent. As the temperature is raised, the number of nematic molecules helically aggregated by the cholesteric would be expected to decrease. This would tend to increase the pitch. On the other hand, the direct effect of temperature on the helical aggregates would be a decreased pitch through increased displacement angle. The net result of these opposing effects would be expected to be composition dependent, as observed, the pitch increasing with temperature for some compositions and decreasing for others.

The second case is more simply rationalized. The only effect of temperature is to increase the displacement angle θ between adjacent molecules in the helix (Fig. 4b). The temperature dependences of the vibrational amplitudes of the two different types of constituent molecules are different. Hence, one contributor, say the right-handed one, undergoes a larger volume change upon heating; its displacement angle relative to its neighbors in the helix will therefore increase more sharply than that of the left-handed compound (Fig. 5b). The resultant average angle along the helix may therefore, in certain composition ranges with certain compounds, initially decrease to zero and then increase as the temperature is raised (Fig. 5c). The net effect, then, on the pitch is to increase to the compensation point (infinite pitch) at T_N, change handedness from left to right, and then decrease (Fig. 5a). This very simple explanation seems never to have been put forward in the literature.

A caveat in the study of pitch as a function of binary mixture composition and temperature is the possible intervention of the smectic mesophase, which causes an increase in pitch. In mixtures of cholesteryl nonanoate (8) and cholesteryl chloride (7), a transition to a smectic phase takes place within a certain range of composition and this results in nonlinear plots of inverse pitch versus composition at 22° C. At 30° C, inverse pitch is linear with composition, since there is no smectic phase. The size of the smectic clusters can be estimated [75].

The pitch and associated optical characteristics of the cholesteric mesophase are also functions of pressure. Little work has been done in this area, however. The pitch of cholesteryl oleyl carbonate (10)-cholesteryl chloride (7) mixtures increases sharply up to 1.5 kbars [86]. A priori one might expect a decrease in pitch with pressure on the grounds of compression of the molecules of the helices, thus decreasing the intermolecular separation, an effect analogous to lowering the temperature (Fig. 4a). On the other hand, it can be argued that increased viscosity due to pressure would lead to an increased pitch by dampening of vibrational modes of the molecules [86], thus reducing the angular displacement between adjacent

$ChOOC(CH_2)_8CH{=}CH(CH_2)_8H$

10

molecules along the helix axis in a manner analogous to lowering the temperature (Fig. 4b). The experimental results [87] indicate that the latter mechanism is dominant. This is then consistent with the effect of temperature and indicates that the angular displacement is more susceptible to change than intermolecular spacing.

The effect of mechanical shear on the cholesteric mesophase is related to these pressure and viscosity effects. Though this area has only recently been scrutinized, a rather large body of data has been accumulated. For a detailed discussion of this subject see Chap. 7 .

V. ELECTRIC AND MAGNETIC PROPERTIES

The molecules that comprise cholesteric mesophases possess inhomogeneous electron distributions; that is, they have dipole moments. These dipole moments are vectors whose orientation is dependent on molecular structure. As a consequence, the electric and magnetic susceptibilities are anisotropic. The dielectric anisotropy is defined as positive when the dielectric constant parallel to the long molecular ($\varepsilon_{||}$) axis exceeds that perpendicular to this axis ($\varepsilon_{\perp}$); when the reverse situation exists it is called negative dielectric anisotropy. The diamagnetic anisotropy is defined analogously based on the diamagnetic susceptibilities $\chi_{||}$ and $\chi_{\perp}$, which derive from the orbital motion of electrons. Cholesteric systems of both signs of dielectric [88,89] and apparently diamagnetic susceptibilities [17,84] are known.

The electric field manipulation of cholesteric mesophases is generally more readily achieved than magnetic field effects. Reorientation of the molecules comprising the mesophase in such a way as to alter the pitch or to cause formation of a nematiclike phase of infinite pitch through parallel molecular alignment at higher fields can be brought about by both electric [61,88,90,91,96] and magnetic fields [17,61,84,92-95,97-99]. These effects agree very well with theoretical expectations [100-102]. Recently, homeotropic nematic states have been converted to cholesteric states by means of an electric field [103]. These orientational changes, of course, result in optical changes. The applications of such effects are discussed in detail in Chap. 12.

The electrical conductivity of cholesteric mesophases in their own right has received only scant attention in the literature. One publication claims that, in contrast to smectics and nematics, cholesterics exhibit an order or magnitude decrease in dc conductivity at the solid-to-mesophase transition [104]. No further results seem to have been published, and the contrasting behavior is not fully rationalized.

In a recent report the dc conduction of the Grandjean and focal conic textures of a mixture of cholesteryl oleyl carbonate (10) and MBBA (6) was examined [105]. Over a wide field range the focal conic texture is about 20% more conductive than the Grandjean. This ratio compares very well with conductivities of the analogous nematic textures. Apparently, the effective conductivity along the molecular axes is greater than that perpendicular to them.

VI. MOLECULAR STRUCTURAL EFFECTS

A. Steroidal Compounds

The original cholesteric liquid crystals were steroidal compounds, and for this reason, a large body of data correlating molecular structure and mesomorphic properties has been gathered. In this section the effect of molecular structure on mesomorphic properties will be discussed.

1. Cholesteryl Derivatives

The most studied sterol system is cholesterol (11). The mesomorphic tendency of cholesterol itself is very low [106], but its esters and halides are mesomorphic and have been studied.

The normal (n) alkanoates (12) are the best characterized. In an extensive investigation of thermal properties, the crystal melting points generally decrease irregularly as the chain length increases, with an

ChOH $\qquad$ ChOOC$(CH_2)_nH$

11 $\qquad$ 12

alteration between odd and even numbers of carbon atoms [46,47,107-112]. In contrast, for those compounds having smectic phases, the smectic-to-cholesteric transition temperature as a function of chain length possesses a maximum at C_{12}, but no apparent odd-even effect [47,108,112]. In contrast to the behavior for these transitions, the cholesteric-to-isotropic transition temperatures after the first mesomorphic member of the series (acetate; formate is not liquid crystalline in the pure state) increase to C_3 and then regularly decrease with a slight odd-even effect [107,109-111]. These findings are depicted in Fig. 6. The odd-even effects are not discernible in composite data for any of the transitions [112]. Taken together, these effects result in an irregular pattern up to C_9 and then a regular decrease in the total cholesteric range as the alkanoate chain is lengthened [110] (see Table 1).

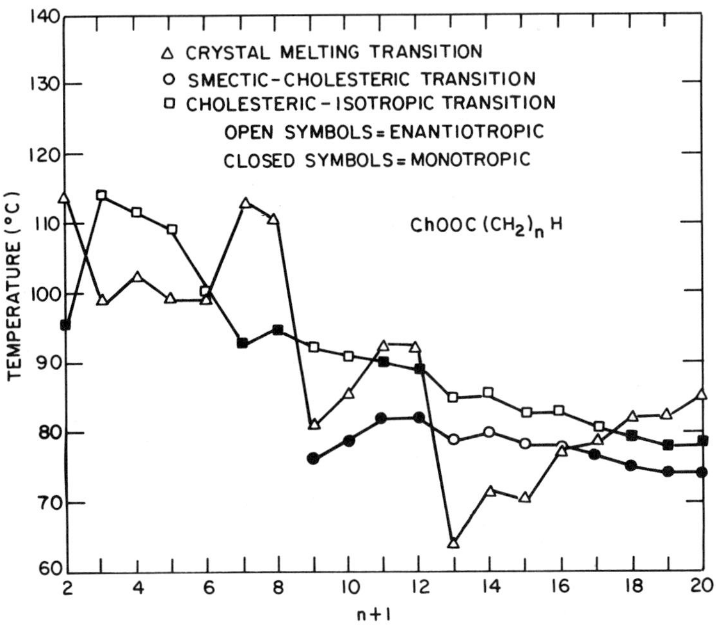

FIG. 6. Transition temperatures as a function of carbon number (n + 1) of cholesteryl n-alkanoates (12) from data of Ref. 110.

TABLE 1. Temperature Ranges and Reflective Wavelengths λ_o of Cholesteric Mesophases of Some Cholesteryl n-Alkanoates (12)

$ChOOC(CH_2)_nH$

12

n	Cholesteric range (°C)	λ_o (nm)
1	95.4 [114][a,b]	540[f]
2	96-112[c]	385[f]
3	98-109[c]	323[f]
4	93-97[d]	286[f]
6	92.7 [113][a,b]	250[f]
8	80.5-92.1[b]	225[f]
13	73.6-85.6[e]	201[g]
15	77.5-82.6[b]	---

[a]Monotropic isotropic-cholesteric temperature, crystal-isotropic temperature in parentheses.
[b]From Ref. 110.
[c]From Ref. 109.
[d]From Ref. 117.
[e]From Ref. 108.
[f]From Ref. 71.
[g]From Ref. 64.

The measurement of the energetics of the mesophase transitions as a function of structure is also a useful technique for understanding the structure of the mesophase. In the n-alkanoates the enthalpy (or the entropy) of fusion of the crystal increases with alkanoate carbon number (see Chap. 9, Fig. 4). A significant feature of this plot is a break at C_8-C_9 which results in two separate lines with an odd-even alternation above C_9 [47,112,113]. This break corresponds to the appearance of the smectic mesophase, but the small additional enthalpy for conversion of the smectic to the cholesteric mesophase is not sufficient to merge the two lines. The break has been ascribed to a change in crystal structure [113]; specifically, the onset at C_9 of chain folding of the alkanoate portion in the crystalline phase has been suggested [54]. The increase in enthalpy per methylene unit averages to 2.18 kcal; this is identical to the value theoretically expected on the basis of the increase in entropy (R ln 3) due to enhanced rotational freedom [47,112].

The enthalpy and entropy for the smectic to cholesteric transition increases sharply and continuously with alkanoate chain length (see Chap. 9 , Fig. 5); there is no odd-even effect [47,108,110-112]. There is an apparent inflection point at about C_{11} or C_{12}, with a greater slope above that carbon number (see Chap. 9 , Fig. 5). No explanation for this behavior has been advanced, but interaction of the side chains in the cholesteric mesophase apparently has a stabilizing effect.

An examination of the dipole moments of the cholesteryl n-alkanoates (12) has revealed a minor, but apparently real, odd-even effect as a function of carbon number [114]. The dipole moment variation corresponds nicely to cholesteric-to-isotropic transition temperatures. The variation is rationalized on the basis that for even carbon atoms the terminal C-C bond lies at an angle to the major chain axis and thus reduces the polarizability of the chain (by a factor of the cosine of this angle) relative to the odd carbon number chains.

Branched cholesteryl alkanoates have also been studied recently in some detail. In the initial investigation the effects of the position, length, and number of branches on the thermal properties of the pure compounds and binary mixtures with cholesteryl nonanoate (8) were studied [106]. Introduction of branching tends to raise the melting point and the isotropic temperature and decrease the temperatures of the mesophase transitions. This leads to monotropic behavior. Branching favors the formation of the smectic over cholesteric mesophase while shrinking the total mesomorphic range. The closer the branch to the carbonyl or the more extensive the branching, the larger these effects. The enthalpy and entropy of fusion increase as branches are introduced and moved closer to the carbonyl group. These results are interpreted in terms of decreased conformational mobility of the alkanoate chain due to steric interactions of the alkyl groups in the branches. This is shown more explicitly in Fig. 7. In Fig. 7a, the three possible noneclipsed conformations of the groups attached to the

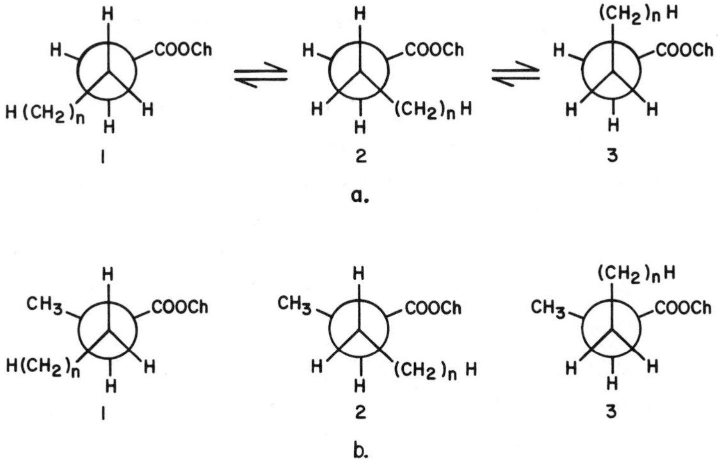

FIG. 7. Newman projections showing the three possible noneclipsed conformations of the C_2-C_3 bond of (a) cholesteryl n-alkanoates and (b) cholesteryl 2-methylalkanoates.

first two carbon atoms adjacent to the carbonyl group are depicted. In this situation conformation a-1 would be preferred since it minimizes interactions of the $(CH_2)_nH$ group; nonetheless, a-2 and a-3 would contribute since the interaction of $(CH_2)_nH$ and the acyloxycholesteryl moiety (COOCh) would probably not be too severe. In Fig. 7b, the effect of introduction of a methyl substituent at the carbon adjacent to the carbonyl group is depicted. Again, conformation b-1 in which the $(CH_2)_nH$ and COOCh groups are trans to one another would probably be most stable. However, conformation b-3, which is analogous to a-3, would now be present in very small amounts due to the steric interaction of the $(CH_2)_nH$ group with both CH_3 and COOCh functions. Therefore, the addition of the methyl group constrains the $(CH_2)_nH$ group. Introduction of more branches leads to even more restriction of the motion of the alkanoate chains. The increased melting point and enthalpy and entropy of fusion are thus a consequence of the increased order, as presumably is the enhanced tendency for smectic behavior.

As stated above, increased chain length in n-alkanoates results in increased enthalpy and entropy of fusion and irregular but decreased crystal melting point. In branched systems the data indicate that increasing the length of one or both carbon chains leads to a slight decrease in the enthalpy and entropy of fusion and lowers the crystal melting point [106]. The mesomorphic transitions are also lowered, but to a lesser extent, so that increased mesomorphic range results. As with the n-alkanoates, the proportion of the mesophase range which is smectic also increases. These effects are attributed to increased conformational mobility of the longer carbon chains which tends to offset the inhibiting effect of branching [106].

The effects of chirality in the alkanoate portion of these esters have also received attention. Chirality in the alkanoate chain results in the possibility of two stereochemical configurations (Fig. 8). These two configurations, of course, alter both intramolecular steric interactions and intermolecular attractions in the cholesteric mesophase. A third possibility is the presence of both configurations in unequal or equal amounts. The transition temperatures are, of course, different for these stereoisomeric modifications. For example, the cholesteryl esters (13) from the 2-methylpentanoic acids have the following melting points: R: 89.7-90.4° C [115], 90.4° [85]; S: 74.3° C [85]; R, S: 78.0-80.0° C [116], 80.1° C [85]. None of these esters shows a cholesteric mesophase in the pure state [85,115,116]; however, extrapolation of transition lines in binary phase diagrams with cholesteryl nonanoate suggests the stereoisomeric modifications have similar cholesteric transition temperatures [116]. Cholesteryl 2(S)-methylbutyrate (14) has an extrapolated cholesteric-isotropic temperature of 74° C [117] compared to 76° C for the ester from the racemic acid [106]; the melting points of the two materials are within 3°C of one another [106,115,117].

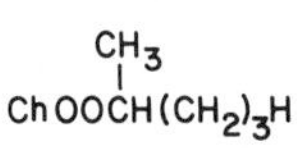

13

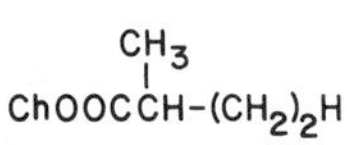

14

In addition to transition temperature differences between chiral forms, there are enthalpy and entropy differences. The 2-methylbutyrates

a. b.

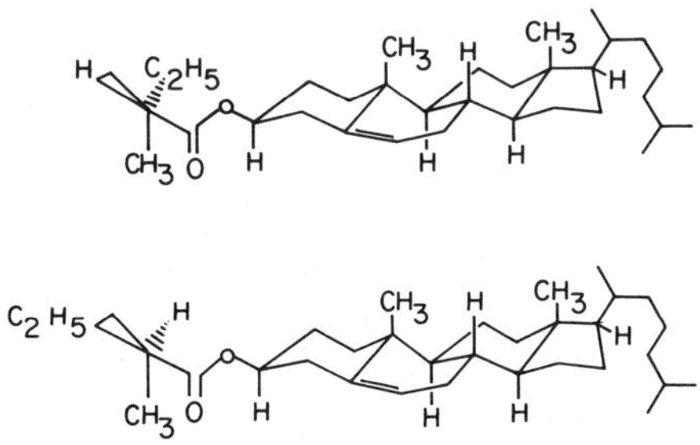

FIG. 8. Configurations of cholesteryl 2(S)-methylbutyrate (a) and cholesteryl 2(R)-methylbutyrate (b).

(14) are an example; ΔH_{fusion} for the S ester is 5.45 kcal/mol, while for the ester from the racemic acid it is 4.98 kcal/mol [54]. Similarly, the 2(R)-methylpentanoate has ΔH_f = 7.17 kcal/mol [54], while the ester from the racemic acid has ΔH_f = 9.26 kcal/mol [116].

Systematic relationships exist among enthalpy of fusion, entropy of fusion, the pitch, and structure. For the n-alkanoates a plot of pitch or reflective wavelength versus ΔH_f (Fig. 9) shows a maximum [54]. This maximum occurs in the formate (point 1). The curve for ΔH_f below that at the maximum includes butyrate (point 2) and pentanoate (point 3), while above $\Delta H_{f,max}$ lie hexanoate (point 4), nonanoate (point 5), and myristate (point 6). This plot is a description of the relationship between the thermal stability of the crystalline phase and the steric interactions in the cholesteric mesophase. If it is assumed that the cholesteryl molecules lie parallel to one another in the crystal, then ΔH_f is a measure of the stability of this arrangement; the more resistance to this arrangement, the lower is ΔH. The pitch of the cholesteric mesophase is a measure of the angular displacement θ between adjacent molecules in the helix (Fig. 1c). Thus, both parameters are related to the steric requirements of the molecules and, more particularly in the present case, the varying steric requirements of the alkanoate chain. In these terms the relationship can be rationalized on the basis of chain folding in the crystalline phase in n-alkanoates of greater than eight or nine carbons [54]. In the crystal phases of the lower alkanoates and in the cholesteric mesophase of all the n-alkanoates the chains are apparently extended [54].

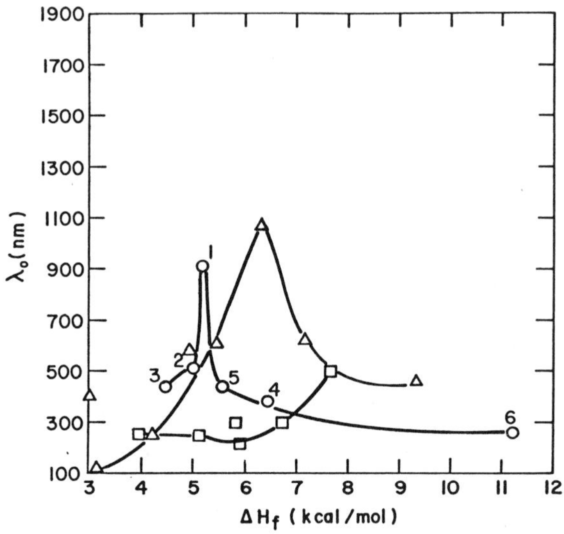

FIG. 9. Reflective wavelength λ_o of cholesteric mesophase versus heat of fusion ΔH_f for cholesteryl alkanoates from data of Ref. 54; ○ n-alkanoates; Δ two-branched alkanoates; □ other branched alkanoates.

Similar effects are observed in the 2-branched alkanoates [54] (see Fig. 9). Apparently, chain folding can occur to some extent in the crystalline state of the longer branched esters.

In alkanoates with branches at the more remote positions of the chain, the plot of reflective wavelength versus ΔH_f (Fig. 9) increases continuously [54]. Apparently, chain folding in the crystal does not occur due to the steric influence of the branches near the chain end, and thus the relationship is continuous because the same extended chain conformation exists in both the crystal and the cholesteric mesophase.

The pitch can also be related to alkanoate structure. In the n-alkanoates the pitch and reflective wavelength decrease as the carbon number increases [10,54,71], as shown in Table 1. This can be rationalized in terms of increasing volume swept by the chain (Fig. 10a) [10,54], which increases the intermolecular angles, and hence θ (Fig. 1), causing a decrease in pitch. The effect is observed in branched systems as chains are lengthened and can be rationalized on this basis, too. The introduction of branches at the 2 position (adjacent to the carbonyl) causes a dramatic increase in pitch; for example, λ_0 of butyrate (12, n = 3) is 529 nm while that of isobutyrate (15) is 1,081 nm [54]. Introduction of branching further out along the alkanoate chain tends to decrease pitch; λ_0 for n-hexanoate (12, n + 5) is 385 nm while 4(S)-methylhexanoate (16) gives λ_0 250 nm [54]. The difference between 2 branching and more remote branches is explained as follows: The introduction of 2 branching results in a restriction of rotation about the CO-C_2 bond, as depicted in Fig. 7b, thus lessening the total volume occupied by the alkanoate chain relative to the unbranched system. This comparison is depicted in Fig. 10. More remote branching does not restrict rotation about this important bond, but does increase the size of the rotating chain, thus increasing the total swept volume and decreasing the pitch (see Figs. 1 and 4).

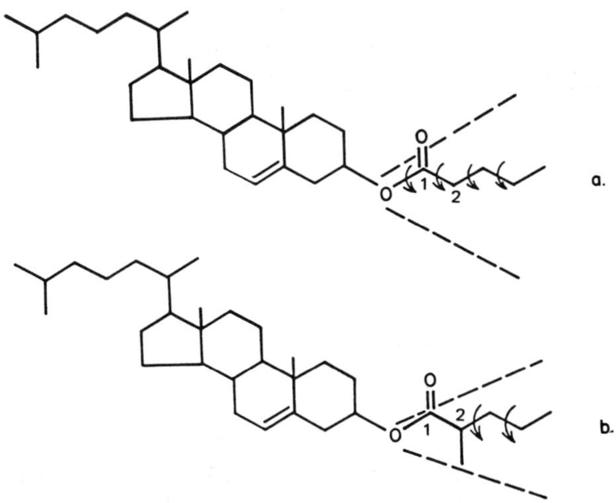

FIG. 10. Displacement volume of alkanoate portion of (a) cholesteryl n-alkanoates and (b) cholesteryl 2-methylalkanoates showing restricted rotation about C_1-C_2 bond of the latter relative to former causing volume swept in latter to be smaller.

$ChOOCCH(CH_3)_2$

15

$ChOOC(CH_2)_2\overset{CH_3}{\overset{|}{C}}H(CH_2)_2H$

16

From the limited data presently available, it is not possible to correlate chiral configuration and pitch. The differing spatial relationships do, of course, affect the pitch. Thus, for example, 2(R)-methylpentanoate (13) has λ_0 = 268 nm and the S derivative, λ_0 = 637 nm [54]. This does seem to be an avenue that ultimately could lead to an understanding of the detailed structure of the cholesteric mesophase.

The effects of chain length on thermal properties have also been examined for thiocholesteryl n-alkanoates (17). The thiocholesteryl n-alkanoates (17) possess generally lower melting points and higher smectic-cholesteric-isotropic transitions [118] than the cholesteryl analogs (12); the total temperature ranges of smectic mesophase increase while those of the cholesteric phase are similar to those of the cholesterol derivatives [110,118]. Compare Table 2 to Table 1. The enthalpy and entropy of fusion are similar for the two series up to C_{12} and above that the thio series values are less [110]. The enthalpy and entropy of the smectic-cholesteric and cholesteric-isotropic transitions are lower for the thio compounds and change very little with chain length [110]. The higher temperature range of the cholesteric phase of the thiocholesterol esters is attributed to increased lateral forces relative to the cholesterol esters deduced from nuclear magnetic resonance (NMR) studies [119].

$ChSC(=O)(CH_2)_nH$	$ChSC(=O)(CH_2)_nC_6H_5$	$ChOOC(CH_2)_nC_6H_5$
17	18	19

The ω-phenylalkanoates of thiocholesterol (18) [120] and cholesterol (19) [121] both show large odd-even decreases in melting point with increasing chain length [110,121,122]. In contrast to the alkanoate case, however, the thio derivatives possess the higher melting points up to n = 5 and lower above that point. The cholesteric-isotropic transitions of series 18 are higher up to n = 6, then higher for even n and lower for odd n than series 19 [120b,122]. The cholesteric ranges of the two series can be compared by examination of Tables 3 and 4. Above C_8 in each series a

TABLE 2. Temperature Ranges of Cholesteric Mesophases of Some Thiocholesteryl n-Alkanoates (17)[a]

$$\mathrm{ChS\overset{\overset{O}{\|}}{C}(CH_2)_nH}$$

17

$\underline{n}$	Cholesteric range (° C)
2	111.1 (111.4)[b]
3	100.6-117.6
4	90.7-104.7
6	73.1-102.1 (107)[b]
8	84.0-97.5
13	85.5-90.4
15	83.0-87.3

[a]From Ref. 118.
[b]Monotropic isotropic-cholesteric temperature or cholesteric range, crystal-isotropic temperature in parenthesis.

smectic phase exists. The results for series 19 are shown in Fig. 11. The entropies of fusion are not related in a regular manner [120b,122]. The entropy of the cholesteric-isotropic transition in the cholesteryl ω-phenylalkanoates is relatively constant to C_3 and then increases smoothly. That of the thiocholesterol derivatives (18) decreases to a minimum at C_3 and then increases smoothly up to n = 10, irregularly above that point; the entropies in the thio compounds are generally lower than the corresponding values in the cholesterol derivatives. The large odd-even effect observed in the ω-phenylalkanoates is attributed to the change in orientation of the terminal phenyl group in the two series [120a]. Like the situation deduced for the cholesteryl alkanoates (12) from dipole moment data [114], on the

TABLE 3. Temperature Ranges of Cholesteric Mesophases of Some Thiocholesteryl ω-Phenylalkanoates (18)[a]

$$\mathrm{ChSC(=O)(CH_2)_nC_6H_5}$$

18

$\underline{n}$	Cholesteric range (° C)
2	136.3-145.6
3	None
4	106.5-109.3
5	None
6	92.4 (95.8)[b]
8	46.9-83.6 (83.5)[b]
13	47.3-52.8 (51.0)[b]
15	49.4-55.3 (54.6)[b]

[a] From Ref. 120b.
[b] Monotropic isotropic-cholesteric temperature or cholesteric range; crystal melting temperature in parentheses.

basis of molecular models the phenyl group with an even number of methylene groups is disposed at an angle to the carbon chain while in the other series it is along the carbon chain axis [120a]. This would be expected to affect the polarizability to a greater extent than in the alkanoates 12 [114], leading to a greater odd-even effect. Steric factors may also play a role [120a].

TABLE 4. Temperature Ranges of Cholesteric Mesophases of Some Cholesteryl ω-Phenylalkanoates (19)[a]

$ChOOC(CH_2)_nC_6H_5$

19

n	Cholesteric range (°C)
2	114.0-114.6
3	26.2 (90.6)[b]
4	91.0 (100.0)[b]
5	44.3 (82.0)[b]
6	83.2 (96.2)[b]
8	76.6 (110.0)[b]
13	48.5-60.3 (64.1)[b]
15	51.2-61.4 (55.0)[b]

[a]From Ref. 121.
[b]Monotropic isotropic-cholesteric temperature or cholesteric range; crystal melting temperature in parentheses.

Cholesteryl alkyl carbonates (20) possess significantly lower melting points than the alkanoate analogs (12), and they decrease irregularly with chain length from C_5 to C_{17} after which they are relatively constant [123]. The cholesteric-isotropic transition temperatures decrease smoothly, rise from C_6 to C_{14}, and then are relatively constant. The enantiotropic cholesteric ranges, with a few exceptions, are relatively narrow (Table 5). The thermodynamic properties do not seem to have been reported.

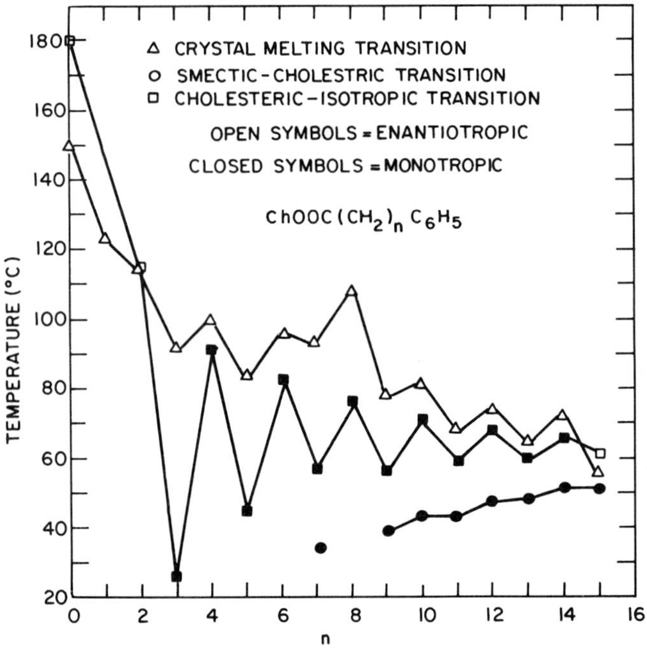

FIG. 11. Transition temperature as a function of methylene number n of cholesteryl ω-phenylalkanoates (19) from data of Ref. 121.

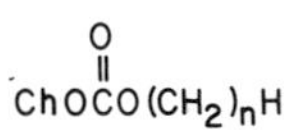

20

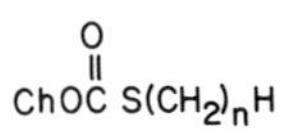

21

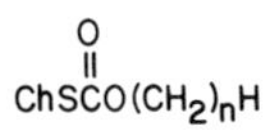

22

The pitches of some of these compounds 20 have been determined and are found to decrease as the carbon number increases [60]. This result is rationalized in the same manner as the similar results with the n-alkanoates, namely the increasing swept volume of the carbon chain (Fig. 10).

The analogous cholesteryl alkyl thiocarbonates' (21) melting points [124a,125] roughly parallel those of oxygen analogs [121], but are

TABLE 5. Temperature Ranges of Cholesteric Mesophases of Some Cholesteryl Alkyl Carbonates (20)[a]

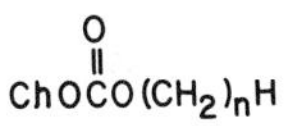

20

n	Cholesteric range (°C)
5	98-102
6	106-108
8	53-83
12	62.5-76
16	69-75
20	70-81

[a]From Ref. 123.

somewhat lower. The cholesteric-isotropic temperatures decrease smoothly with increasing chain length. The smectic phase begins at C_5 and the smectic-cholesteric temperature increases up to C_9, after which it declines in parallel with the clearing point. Thus, the total cholesteric range narrows up to C_9 and then is constant. The cholesteric and smectic phases are monotropic from C_2 to C_6, for C_{17}, C_{19}, and C_{20} (see Table 6). The entropies associated with fusion and smectic-cholesteric and cholesteric-isotropic transitions increase with chain length [125] in a manner analogous to the cholesteryl alkanoates. The values for the fusion entropies are very similar; those for the mesophase transitions are significantly lower in the thiocarbonates 21.

TABLE 6. Temperature Ranges of Cholesteric Mesophases of Some Cholesteryl Alkyl Thiocarbonates (21)[a]

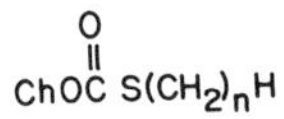

21

n	Cholesteric range (° C)
2	97.1 (114.7)[b]
3	90.3 (97.9)[b]
4	93.3 (97.5)[b]
5	40.3-87.3 (92.9)[b]
6	58.0-85.4 (94.1)[b]
8	69.8-82.3 (80.9)[b]
12	65.2-71.6
15	61.9-67.6 (62.7)[b]
20	54.2-60.4 (64.8)[b]

[a]From Ref. 125.
[b]Monotropic isotropic-cholesteric temperature or cholesteric range; crystal melting temperature in parentheses.

The transition temperatures of the isomeric alkyl carbonates (22) from thiocholesterol are roughly parallel to, but approximately 10°C higher [124b] than 20. Only n = 9, 10, 17, 18 exhibit enantiotropic cholesteric phases; the rest are all monotropic cholesteric. Again, this is probably due to increased lateral interactions in the sulfur compounds. The cholesteric ranges of some of these compounds are listed in Table 7. The entropies of fusion for 21 and 22 are virtually identical; the entropy associated with the cholesteric-isotropic transition is slightly higher in 21 than in analogous members of series 22 [124b, 125].

TABLE 7. Temperature Ranges of Cholesteric Mesophases of Some Thiocholesteryl Alkyl Carbonates (22)[a]

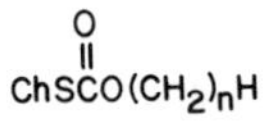

22

n	Cholesteric Range (° C)
2	109.0 (132.2)[b]
3	96.1 (109.2)[b]
4	35.0-95.8 (101.7)[b]
5	56.0-90.3 (96.0)[b]
6	65.9-91.6 (95.3)[b]
8	66.6-82.8 (87.7)[b]
12	69.1-80.2 (92.5)[b]
15	66.3-75.9 (84.4)[b]
20	60.4-70.2 (70.1)[b]

[a]From Ref. 124b.
[b]Monotropic isotropic-cholesteric temperature or cholesteric range; crystal melting point in parentheses.

Aromatic acid esters of cholesterol have also been studied, though not as extensively or intensively as alkanoates. The behavior of cholesteryl p-n-alkoxybenzoates (23) have been investigated as a function of the length of the n-alkoxy substituent [126]. The melting points decrease in an irregular fashion as the carbon number increases. All members yield enantiotropic cholesteric phases, but above C_7 the crystal initially melts to a smectic state. The smectic to cholesteric transition temperature is a maximum at C_{12}. The cholesteric to isotropic transition temperature has a real odd-even effect and decreases with increasing chain length. As a

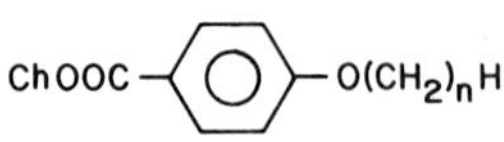

23

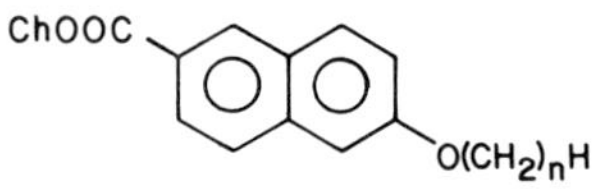

24

result, the total cholesteric range decreases from 90° C at C_1 to a few degrees at C_{18} (see Table 8). These effects then are comparable to those observed with the n-alkanoates (12), with the exception that the transition temperatures are much (≈ 100°C up to C_{10}) higher (compare Tables 1 and 8). This is apparently due to the increased bonding forces arising from the alkoxy and aromatic moieties. No enthalpic data is available for this series of compounds.

TABLE 8. Temperature Ranges of Cholesteric Mesophases of Some Cholesteryl p-n-Alkoxybenzoates (23)[a]

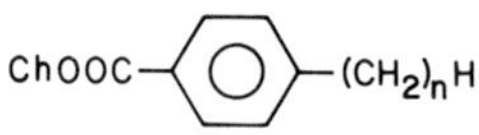

n	Cholesteric range (° C)
1	180-268
2	149.5-265
3	141-253
4	134-248
5	148.5-236.5
6	150-234.5
8	171.5-220.5
12	179.5-200.5
16	170.5-179.5
18	161-163

[a]From Ref. 126.

In a closely related investigation, the cholesteryl 6-(n-alkoxy)-2-naphthoates (24) were examined [127]. As might be anticipated on the basis of the larger aromatic system, all transition temperatures are raised relative to 23. The melting points are 10-60° C higher. More striking is the effect on the mesophase transitions, which occur about 50° C above those in the corresponding alkoxybenzoates (23). Therefore, the cholesteric range in 24, n = 1 extends 120° C from 183-303° C (see Table 9). The effects of chain length are virtually identical in the two systems, as might be expected.

TABLE 9. Temperature Ranges of Cholesteric Mesophases of Some Cholesteryl 6-(n-Alkoxy)-2-Naphthoates (24)[a]

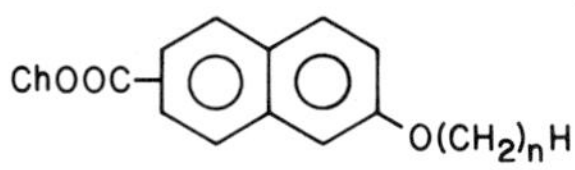

n	Cholesteric range (° C)
1	183.0-302.5
2	198.0-310.0
3	184.0-287.5
4	193.0-290.5
5	179.5-278.0
6	198.0-277.5
8	225.0-268.0
12	230.5-248.0
16	219.0-227.5
18	208.0-221.0

[a]From Ref. 127.

The interposition of a vinyl group between the aromatic moiety and the carbonyl group of such compounds was accomplished by the synthesis of the cholesteryl trans-p-(n-alkoxy)cinnamates (25) [128]. The effect of this double bond is to raise all the mesophase transition temperatures relative to the benzoates 23, the cholesteric mesophase being affected to a much greater extent than the smectic. The general effect has been attributed to an enhanced polarizability and the difference between mesophases to the greater sensitivity of the smectic phase to the breadth increase accompanying the introduction of the double bond [128]. The effect of carbon number roughly parallels that observed in the benzoates 23. As can be seen from Table 10, from C_2 to C_6 the average range of the cholesteric phase is 136° C. These compounds provide some of the broadest cholesteric ranges reported for derivatives of cholesterol.

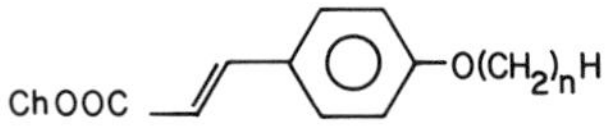

25

$ChOOCC_6H_5$

26

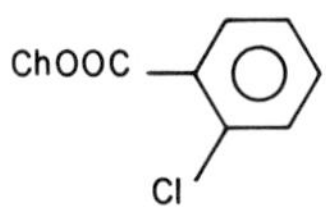

27

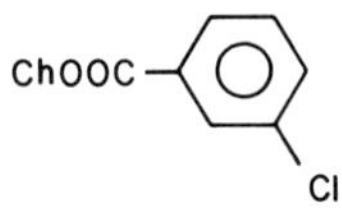

28

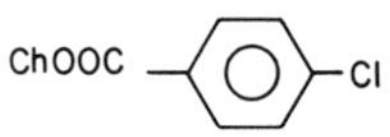

29

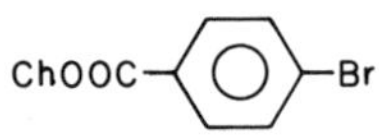

30

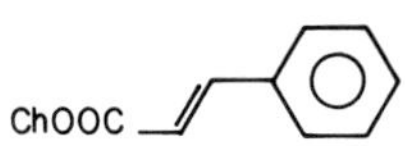

31

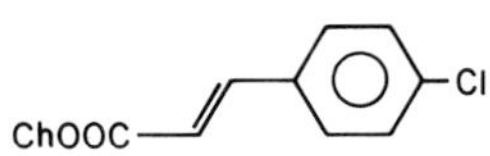

32

TABLE 10. Temperature Ranges of Cholesteric Mesophases of Some Cholesteryl Trans-p-(n-Alkoxy)-Cinnamates (25)[a]

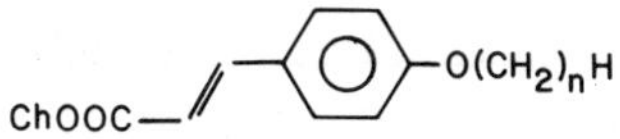

n	Cholesteric range (°C)
1	173.0-292.5
2	152.0-293.5
3	146.0-280.0
4	134.0-281.0
5	128.0-268.0
6	128.5-262.5
8	144.0-249.0
12	181.5-226.5
16	181.5-200.0
18	167.0-186.0

[a]From Ref. 128.

The interrelationship of the electronic and steric effects of substituents on aromatic rings of aromatic esters of cholesterol has recently been examined [129]. The transition temperatures, enthalpies, and entropies of cholesteryl benzoate (26), 2- (27), 3- (28), and 4-chlorobenzoates (29), 4-bromobenzoate (30), cinnamate (31), and 4-chlorocinnamate (32) were determined. From the results the following conclusions were drawn. First, the steric effect of 2 substitution causes decreases in the transition temperatures and entropies of transitions from the crystalline and the cholesteric phases. 3 Substitution does not affect the melting point but lowers the cholesteric-to-isotropic temperature and increases the entropies of both transformations. 4 Substitution raises both the melting point and cholesteric-to-isotropic temperature; the melting entropy is increased, but that of the transition to the isotropic phase is unaltered. These conclusions and others were based on a limited number of examples and their generality is not known. Nonetheless, the influence of both steric and electronic effects were demonstrated, as well as the fact that the crystalline solid and mesophases respond differently to these effects. The cholesteric mesophase ranges for these compounds appear in Table 11.

2. Other Steroidal Nuclei

Modifications of the sterol nucleus have been studied in addition to the 3β substituent variations mentioned above. It has been concluded that three basic structural features are necessary for the formation of the cholesteric phase: (1) a trans A/B ring juncture, (2) a 3β-acyloxy group, and (3) no double bond at the 1-2, 4-5, or 8-9 positions.

Wiegand demonstrated the lack of mesomorphism in derivatives of epicholesterol (33), in which the 3-hydroxy function occupies the α position out of the plane of the A ring [130]. Likewise, he showed that esters of coprostanol (34), which features a cis A/B ring junction, are nonmesomorphic, apparently due to the sharp angle that the A ring makes with the rest of the molecule [130]. Derivatives of $\Delta^{14(15)}$-cholestan-3β-ols are nonmesomorphic [130]. This has been ascribed to the tilting of the D ring away from the rest of the nucleus due to the constraints of the bridgehead double bond at C_{14}, C_{15} [131]. Other double bond isomers and nuclear dienes related to cholesterol were claimed to be mesomorphic by Wiegand

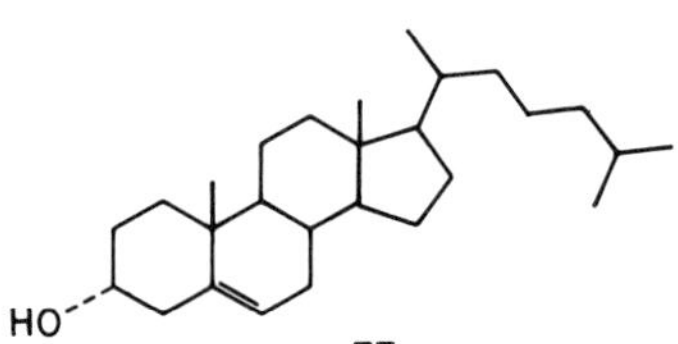

33

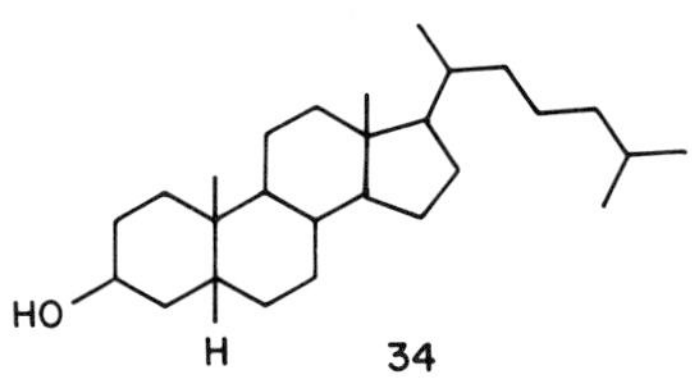

34

TABLE 11. Temperature Ranges of Cholesteric Mesophases of Selected Compounds

Compound	Cholesteric range (°C)	Reference
$ChOOC(OCH_2CH_2)_3H$ (4)	26-32	58a
ChCl (7)	64 (96)[a]	134
$ChOOC(CH_2)_8CH{=}CH(CH_2)_8H$ (10)	17.5-33	
26	145.8-180.7	129
27	96.0-144.8	129
28	142.0 (144.4)[a]	129
29	165.0-240.0	129
30	175.3-245.3	129
31	162.6-215.2	129
32	146.2-262.4	129
35, n = 12	55-60	124a
35, n = 19	53-57	124a
37, n = 8	58.4 (58.5)[a]	133
37, n = 10	50.0-55.1	133
38, R = C_6H_5	137-155	130
38, R = p-$C_6H_4OCH_3$	160-239	130
38, R = p-$C_6H_4C_6H_5$	172-291	130
38, R = p-$C_6H_4CH_3$	172-225	130
38, R = o-$C_6H_4CH_3$	108 (123)[a]	130
40, R = C_6H_5	140-189	134
40,R = CH_3	120 (129.5)[a]	134
47, R = n-C_5H_{31}	52-72	144
50, R = n-C_7H_{15}	61 (78)[a]	145
50, R = n-C_9H_{19}	56 (76)[a]	145
2	90-133	30
52	15-60	20
53	9 (53.5)[a]	21
55	47-95	27

[a]Monotropic isotropic-cholesteric temperature; crystal-isotropic temperature in parentheses.
[b]G.D. Dixon and L.C. Scala, Mol. Cryst. Liq. Cryst., 10:327 (1970).

[130]. However, recent work has demonstrated that introduction of double bonds at the C_1-C_2 [132], C_4-C_5 [132], or C_8-C_9 [131] positions results in loss of all the mesomorphic properties. Steric alterations as a result of these structural changes relative to cholesterol appear to be minimal. NMR studies of the C_1-C_2 and C_4-C_5 compounds support the suggestion that there are significant polarizability alterations relative to cholesteryl derivatives and that these alterations are reflected in the loss of mesomorphism [132]. The same rationale has been put forward for the C_8-C_9 unsaturated compound [131].

The behavior of the 5α-cholestan-3β-yl alkyl thiocarbonates (35) closely resembles that of the cholesteryl analogs 21, with the exception that all transitions are depressed about 10° C [124]. Contrarily, unlike the cholesteryl analogs, the 5α-cholestanyl alkyl carbonates (36) do not show a cholesteric mesophase at all, presumably due to high melting points [124a].

35, X = $O\overset{\overset{O}{\|}}{C}S(CH_2)_nH$

36, X = $O\overset{\overset{O}{\|}}{C}O(CH_2)_nH$

37, X = $OOC(CH_2)_nC_6H_5$

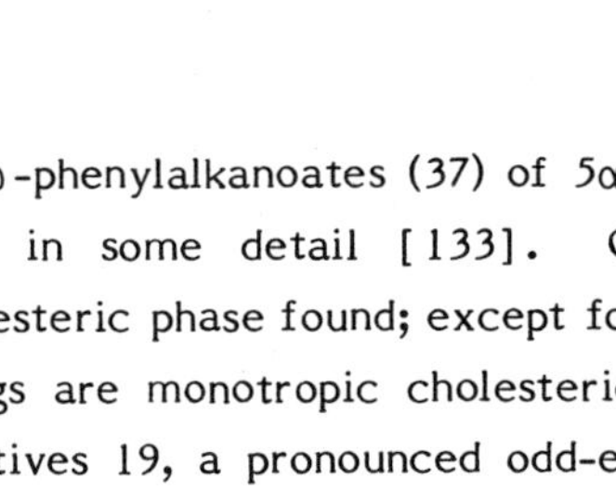

A series of ω-phenylalkanoates (37) of 5α-cholestan-3β-ol have also been investigated in some detail [133]. Only when n = 10 is an enantiotropic cholesteric phase found; except for n = 1, 3, and 5, however, the other homologs are monotropic cholesteric. As with the analogous cholesteryl derivatives 19, a pronounced odd-even effect is manifest for melting point and cholesteric-isotropic transitions, but not for smectic-cholesteric temperatures. The transition temperatures are 15-28° C lower in 37 than in 19. This is attributed to the loss of the bonding forces of the

C_5-C_6 double bond present in 19 [133]. The entropy of fusion increases in an odd-even fashion as n increases in 37 [133], in contrast to the relatively constant level in 19 [110]. The entropy associated with the cholesteric-to-isotropic transition is significantly higher in 37 [133] than in 19 [110], but has the same dependence on n and a definite breakdown into two lines, one for even carbon number and one for odd, as for 19. Again, this is presumably due to the different orientations of the phenyl group in the two series [120a].

The mesomorphic properties of the cholestanyl alkanoates 38, R = n-alkyl, have not been systematically studied, so they cannot be compared to the cholesteryl analogs 12. The substituted benzoates 38, R = aromatic, have been examined [130]. They have lower melting and cholesteric-isotropic temperatures [130] than the corresponding cholesteryl derivatives 26-30 [129] (see Table 11). Again, the data indicate that the C_5-C_6 double bond of cholesteryl compounds stabilizes the cholesteric mesophase.

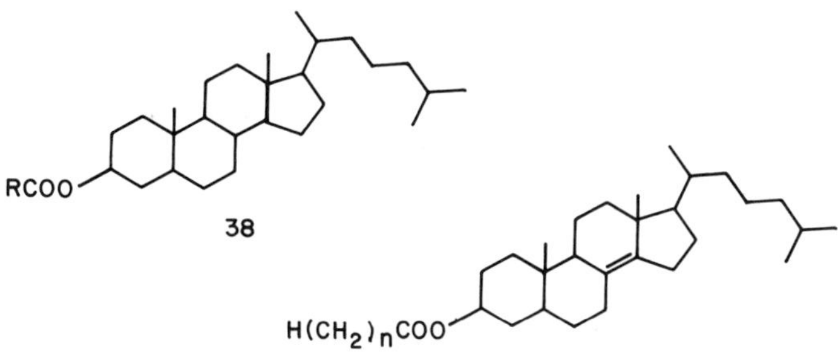

38

39

The n-alkanoates (C_1 to C_{18}) of 5α-cholest-8(14)-en-3β-ol (39) have been synthesized and studied in detail recently [134,135]. These compounds are isomeric to the cholesteryl n-alkanoates (12), differing only in the position of the double bond. In this series the melting point decreases irregularly up to C_{10} and then increases, manifesting a minor odd-even effect from C_{12} to C_{18}. The cholesteric-isotropic transition decreases from C_3 to C_5 (no mesophases for C_1 and C_2) and then remains relatively constant to C_{15}, above which it no longer exists. This combination of effects results in a monotropic cholesteric phase from C_4 to C_6, while C_3 and C_7 to C_{15} are enantiotropic cholesteric. A smectic phase is observed for C_9 through C_{18}, enantiotropic only for C_{10} through C_{15}. Thus, as shown in Table 12, the cholesteric ranges of these compounds are considerably lower (10-40° C) than the analogous cholesteryl compounds (Table 1). A unique feature is the fact that they form left-handed helices in the cholesteric mesophase, so that in contrast to the cholesteryl n-alkanoates, right-handed circularly polarized light is reflected [134].

TABLE 12. Temperature Ranges[a] and Reflective Wavelengths λ_o[b] of Cholesteric Mesophases of Some 5α-Cholest-8(14)-en-3β-yl n-Alkanoates (39)

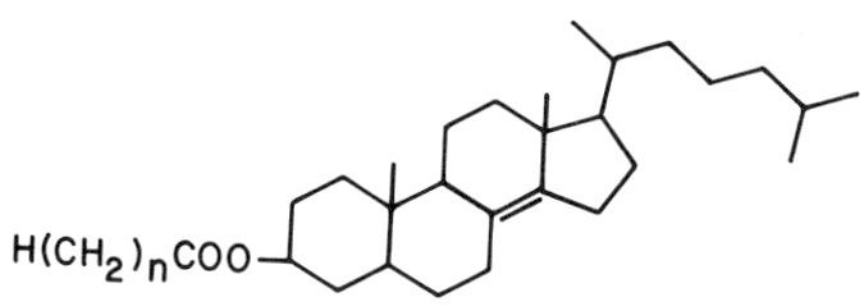

n	Cholesteric range (° C)	λ_o (nm)
1	None	--
2	None	121
3	82.3-85.2	133
4	77.3 (85.3)[c]	141
6	68.0 (85.9)[c]	158
8	48.7-68.4	--
13	68.1-70.0	--
15	68.1-68.5	--

[a]From Ref. 135.
[b]From Ref. 134.
[c]Monotropic isotropic-cholesteric temperature; crystal-isotropic temperature in parentheses.

Two lines result when the total transition entropy is plotted versus carbon number for 39 [136], as shown in Fig. 12a. The break here occurs between C_6 and C_7. As in the isomeric cholesteryl series, this may be due to chain folding of the alkanoate chain in the crystalline phase of esters above C_6. The values are very similar to those of the cholesteryl series (Chap. 9, Fig. 4). The entropy change associated with the interconversion of the smectic and cholesteric phases increases sharply from C_8 to C_{15} and is higher in each case than for the isomeric esters (12) of cholesterol (Chap. 9 , Fig. 5). The cholesteric-isotropic transition entropy starts at a higher level than the cholesteryl analog, remains fairly constant as does that of the isomeric system, and then increases rapidly with chain length to a maximum value six to seven times that of isomeric system (Fig. 12b). This effect has been interpreted to mean that in both the systems the short

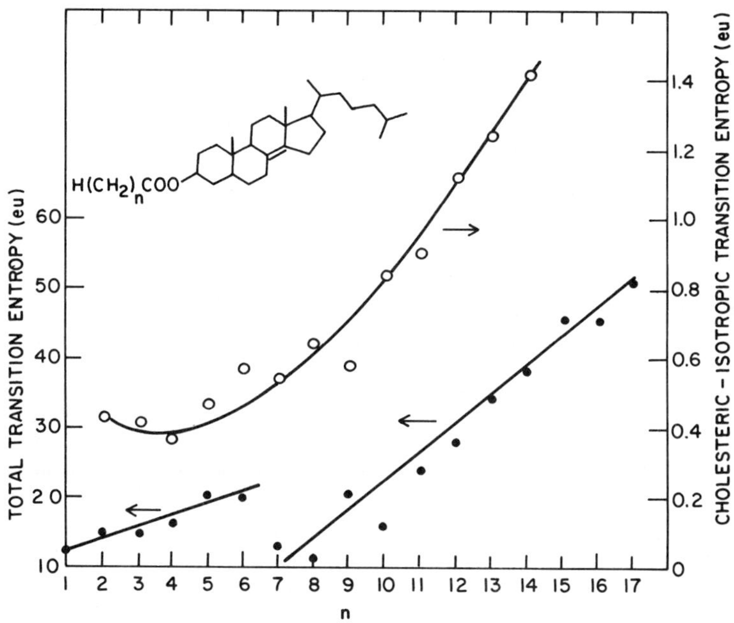

FIG. 12. Entropy of transition versus carbon number n for 5α-cholest-8(14)en-3β-yl alkanoates (39): (a) sum of all transitions (•) and (b) cholesteric-isotropic transition (o) from data of Ref. 136.

chain alkanoate cholesteric-to-isotropic transition primarily involves dissociation of the sterol nuclei, while in the longer chain analogs the dissociation of the alkanoate chains becomes increasingly important [136]. The reason for the apparent increased association of the chains in the 5α-cholest-8(14)-en-3β-yl alkanoates (39) is not clear.

To pursue this question, the dipole moments of series 39 were measured [137]. The results plotted as a function of alkanoate carbon number are very similar to those for the cholesteryl series [114]. No odd-even effect is discernible up to C_6, but from C_7 to C_{13} a relatively large odd-even effect seems to exist. Up to C_6 the dipole moments of series 39 are somewhat lower than those of series 12, but above C_6 the averages for the two are nearly the same. These results then do not appear to illuminate the question of chain association in mesophases. Again, the subtlety of structural effects is underscored.

The pitches of a number of derivatives (39) of 5α-cholest-8(14)-en-3β-ol have been measured [134]. As noted above, these compounds tend to be right-handed (reflect right-handed circularly polarized light). In this system, however, in contrast to the cholesteryl derivatives (12), the effect of increasing chain length is to increase the pitch of the mesophase (see Table 12). In unbranched systems the inverse pitch or displacement angle θ decreases in direct proportion to the number of carbon atoms in the 3β substituent [134]. In all cases studied to date (12,20, 38, 39, 40), increased chain length causes the rotatory sense to become more left-handed (based on helicity of reflected light), i.e., increase pitch of right-handed

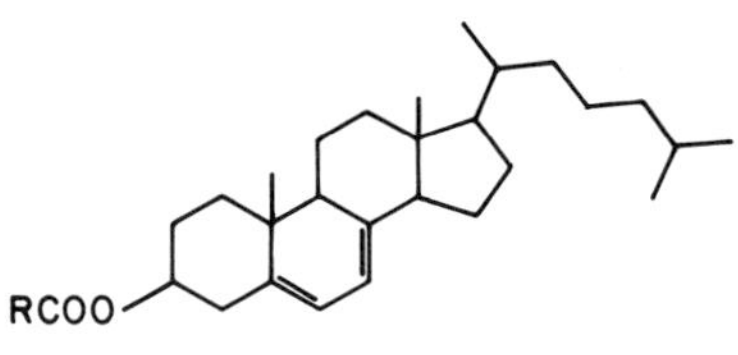

40

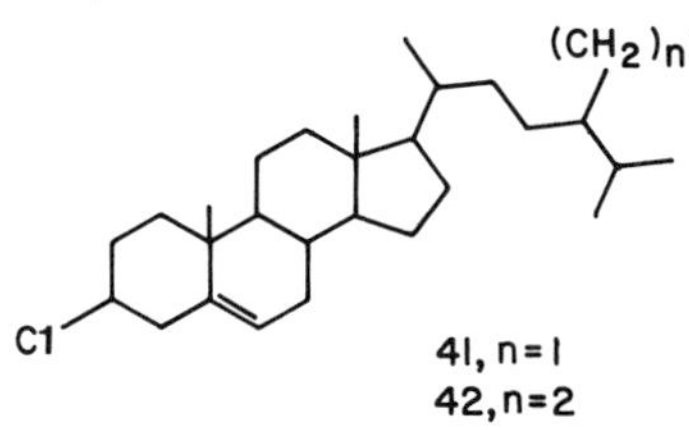

41, n=1
42, n=2

compounds and decrease the pitch in left-handed compounds. This suggests that the disposition of the 3β group is the same in all these compounds. The conformation of these substituents in the cases of esters still is not known. Nonetheless, it is possible to rationalize these results, at least in a general way, as follows. Construction of molecular models of 12, 38, 39, and 40 shows that structural changes in the steroid nucleus cause changes in the shape. The stacking of the molecules in the cholesteric mesophase is dependent on the shape of the nucleus as well as the shape and effective size of the 17 and 3β-substituents. Comparison of molecular models of 12, n = 1, and 39, n = 1 (Fig. 13), reveals two salient differences: (1) the D ring in 12 is up while the D ring of 39 is down, (2) the 17-alkyl chain of 39 is much more rigid due to interaction of the 18- and 20-methyl groups. (This may also play a role in the increased chain associations reflected in the higher entropies associated with mesophase transitions noted above.) The effects of these two parts (nucleus and substituent) may be reinforcing or opposing, depending on their geometrical relationship. In the cholesteryl alkanoates (12) the chirality tendencies of the cholesteryl nucleus and the alkanoate chain are the same. Thus, as the volume occupied by the alkanoate chain increases (Fig. 10), the minimization of interaction energies for interaction of adjacent molecules in the helix for both the steroid nucleus and the alkanoate lead to an increase in the displacement angle θ, as described above. In the case of 39, however, the energy minimization processes are in opposition: the angle for the alkanoate portion tends to increase as the length of the chain increases in the direction of a right-handed helix just as in the cholesteryl derivatives 12, while the nucleus prefers to form a left-handed helix. The displacement angle θ, such that the total interaction energies between adjacent molecules are minimized, will therefore tend to decrease from a value for a left-handed helix (right-handed cholesteric, reflecting right-handed circularly polarized light) to zero and then increase as the chirality of the helix becomes right-handed. Thus, the situation is somewhat similar to the temperature dependence of pitches of mixtures of right- and left-handed components discussed in Sec. IV and Fig. 5.

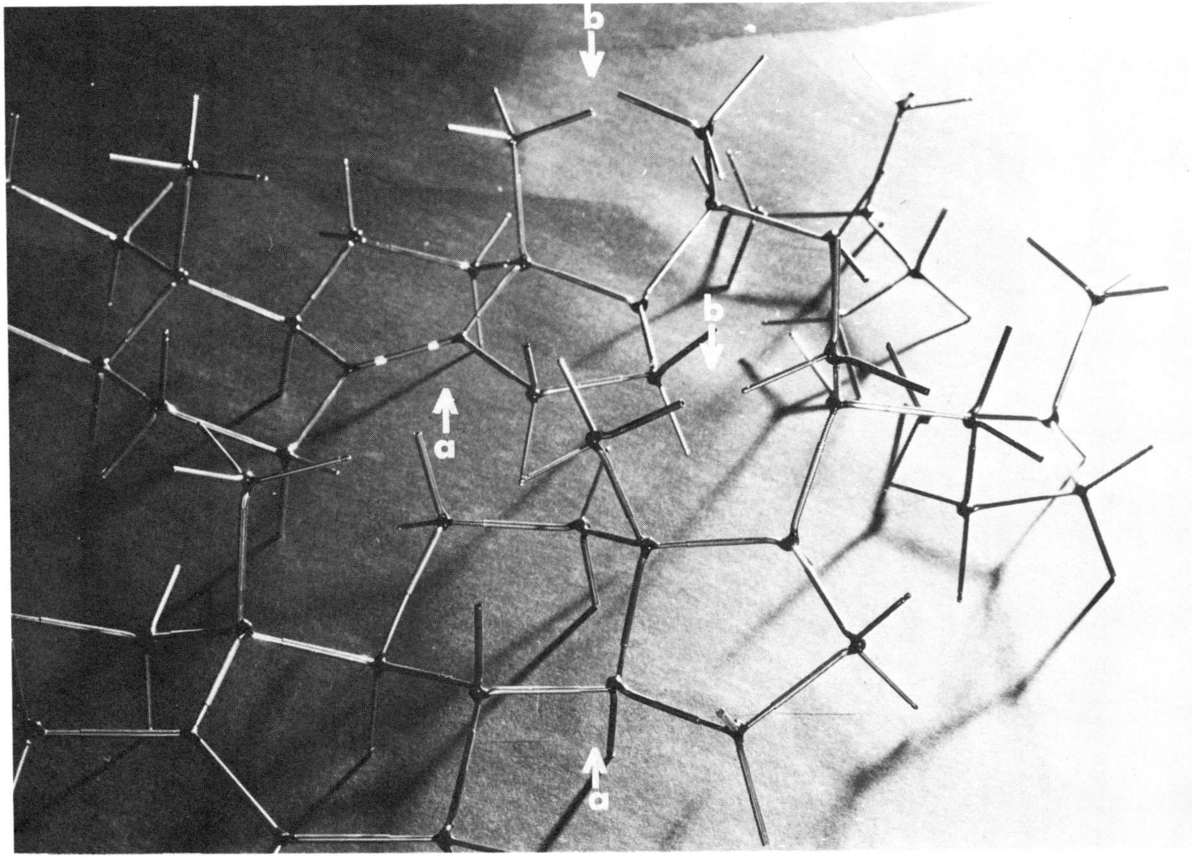

FIG. 13. Photographs of Dreiding models of cholesteryl acetate (12, n = 1) (bottom) and 5α-cholest-8(14)-en-3β-yl acetate (39, n = 1) (top) showing change in shape of steroid nucleus with change in position of double bond, particularly the shape of the D ring (a) and the interaction between C_{18} and C_{20} (b).

Modification of the alkyl chain at the C_{17} position influences the liquid crystalline behavior of steroidal derivatives. The effects of branching of this alkyl chain parallel those for the alkanoate moiety at C_3 of cholesteryl compounds 12. The same principles regarding pitch apply to the steryl chlorides cholesteryl (7), campesteryl (41), and sitosteryl (42). Introduction of a methyl or an ethyl group on the 17β-alkyl chain leads to a large decrease in pitch from 488 nm (7) to 256 nm (41) to 238 nm (42) [138], ostensibly due to an increase in the total volume occupied by the 17β-alkyl chain in a manner similar to that depicted for the 3β group of 12 in Fig. 10a.

Inclusion of double bonds in the side chain has received some attention. Stigmasteryl derivatives (43), which differ from cholesteryl compounds only by the presence of a double bond at C_{22} and an additional ethyl group at C_{24}, do not form cholesteric mesophases [139,140], though a few have smectic phases [140]. β-Sitosteryl esters (44), which are the dihydro derivatives of 43 and which differ from cholesteryl esters 12 only by the presence of the C_{24} ethyl group, are smectic but not cholesteric [141]. Ergosteryl esters (45) which possess the C_{22} double bond and a methyl substituent at C_{24} are smectic, but not cholesteric [142].

In pursuit of this question of the influence of sterol structure on mesomorphism, a number of triterpenol esters have been studied. For

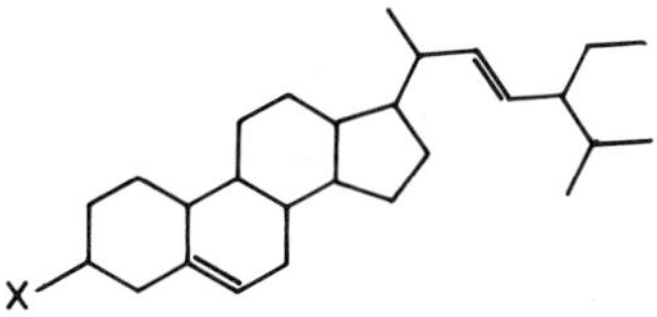

43

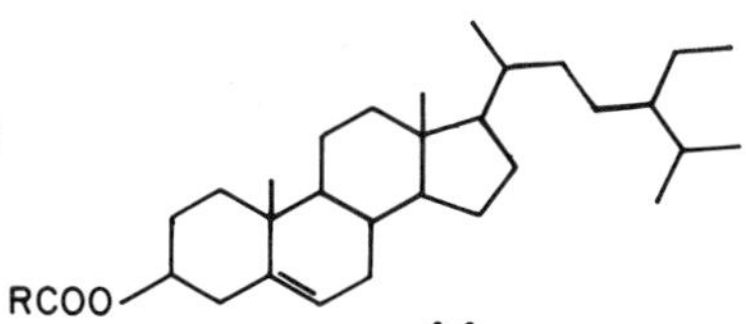

44

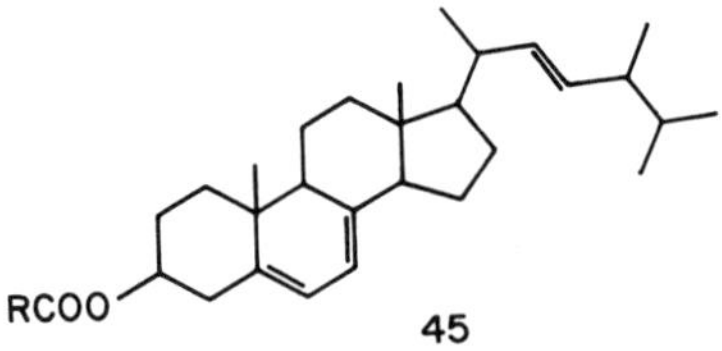

45

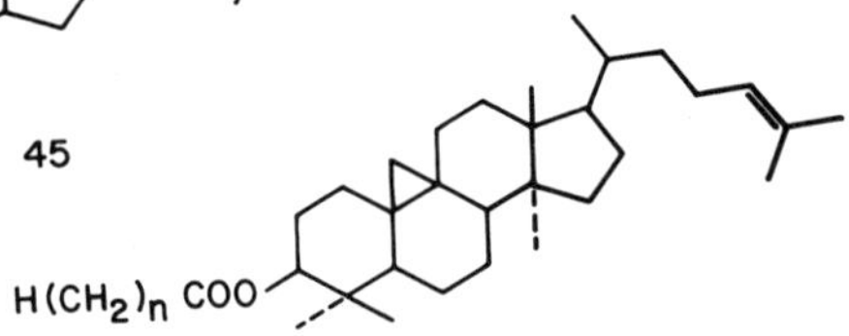

46

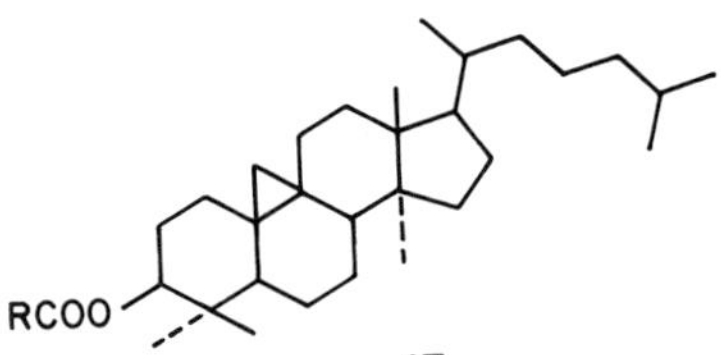

47

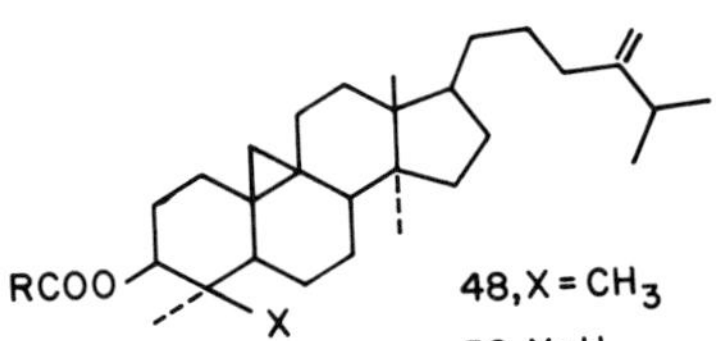

48, X = CH_3

50, X = H

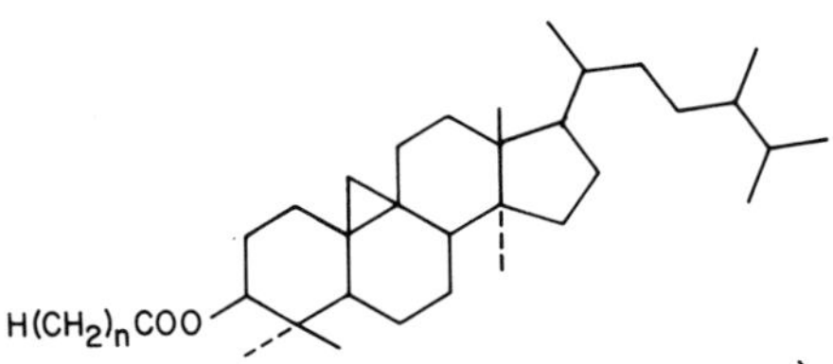

49

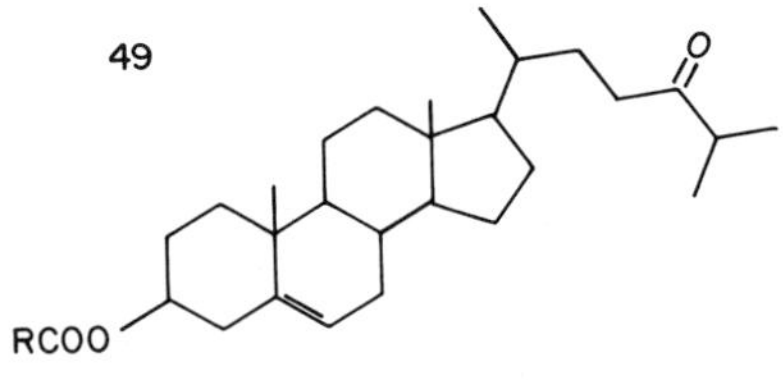

51

example, esters (46) of cycloartenol, which possess a C_{24}-C_{25} double bond, are cholesteric (Table 13), as are the esters of the reduced analog, cycloartanol (47) (Table 1) [143-145]. 24-Methylene cycloartenol esters (48), on the other hand, are not cholesteric, but the saturated derivatives (49) are cholesteric [144,145] (Table 14). The effect of this 24-methylene substituent can be ameliorated by other substituents, however; cycloeucalenol (50) and its dihydro derivatives both generally yield monotropic cholesteric mesophases [144,145]. The corresponding 24-keto cholesterol esters (51) may be cholesteric [145], but need to be examined in more detail.

TABLE 13. Temperature Ranges of Cholesteric Mesophases of Some Cycloartenyl n-Alkanoates (46)[a]

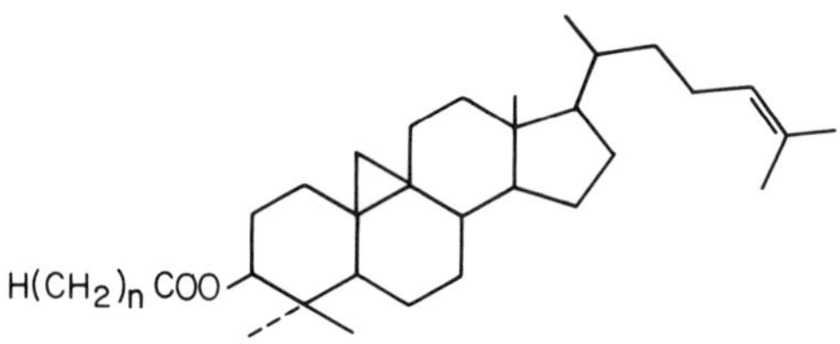

n	Cholesteric range (° C)
3	None
7	58-59
9	59 (71)[b]
11	53.5-58
13	50-56
15	51 (64)[b]

[a]From Ref. 145.
[b]Monotropic isotropic-cholesteric temperature, crystal-isotropic temperature in parentheses.

TABLE 14. Temperature Ranges of Cholesteric Mesophases of Some 24ξ-Methylcycloartanyl n-Alkanoates (49)[a]

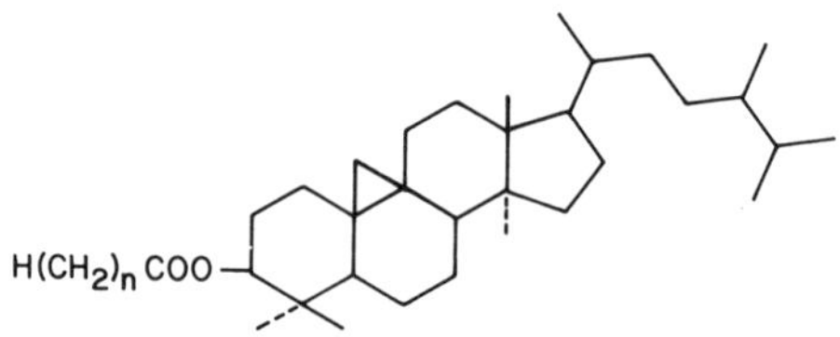

n	Cholesteric range (°C)
3	None
9	None
11	60-65.5
13	66-67.5
15	(64) 65[b]

[a]From Ref. 145.
[b]Monotropic isotropic-cholesteric temperature; crystal-isotropic temperature in parentheses.

All of the foregoing results with unsaturated C_{17} side chains appear to be based on steric grounds. The same general arguments regarding total volume swept by the chain as are applicable to the alkanoate portion of the molecule can be made, though the pitches of these compounds have not been measured. One would predict that the introduction of unsaturation close to the nucleus (e.g., C_{22}-C_{23}) would lead to a larger increase in pitch than unsaturation further away (e.g., C_{24}-C_{25}). This would be a consequence of the lower swept volume due to a more rigid chain in the former case. As previously discussed for branching, however, in addition to this effect on swept volume, the introduction of the double bond may alter the configuration of the side chain in such a way that it does not lie in the requisite attitude, and mesomorphism may be lost. At present our knowledge of these requirements is too limited to rationalize these results in detail.

Ester groups on the C_{17} side chain are compatible with cholesteric behavior. A series in which a carbomethoxy substituent is placed on this chain yields cholesteric mesophases [131]. Even when the alkyl chain is shortened cholesteric behavior is retained as long as there is a saturated carbon atom at C_{20} [131].

The presence of a keto group at C_{17} causes loss of all mesomorphic behavior, as does a hydroxyl or esterified hydroxyl [131]. While the loss of the chiral center at C_{20} may be a factor in these results, the polar character of the substituents is probably determinant [131].

The cholesteric ranges of a number of compounds mentioned above are collected in Table 11.

B. Chiral Nematic Compounds

The relationship of structure to mesomorphic properties in chiral nematics bears a close relationship to that for the normal nematic mesomorphic materials [29]. For a discussion of this mesophase see Chap. 2.

Introduction of chirality into molecules which are nematic gives rise to compounds with cholesteric properties [7,15-30]. Enantiomeric compounds give rise to enantiomeric (opposite-handed) mesophases [3,11]. The temperatures and heats associated with melting of the individual enantiomers should be identical to each other, but may in theory vary from those of the racemic mixture [8]. In practice this variation has been small [7,15,19,20], so that racemic compounds provide a good means of prediction of melting temperatures and heats. On theoretical grounds, the transition temperature and enthalpy associated with the chiral nematic-to-isotropic liquid transition are nearly identical to those of the racemic nematic compounds [20]. The viscosities of the chiral nematics are generally lower than those of the sterol derivatives and are very similar to those of the ordinary nematics. The dielectric anisotropies of the chiral nematics are two orders of magnitude greater than those of cholesteryl derivatives [20], making them superior for some applications (see Chap. 12).

The measured pitches of the pure chiral nematics cover a range from the near ultraviolet to the infrared. For example, the pitch of anil 52 is 223 nm [20]; that of 4'-(2"-methylbutoxy)-4-cyanobiphenyl (53) is about 1,000 nm [21]. Some other representative chiral nematic structures are

54 [26] and 55, with pitches about 350 nm [27]. The cholesteric ranges of a number of chiral nematic compounds are listed in Table 11.

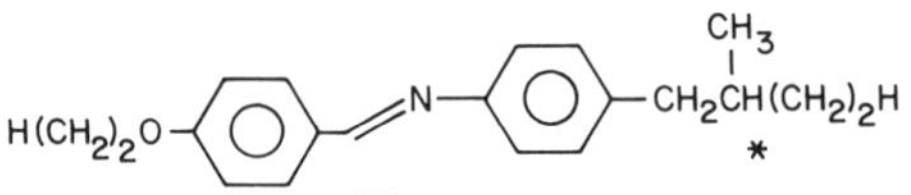

52

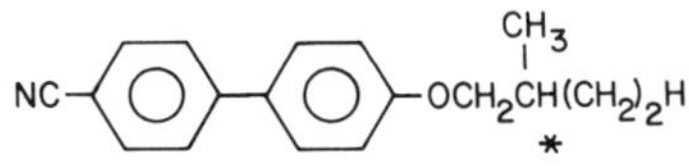

53

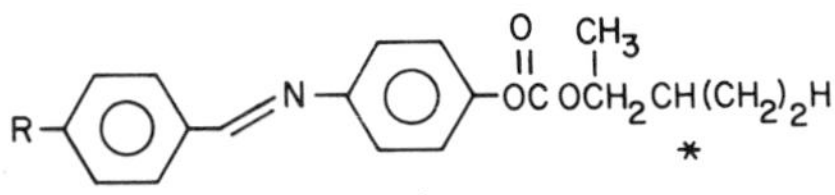

54

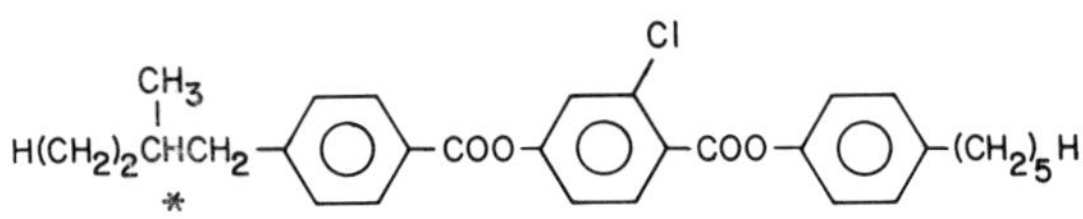

55

VII. APPLICATIONS

The commercial utility of cholesteric liquid crystals derives primarily from the response of their unique optical properties to applied electrical, thermal, and mechanical stimuli. Commercial applications of cholesteric liquid crystals are discussed in greater detail in Chap. 12.

Other applications which are of theoretical importance utilize the organizing properties of the cholesteric mesophase. Solutes oriented in such matrices may thus be spectroscopically characterized by infrared and ultraviolet absorption (polarized radiation included), fluorescence (again polarized), and in favorable cases NMR. Such studies afford fundamental information with regard to molecular and electronic parameters.

Chemical reactions in the cholesteric mesophase with the expectation of enhanced rates and stereoselective control have received some attention [146-153]. The synthesis of optically active 57 by rearrangement of the achiral 56 shows the promise of this area [153].

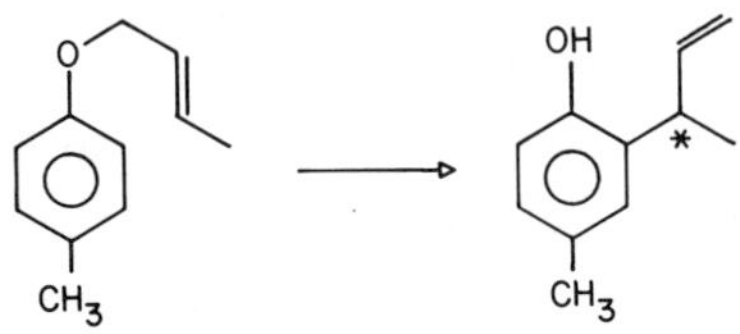

56 57

Another promising field is the use of cholesteric mesophases in chromatography for the resolution of enantiomers [154]. The enantiomeric compounds interact with the helical structure with differing energetics. In theory, this could be large enough for practical resolution.

VIII. THE FUTURE

At this point in time the fundamental properties of the cholesteric mesophase appear to have been fully exploited; ongoing work seems, to this author at least, to involve optimization of variables, which is not a small task, and creative conjoining of these effects with other technologies; this effort is as vast as human imagination makes it. In the near future cholesteric liquid crystalline devices should become commercially important on a number of fronts.

On the theoretical side, however, there is much to be learned. We do not really understand the molecular details of the helical structure of the mesophase. This appears to be the direction of present and future efforts in this field. A number of powerful experimental techniques can be brought to bear on this point; spectroscopic study of the host mesophase and guest solutes (Chap. 6) seems one of the most promising. The combination of this empirical data base and energy minimization-type theoretical calculations similar to those fruitfully employed for polymer conformational analysis [155] should lead to an appreciation of the molecular factors important in determining and controlling the interactions in the mesophase. This will encompass both the host-host and guest-host interactions.

Inasmuch as the formation of the mesophase minimizes the sums of the individual intermolecular interaction energies over the length of the entire helix, relatively small changes in intermolecular interactions affect the pitch. These intermolecular interaction energies in turn are dependent upon the conformations (and configurations) of the molecules. As a consequence, the properties of the cholesteric mesophase are very sensitive probes of the conformations of the molecules comprising the phase. At an advanced point of knowledge, the cholesteric mesophase could in the future be of tremendous utility in conformational analysis of guest molecules.

The utility of the cholesteric mesophase as a medium for asymmetric synthesis has received scant attention. Fruitful results from this area are likely, and these would be a boon to the organic chemist. Polymerizations in cholesteric media have received attention in the past with only limited success as far as stereochemically directed polymer synthesis [146-152]. It is likely, in view of the potential value of such a technique, that such studies will continue.

There is increasing interest in the role of cholesteric mesophases in biological systems (see also Chap. 8). Deeper understanding of the properties of the mesophase would undoubtedly have a great impact on the treatment of certain human disorders.

In the future, then, it appears that the results of ongoing investigations of the cholesteric mesophase will have an impact on a number of areas: new and more diversified commercial products, conformational analysis and stereochemistry, asymmetric synthesis, and medicine. In short, the future of the cholesteric mesophase seems to be filled with a myriad of exciting possibilities.

ACKNOWLEDGMENT

The author is indebted to the following coworkers who took the time to read this chapter and offer many helpful suggestions: Mr. F.C. Bailey, Drs. J.Y.C. Chu, W.H.H. Gunther, J.M. Pochan, and F.D. Saeva.

REFERENCES

1. F. Reinitzer, Monatsh. Chem., 9:421 (1888).
2. G.W. Gray, Molecular Structure and the Properties of Liquid Crystals, Academic, New York, 1962.
3. F.D. Saeva, Mol. Cryst. Liq. Cryst., 23:171 (1973).
4. G.H. Brown, Amer. Scientist, 60:64 (1972).
5. T. Nakagiri, H. Kodama, and K.K. Kobayashi, Phys. Rev. Lett., 27:564 (1971).
6. I.G. Chistyakov, Sov. Phys. Upsekki, 9:551 (1967).
7. G. Friedel, Ann. Phys., 18:273 (1922); C.R. Acad. Sci., Paris, Ser. A.B., 176:475 (1923).
8. E.L. Eliel, Stereochemistry of Carbon Compounds, McGraw Hill, New York, 1962.
9. A.D. Buckingham, G.P. Ceasar, and M.B. Dunn, Chem. Phys. Lett., 3:540 (1969).
10. H. Baessler and M.M. Labes, J. Chem. Phys., 52:631 (1970).
11. H. Stegemeyer and K.J. Mainusch, Chem. Phys. Lett., 6:5 (1970); 8:425 (1971); Naturwissenschatten, 58:599 (1971).
12. W.J.A. Goossens, Mol. Cryst. Liq. Cryst., 12:237 (1971).
13. P.E. Cladis, M. Kleman, and M.A. Guiner, C.R. Acad. Sci. Paris, Ser. A.B., 273:275 (1971).
14. I.G. Chistyakov, Kristallografiya, 7:764 (1962).
15. R. Cano and P. Chatelain, C.R. Acad. Sci., Paris, Ser. A.B, 253B:1815 (1961); 259B:252 (1964).
16. R. Cano, Bull. Soc. Fr. Mineral. Cristallogr., 90:333 (1967).
17. G. Durand, L. Leger, F. Rondelez, and M. Veyssie, Phys. Rev. Lett., 22:227 (1969).
18. W. Haas, J. Adams, and G. Dir, Chem. Phys. Lett., 14:95 (1972).
19. M. Leclercq, J. Billard, and J. Jacques, C.R. Acad. Sci., Paris, Ser. C, 266C:654 (1968); Mol. Cryst. Liq. Cryst., 8:367 (1969).
20. D. Dolphin, Z. Muljiani, J. Cheng, and R.B. Meyer, J. Chem. Phys., 58:413 (1973).
21. G.W. Gray, K.J. Harrison, J.A. Nash, and E.P. Raynes,Electronic Lett., 9:616 (1973).

22. G.W. Gray, Mol. Cryst. Liq. Cryst., 7:127 (1969).

23. C.S. Oh and E.F. Pasierb, U.S. Patent 3,792,915 (1974).

24. J.A. Castellano, et al., Liquid Crystal Systems for Electro-Optic Storage Effects, Final Report to U.S. Air Force under Contract No. F 33615-70-C-1590, Project 7360, December 1971.

25. D. Vorlander, Ber. Dtsch. Chem. Ges., 41:2033 (1901).

26. R.J. Cox, E.M. Barrall II, A.J. Logan, and B.L. Dawson, Abstracts, International Liquid Crystal Conference, 1974, p. 51.

27. B. Klanderman and T.R. Criswell, presented at International Liquid Crystal Conference, 1974.

28. G.W. Gray, presented at International Liquid Crystal Conference, 1974.

29. Y.Y. Hsu and D. Dolphin in Liquid Crystals and Ordered Fluids, Vol. 2 (J.F. Johnson and R.S. Porter, eds.), Plenum, New York, 1974, p. 461.

30. D. Coates and G.W. Gray, Mol. Cryst. Liq. Cryst., 24:163 (1973); G.W. Gray, Mol. Cryst. Liq. Cryst., 21:161 (1973).

31. C. Mauguin, Bull. Soc. Fr. Mineral. Cristallogr., 34:71 (1911).

32. M. Schadt and W. Helfrich, Appl. Phys. Lett., 18:127 (1971).

33. H. Zocher, Z. Phys. Chem. (Leipzig), 132:295 (1928).

34. P. Chatelain, C.R. Acad. Sci., Paris, 213:875 (1941); Bull. Soc. Fr. Minal. Cristallogr., 66:105 (1943).

35. L.T. Creagh and A.R. Kmetz, Mol. Cryst. Liq. Cryst., 24:59 (1973).

36. D.W. Berreman, Mol. Cryst. Liq. Cryst., 23:215 (1973); Phys. Rev. Lett., 28:1683 (1972).

37. H. Sackmann, Z. Elektrochem., 63:1171 (1959), H. Sackman and D. Demus, Mol. Cryst., 2:81 (1966); H. Arnold, K.H. Kolz, and H. Sackman, Z. Phys. Chem. (Leipzig), 252:93 (1973).

38. C.W. Osien, Trans. Faraday Soc., 29:883 (1933); H. Zocher, Z. Phys. 28:790 (1927).

39. F.C. Frank, Disc. Faraday Soc., 25:19 (1958).

40. A. Anzelius, Uppsala Univ. Arsskr. Mat. Naturvet., 1 (1931).

41. A.C. Erigen and J.D. Lee, in Liquid Crystals and Ordered Fluids, Vol. 2 (J.F. Johnson and R.S. Porter, eds.), Plenum, New York, 1974, p. 383.

42. J.L. Ericksen, Mol. Cryst. Liq. Cryst., 7:153 (1969).
43. F.M. Leslie, Mol. Cryst. Liq. Cryst., 7:407 (1969).
44. J.D. Lee and A.C. Eringen, in Liquid Crystals and Ordered Fluids, Vol. 2 (J.F. Johnson and R.S. Porter, eds.), Plenum, 1974, p. 315.
45. F.P. Price and C.S. Bak, In Liquid Crystals and Ordered Fluids, Vol. 2 (J.F. Johnson and R.S. Porter, eds.), Plenum, 1974, p. 411.
46. J.F. Johnson, R.S. Porter, and E.M. Barrall II, Mol. Cryst. Liq. Cryst., 8:1 (1969).
47. G.J. Davis, R.S. Porter, and E.M. Barrall II, Mol. Cryst. Liq. Cryst., 10:1 (1970).
48. I.G. Chistyakov and E.A. Kosterin, Rost. Kristallov. Akad. Nauk SSSR, 4:6 (1964); S.E.B. Petrie, Abstracts, 162nd National Meeting of American Chemical Society, 1970, Phys. 210.
49. K. Tsuji, M. Sorai, and S. Seki, Bull. Chem. Soc. Jap., 44:1452 (1971).
50. L.B. Leder, U.S. Patent 3,789,225 (1974).
51. W. Mahler and M. Panar, J. Amer. Chem. Soc., 94:7195 (1972).
52. J.M. Pochan and H.W. Gibson, J. Amer. Chem. Soc., 93:1279 (1971); 94:5573 (1972).
53. F.P. Price and A.K. Fritzsche, J. Phys. Chem., 77:396 (1973).
54. H.W. Gibson, J.M. Pochan, and D. Hinman, in Liquid Crystals and Ordered Fluids, Vol. 2, (J.F. Johnson and R.S. Porter, eds.), Plenum, New York, 1974, p. 593.
55. F.P. Price and J.H. Wendorff, J. Phys. Chem., 75:2389 (1971).
56. S.A. Jabarin and R.S. Stein, J. Phys. Chem., 77:399,409 (1973).
57. L.B. Leder, Chem. Phys. Lett., 6:285 (1970).
58. H. Baessler, P.A.G. Malya, W.R. Ness, and M.M. Labes, Mol. Cryst. Liq. Cryst., 6:329 (1970).
59. J.P. Matheiu, Bull. Soc. Fr. Mineral. Cristallogr., 61:174 (1938).
60. K. Ko, I. Teucher, and M.M. Labes, Mol. Cryst. Liq. Cryst., 22:203 (1973).
61. H. Baessler, F.M. Laronge, and M.M. Labes, J. Chem. Phys., 51:3213 (1969).

62. W. Maier, in Zahlenwerk and Funktionen aus Physik, Chemie, Astronomie, Geophysik and Technik, 6th ed. (Landolt-Bornstein, ed.), Springer, Berlin, Vol. II, Part 6, p. 607.
63. R. Cano and P. Chatelain, C.R. Acad. Sci., Paris, Ser. A.B, 251:1815 (1961); 259:352 (1964).
64. H. Baessler and M.M. Labes, J. Chem. Phys., 52:631 (1970).
65. F. Strumpf, Phys. Z., 11:780 (1910).
66. J.L. Fergason, in Liquid Crystals (G.H. Brown, G.J. Dienes, and M.M. Labes, eds.), Gordon and Breach, New York, 1966, p. 89.
67. H. de Vries, Acta Crystallogr., 4:219 (1951).
68. S. Chandrasekhar and J.S. Prasad, in Liquid Crystals, (G.H. Brown and M.M. Labes, eds.), Gordon and Breach, New York, 1972, p. 209.
69. G.S. Ranganath, S. Chandrasakhar, U.D. Kini, K.A. Suresh, and S. Ramaseshan, Chem. Phys. Lett., 19:556 (1973).
70. J.E. Adams, W. Haas, and J. Wysocki, Bull. Amer. Phys. Soc., 14:6 (1969).
71. J.E. Adams, W.E. Haas, and J.J. Wysocki, in Liquid Crystals and Ordered Fluids (J.F. Johnson and R.S. Porter, eds.), Plenum, New York, 1970, p. 463.
72. J.E. Adams and W. Haas, Mol. Cryst. Liq. Cryst., 15:27 (1971).
73. F.D. Saeva and J. Wysocki, J. Amer. Chem. Soc., 93:5928, (1971).
74. H. Finkelmann and H. Stegemeyer, Z. Naturforsch., 28a:1046 (1973).
75. J. Voss and E. Sackmann, Z. Naturforsch., 28a:1496 (1973).
76. H. Stegemeyer and H. Finkelmann, Chem. Phys. Lett., 23:227 (1973).
77. J. Adams, G. Dir, and W. Haas, in Liquid Crystals and Ordered Fluids, Vol. 2 (J.F. Johnson and R.S. Porter, eds.), Plenum, New York, 1974, p. 421.
78. J.E. Adams, W.E. Haas, and G.A. Dir, Abstracts, V International Liquid Crystal Conference, 1974, p. 166.
79. J.E. Adams and L.B. Leder, Chem. Phys. Lett., 6:90 (1970).
80. J.M. Pochan and D.D. Hinman, J. Phys. Chem., 78:1206 (1974).
81. P.N. Keating, Mol. Cryst. Liq. Cryst., 8:315 (1969).
82. J.L. Fergason, N.N. Goldberg, and R.J. Nadalin, in Liquid Crystals (G.H. Brown, G.J. Dienes, and M.M. Labes, eds.), Gordon and Breach, New York, 1966, p. 105.

83. G. Durand, C.R. Acad. Sci., Paris, Ser. A.B, 264:1397 (1967).
84. E. Sackmann, S. Meiboom, L.C. Snyder, A.E. Meixner, and R.E. Dietz, J. Amer. Chem. Soc., 90:3567 (1968).
85. H. Hakemi and M.M. Labes, J. Chem. Phys., 58:1318 (1973).
86. P. Pollmann and H. Stegemeyer, Chem. Phys. Lett., 20:87 (1973); P. Pollman, Ber. Bunsenges. Phys. Chem., 78:374 (1974).
87. B. Boettcher, Chem. Z., 96:214 (1972).
88. J.J. Wysocki, J. Adams, and W. Haas, Phys. Rev. Lett., 20:1024 (1968).
89. W. Haas, J. Adams, and J.B. Flannery, Phys. Rev. Lett., 24:577 (1970).
90. J. Wysocki, J. Adams, and W. Haas, Mol. Cryst. Liq. Cryst., 8:471 (1969).
91. G.H. Heilmeier and J.E. Goldmacher, J. Chem. Phys., 51:1258 (1969).
92. F. J. Kahn, Phys. Rev. Lett., 24:209 (1970).
93. H. Baessler and M.M. Labes, Phys. Rev. Lett., 21:1791 (1968); J. Chem. Phys., 51:1846 (1969).
94. J.R. Hansen and R.J. Schneeberger, IEEE Trans. Elec. Dev. ED-15:896 (1968).
95. J.L. Fergason, Brit. Patent, 1,123,117 (1968).
96. E.F. Carr, J.H. Parker, and D.P. McLemore, in Liquid Crystals and Ordered Fluids (J.F. Johnson and R.S. Porter, eds.), Plenum, New York, 1970, p. 201.
97. R.B. Meyer, Appl. Phys. Lett., 14:208 (1969).
98. E. Sackmann, S. Meiboom, and L. Snyder, J. Amer. Chem. Soc., 89:5981 (1967).
99. H. Gasparoux, G. Sigaud, and J. Prost, Mol. Cryst. Liq. Cryst., 22:189 (1973).
100. P.G. de Gennes, Solid State Commun., 6:163 (1968).
101. R.B. Meyer, Appl. Phys. Lett., 12:281 (1968).
102. P.G. de Gennes, Mol. Cryst. Liq. Cryst., 7:325 (1969).
103. W.E. Haas and J.E. Adams, Abstracts, V International Liquid Crystal Conference, 1974, p. 239.
104. S. Kusabaysahi and M.M. Labes, Mol. Cryst. Liq. Cryst., 7:395 (1969).

105. G. Dir, J. Adams, and W. Haas, in Liquid Crystals and Ordered Fluids, Vol. 2 (J.F. Johnson and R.S. Porter, eds.), Plenum, New York, 1974, p. 429.
106. H.W. Gibson and J.M. Pochan, J. Phys. Chem., 77:837 (1973).
107. G.W. Gray, J. Chem. Soc., 3733 (1956).
108. E.M. Barrall II, R.S. Porter, and J.F. Johnson, J. Phys. Chem., 71:1224 (1967).
109. P.J. Sell and A.W. Neumann, Z. Phys. Chem. Neue Folge, 65:13 (1969).
110. R.D. Ennulat, Mol. Cryst. Liq. Cryst., 8:247 (1969).
111. H. Arnold, D. Demus, H.J. Koch, A. Nelles, and H. Sackmann, Z. Phys. Chem. (Leipzig), 240:185 (1969).
112. G.J. Davis, R.S. Porter, E.M. Barrall II, Mol. Cryst. Liq. Cryst., 11:319 (1970).
113. E.M. Barrall II, J.F. Johnson, and R.S. Porter, Mol. Cryst. Liq. Cryst., 8:27 (1969).
114. S. Chandra and B. Bahadur, J. Chem. Phys., 70:605 (1973).
115. H.W. Gibson, Mol. Cryst. Liq. Cryst., 27:43 (1974).
116. H.W. Gibson, J. Phys. Chem., 80:1310 (1976).
117. J.M. Pochan and H.W. Gibson, J. Phys. Chem., 78:1740 (1974).
118. W. Elser, J.L.W. Pohlmann, and P.R. Boyd, Mol. Cryst. Liq. Cryst., 11:279 (1970).
119. L.M. Cameron, R.E. Callender, and A.J. Kramer, Mol. Cryst. Liq. Cryst., 16:75 (1972).
120a. W. Elser, J.L.W. Pohlmann, and P.R. Boyd, Mol. Cryst. Liq. Cryst., 15:175 (1971).
120b. W. Elser, J.L.W. Pohlmann, and P.R. Boyd, Mol. Cryst. Liq. Cryst., 27:325 (1974).
121. J.L.W. Pohlmann, W. Elser, and P.R. Boyd, Mol. Cryst. Liq. Cryst., 20:87 (1973).
122. R.D. Ennulat and A.J. Brown, Mol. Cryst. Liq. Cryst., 12:367 (1971).
123. W. Elser, Mol. Cryst., 2:1 (1966).
124a. W. Elser, Mol. Cryst. Liq. Cryst., 8:219 (1969).
124b. W. Elser, R.D. Ennulat, aNd J.L.W. Pohlmann, Mol. Cryst. Liq. Cryst., 27:375 (1974).

125. W. Elser and R.D. Ennulat, J. Phys. Chem., 74:1545 (1970).
126. J.S. Dave and R.A. Vora, in Liquid Crystals and Ordered Fluids (J.F. Johnson and R.S. Porter, eds.), Plenum, 1970, p. 477.
127. J.S. Dave and G. Kurian, Indian J. Chem., 11:833 (1973).
128. J.S. Dave and R.A. Vora, Mol. Cryst. Liq. Cryst., 14:319 (1971).
129. E.M. Barrall II, K.E. Bredfeldt, and M.J. Vogel, Mol. Cryst. Liq. Cryst., 18:195 (1972).
130. C. Wiegand, Z. Naturforsch., 4b:249 (1949).
131. A.M. Attallah and H.J. Nicholas, Mol. Cryst. Liq. Cryst., 18:339 (1972).
132. J.L.W. Pohlmann, W. Elser, and P.R. Boyd, Mol. Cryst. Liq. Cryst., 13:243 (1971).
133. J.L.W. Pohlmann, W. Elser, and P.R. Boyd, Mol. Cryst. Liq. Cryst., 26:59 (1974).
134. L.B. Leder, J. Chem. Phys., 55:2649 (1971); 58:1118 (1973).
135. J.Y.C. Chu, J. Chem. Soc. Chem. Commun., 374 (1974).
136. J.Y.C. Chu, J. Phys. Chem., 79:119 (1975).
137. J.M. Pochan, J.E. Kuder, J.Y.C. Chu, and D. Wychick, Can. J. Chem., 53:578 (1975).
138. L.B. Leder, J. Chem. Phys., 54:4671 (1971).
139. A. Kuksis and J.M.R. Beveridge, J. Org. Chem., 25:1209 (1960).
140. J.L.W. Pohlmann, Mol. Cryst., 2:15 (1966); G.G. Maidachenko and I.G. Chistyakov, Zhur. Obshch. Khim., 37:1730 (1967).
141. J.L.W. Pohlmann, Mol. Cryst. Liq. Cryst., 8:417 (1969).
142. F. Knapp and H.J. Nicholas, Mol. Cryst. Liq. Cryst., 10:173 (1970).
143. F.F. Knapp and H.J. Nicholas, J. Org. Chem., 33:3995 (1968).
144. F.F. Knapp and H.J. Nicholas, J. Org. Chem., 34:3328 (1969).
145. F.F. Knapp and H.J. Nicholas, in Liquid Crystals and Ordered Fluids (J.F. Johnson and R.S. Porter, eds.), Plenum, New York, 1970, p. 147.
146. Y.B. Amerik and B.A. Krentzel, J. Polym. Sci., Part C, 16:1382 (1967).
147. W.J. Toth and A.V. Tobolsky, Polym. Lett., 8:289 (1970).
148. H. Kamogawa, Polym. Lett., 10:7 (1972).
149. A.C. de Visser, K. de Groot, J. Feyen, and A. Bantjes, Polym. Lett., 10:851 (1972).

150. L. Strzelecki and L. Liebert, Bull. Soc. Chim. Fr., 2:597 (1973).

151. L. Liebert and L. Strzelecki, Bull. Soc. Chim. Fr., 2:603 (1973).

152. W.E. Bacon, Abstracts, V International Liquid Crystal Conference, 1974, p. 109.

153. F.D. Saeva and P.E. Sharpe, J. Amer. Chem. Soc., 97:204 (1975).

154. C.H. Lochmüller and R.W. Souter, J. Chromatogr., 87:243 (1973); 88:41 (1974).

155. H.A. Scheraga, Curr. Top. Biochem., 1 (1974).

Chapter 4

SMECTIC LIQUID CRYSTALS

S. E. B. Petrie

Eastman Kodak Company
Research Laboratories
Rochester, New York

I. INTRODUCTION

Many organic compounds that show liquid crystallinity on heating (mesogenic compounds or mesogens) give a series of mesophases, with order being lost stepwise over a range of temperature [1-10]. A given compound may give one or more smectic phases or one or more smectic phases in addition to a nematic or cholesteric (skewed nematic) phase. In contrast to the nematic phase, the smectic phases have stratified structures, with the long axes of the rod- or lathlike molecules parallel to each other in the layers. In most smectic phases, the molecules are mobile in two directions and can rotate about one axis. The interlayer attractions are weak compared with the lateral forces between the molecules, and the layers are able to slide over one another relatively easily. Consequently, the smectic phases have fluid properties, but are much more viscous than the other mesophases. Consistent with their higher order, the smectic phases always occur at lower temperatures.

The lower order smectic phases are unstructured, with the molecules in the layers being randomly distributed; the layers of higher order mesophases, however, are structured, the molecules being arranged in a regular two-dimensional lattice. The long axes of the molecules may be approximately normal to the plane of the layers or inclined, regardless of whether the smectic layers are structured or unstructured.

As demonstrated by Friedel [11] and others [5,8,12-15] the well-defined interlayer spacings of the characteristic periodic packing of molecular layers can be measured by x-ray diffraction. The x-ray diffraction pattern of nonoriented samples (powder pattern) consists essentially of two rings: (1) an inner ring at small Bragg angles that corresponds to the layer spacing and is very well defined and (2) an outer ring at larger angles that is characteristic of the lateral distribution of parallel molecules within each layer and that generally looks like a liquid diffraction ring [11-15]. For further detail, see the chapter by de Vries on x-ray studies of mesophase structures (Chap. 1).

In this chapter, an attempt is made to integrate the information available from the thermal data, x-ray diffraction patterns, and textural phenomena in a discussion of the specific features of the various smectic phases known.

II. CLASSIFICATION AND CHARACTERISTICS OF SMECTIC PHASES

As many as eight thermotropic smectic phases have been identified [8,11,12]. After the classification system developed by Sackmann, Demus, and coworkers (e.g., see Ref. 8 and papers cited therein), they are designated S_A, S_B, S_C, S_D, S_E, S_F, S_G, and on the basis of recent x-ray evidence [14], S_H. The seven smectic phases S_A through S_G, as well as the nematic (N) and cholesteric (Ch) phases, have been classified on the basis of the miscibility relationships of liquid crystal modifications in binary systems, i.e., on isobaric temperature-concentration diagrams, supported by characteristic microscopic textures and x-ray diffraction patterns. According to the rule of selective miscibility, "all liquid crystalline modifications which exhibit an uninterrupted series of mixed crystals in binary systems without contradiction can be marked with the same symbol" [8]. In this selective miscibility rule, "without contradiction" means that those liquid crystalline modifications that have the same designation in no case exhibit an uninterrupted series of mixed liquid crystals with the liquid crystalline states of another designation.

The temperature-composition diagram of a typical binary system composed of components both exhibiting smectic A and C phases, in addition to the crystal (C) and isotropic liquid (Is) phases, is illustrated in Fig. 1 [16]. The various liquid crystalline states show mutually selective miscibility; that is, they form an uninterrupted series of mixed liquid crystals between corresponding smectic phases. A region of immiscibility is observed in the phase diagram illustrated in Fig. 2, however, since one component gives nematic and smectic C phases while the other gives smectic phases A and C.

Temperature-composition phase diagrams for liquid crystalline mixtures can be generated from thermal data [17-22] or, because of the various optical features characteristic of each mesophase structure, from observations of microscopic textures of the mesophases between crossed polarizers [1-10]. For example, the temperature-composition phase diagram illustrated in Fig. 3 for bis-(4'-n-octyloxybenzal)-1,4-phenylenediamine (OOBPD) and the corresponding heptyl derivative (HOBPD), both of which exhibit five smectic phases and a nematic phase, can be generated either from observed changes in the textures exhibited by the birefringent liquid crystalline film as a function of temperature or from differential scanning calorimetry (DSC) data [20,22].

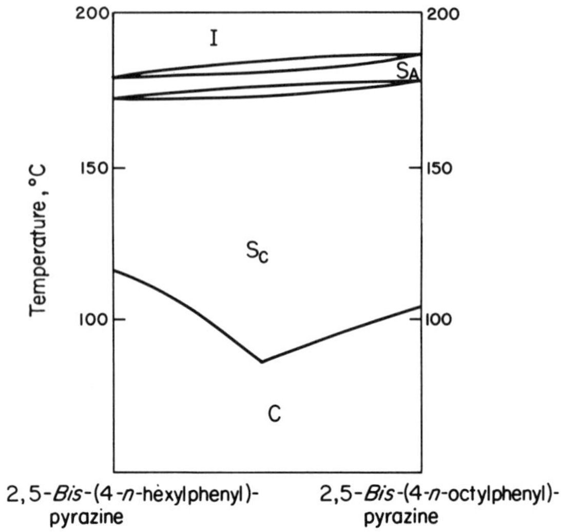

FIG. 1. Composition-temperature phase diagram for the binary system of the n-hexyl and n-octyl derivatives of 2,5-bis-(4-n-alkylphenyl)pyrazines [16].

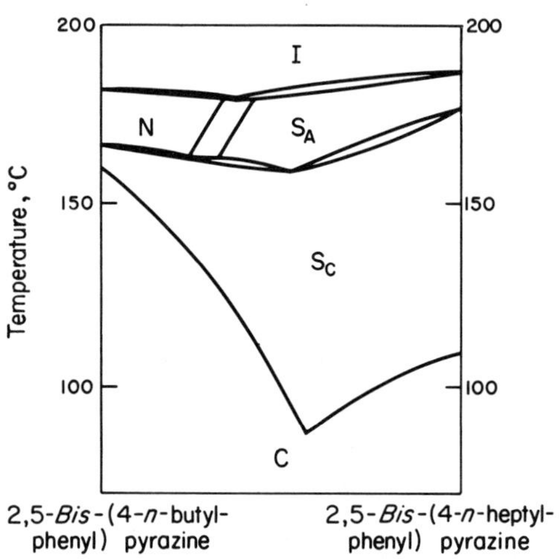

FIG. 2. Composition-temperature phase diagram for the binary system of the n-butyl and n-heptyl derivatives of 2,5-bis-(4-n-alkylphenyl)pyrazines [16].

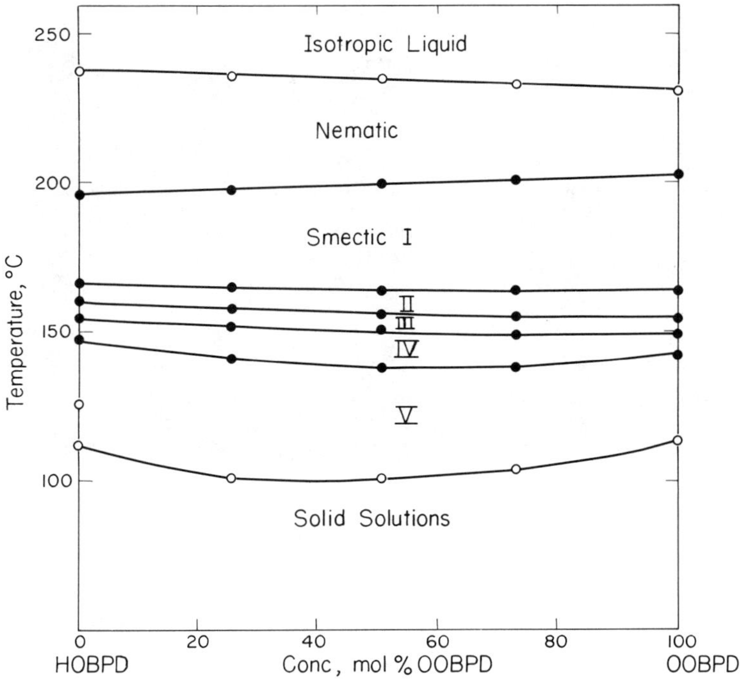

FIG. 3. Composition-temperature phase diagram for the binary system of HOBPD and OOBPD.

Microscopic textures [4-8] corresponding to various mesophases, ranging from the typical schlieren texture of the nematic phase to the mosaic textures characteristic of highly structured, lower temperature smectic phases, are illustrated in Fig. 4 for N,N'-terephthalylidene-bis-(4-n-butylaniline) (TBBA) [23]. The textural changes at some of the transitions are quite dramatic.

The series of microscopic textures observed between the melting points and the clearing points for the two phenylenediamine derivatives are comparable, supporting the conclusion that the two compounds give equivalent series of mesophase structures deduced from the observation that all the mesophases are continuous over the complete composition range in the binary temperature-composition phase diagram.

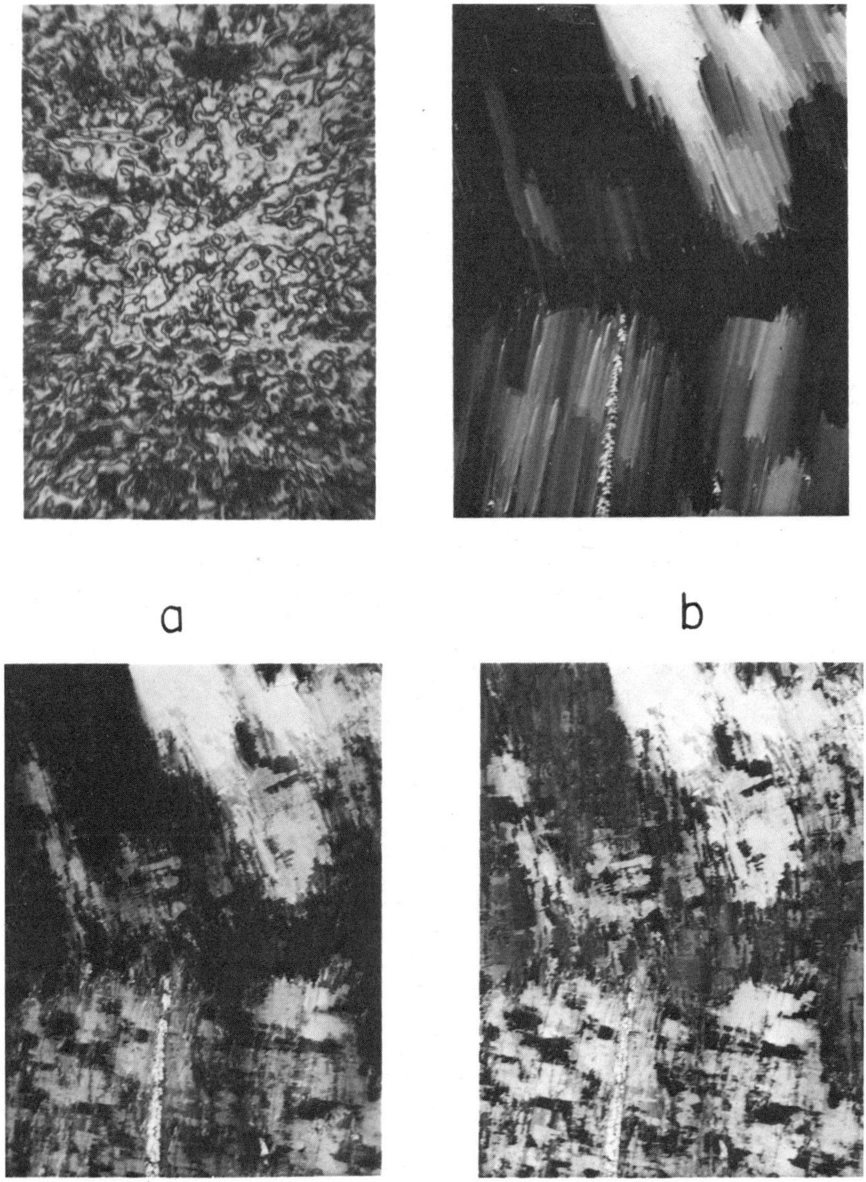

FIG. 4. Photomicrographs of TBBA between crossed polarizers, magnification about 100 X: (a) 200 °C, (b) 195 °C, (c) 165 °C, (d) 145 °C, (e) 137 °C, (f) 80 °C, and (g) 70 °C.

e

f

g

The differences between the textures exhibited by some mesophases are subtle, and some transitions may go undetected [23-25]. The absence of an observable change in optical properties between mesophases has been reported by Arora et al. for a number of 4-n-alkoxybenzylidene-4'-aminopropiophenones [25]. This effect is illustrated in Fig. 4, in which the changes in texture among the highly structured smectic phases of TBBA are minor. It is advisable, therefore, to use a combination of detection techniques including both optical and thermal measurements in order to ensure that characterization is as complete as possible. Even small changes in thermodynamic quantities at transitions between mesophases can be detected by DSC. Typical thermodynamic data for various thermal transitions between mesophases are given in Table 1 for OOBPD [23].

The ease and reliability with which mesophase transitions can be detected by DSC are illustrated by the DSC thermograms reproduced in Figs. 5a and 5b for HOBPD and OOBPD [22], respectively. The lowest temperature transition in HOBPD is concluded to be crystal-to-crystal, since supercooling is observed for the two lowest temperature transitions in cooling cycles, and mesophase-mesophase and mesophase-isotropic liquid transitions characteristically exhibit little or no supercooling when the transition temperatures are reached. The transitions between the highly structured phases are readily detectable by DSC.

TABLE 1. Transition Enthalpies and Entropies for OOBPD

Transition from preceding phase to	Temperature (°C)	Enthalpy (kcal/mol)	Entropy (cal/mol/K)
S_5	115	4.2	10.9
S_4	142	1.6	3.9
S_3	149	0.50	1.2
S_2	155	0.10	0.23
S_1	164	0.76	1.7
N	203	0.85	1.8
I	231	0.54	1.1

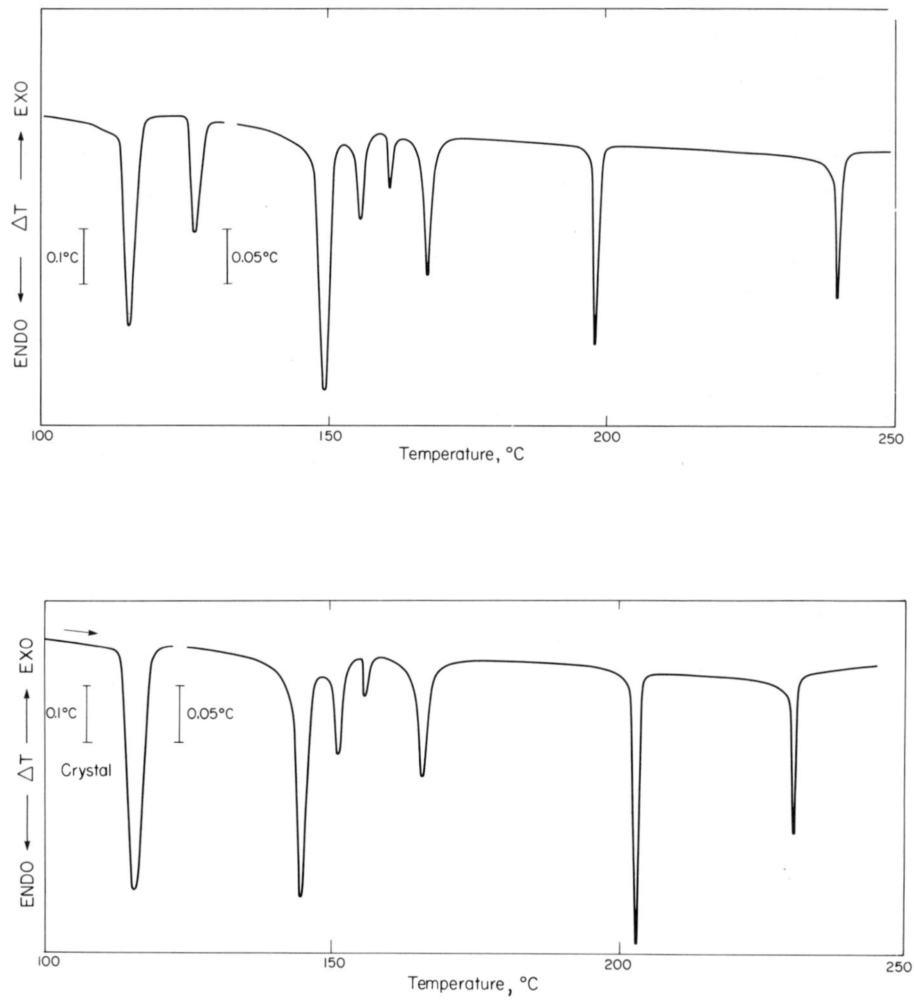

FIG. 5. DSC thermograms obtained at a heating rate of 5°C/min for (a) HOBPD and (b) OOBPD.

The similarities in the enthalpy changes and pretransition phenomena associated with the series of mesophase transitions for the two compounds further substantiate the conclusion reached in the microscopic texture and binary mixture studies that the two compounds have equivalent mesomorphic behavior. Also consistent with this conclusion is the observation that the thermal behavior for the 50:50 mixture, aside from slight temperature shifts, is not noticeably different from that of the pure components (Fig. 6). No detectable broadening of the transition temperature intervals associated with the mesophase transitions is observed [23,24]. It is recognized, of course, that these compounds are sufficiently similar structurally that they form a continuous series of solid solutions throughout the complete composition range. For compounds less similar in molecular structure, broadening of the transition temperature intervals may be observed [20,23]. These effects are indicated by the transition envelopes depicted in Figs. 1 and 2.

The objective in the classification of liquid crystalline modifications is to group the great number of modifications into a few types based on

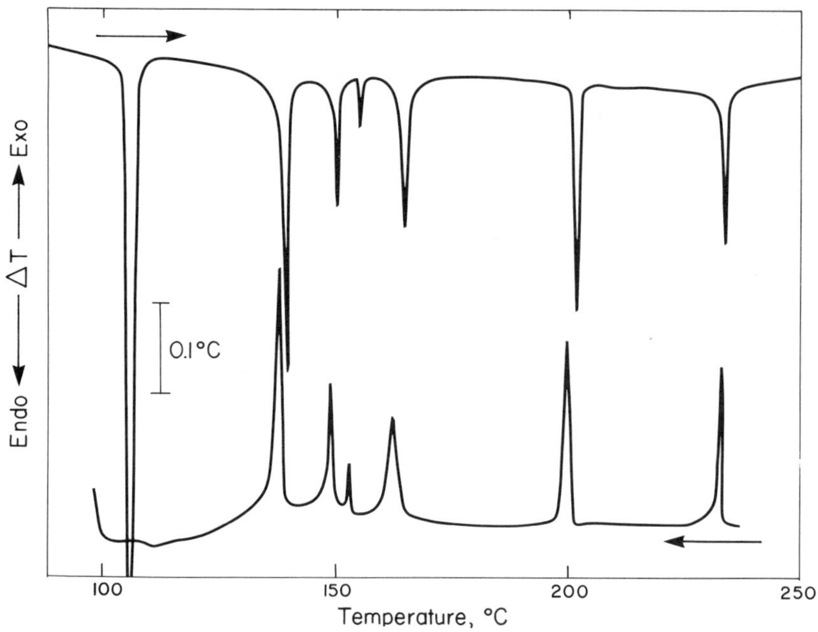

FIG. 6. DSC thermogram obtained at a heating rate of 5°C/min for 50:50 mixture of HOBPD and OOBPD.

general ordering features. It is recognized that the classification by means of miscibility investigations is quite empirical and is limited in some cases by molecular characteristics. It is possible that complete miscibility between similar modifications may be prevented because of great differences of intermolecular forces, as expressed by great differences in thermal transition temperatures and molecular size. Moreover, organic compounds with strong polar groups may be incompatible with nonpolar compounds having similar liquid crystal modifications.

The latitude for variations in molecular structural features of the components in systems with complete miscibility appears to decrease generally in the sequence: isotropic liquid, nematic phase, smectic phases, crystal. This trend is observed in the binary system of OOBPD and the p-octyl derivative of phenylene dibenzoate (PDB) [20,22], whose temperature-composition phase diagram is illustrated in Fig. 7. A well-defined envelope for the smectic-nematic phase transition is observed, while the nematic-isotropic liquid transition remains sharp throughout the complete composition range.

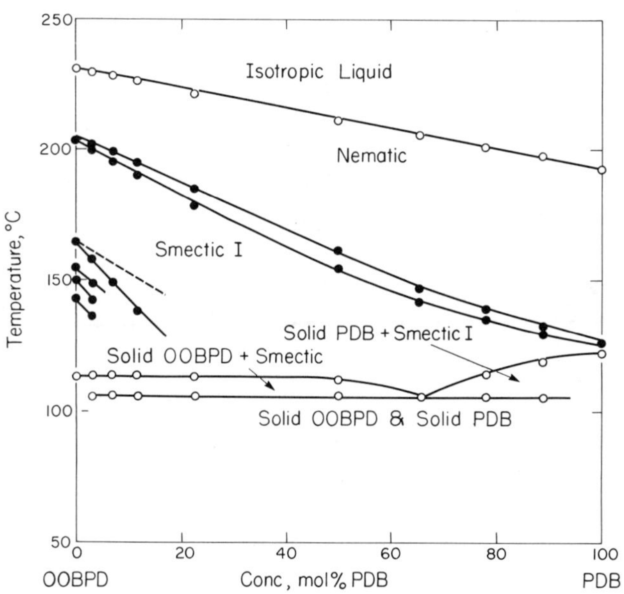

FIG. 7. Composition-temperature phase diagram for OOBPD and PDB.

The classification of liquid crystal modifications on the basis of miscibility relationships, however, has been effective in permitting reasonable selection of substances for further physical studies that are basic to definitive characterization of the main structural types. As the various smectic phases become more adequately characterized, a more rigorous classification system based on the structural features of the various smectic modifications should be realized. Such a classification system would bring better correspondence between the structure of a phase and its identification symbol. The original assignment of phase types was done on the basis of the order in which they were discovered, rather than according to their structures.

With this objective, de Vries [26] has proposed a classification system (Table 2) in which the smectic phases are divided into three main classes according to the degree of order in the molecular packing, the degree of order being judged by the appearance of the outer ring(s) in the diffraction patterns (diffraction rings with Bragg angles of about 10° or more). The viability of this classification scheme will become more apparent as detailed structural information for the various liquid crystal modifications becomes available.

TABLE 2. de Vries Classification of Smectic Phases

Characteristics of outer ring in diffraction pattern	Phase type by original classification scheme	Proposed classification scheme	
		Class symbol	Phase type
Diffuse	$S_A, S_C, S_F, (S_D)$[a]	$S(\alpha)$	$\alpha_1, \alpha_2, \alpha_3$
Sharp, one	S_B	$S(\beta)$	β_1
Sharp, several	S_E, S_G, S_H	$S(\gamma)$	$\gamma_1, \gamma_2, \gamma_3$[b]

[a]More characterization of the smectic D phase is required before unequivocal assignment can be made. If the S_D phase is to be included in the $S(\alpha)$ class, its place would be between α_1 and α_2 based on current data.

[b]Since sequence information is not available for the smectic modifications in this class, the assignment of γ_1 to S_E, etc., is rather arbitrary.

A. Smectic A Phase

Within the layers of a smectic A phase, the long axes of the molecules lie parallel to one another, their direction being normal to the plane of the layer. The molecules are free to rotate about their long axis, and the distribution of the molecules within the layers is random [27-29]. Thus, the layers have a low degree of order, each layer being a two-dimensional liquid. The layer thickness, determined from x-ray scattering data, is in most cases close to the full molecular length. The molecular arrangement is illustrated schematically in Fig. 8.

The smectic A phase is the most common type encountered [1,3-9,30] and is the highest temperature smectic form in compounds exhibiting smectic polymorphism. On heating, it transforms accordingly to a nematic or cholesteric phase, or to the normal amorphous liquid. The enthalpy change associated with S_A-N or S_A-Ch transitions is significantly less than that of S_A-Is transitions, as the layered structure disappears along with the orientational order [31].

As a consequence of infinite-fold rotational symmetry about an axis parallel to the direction normal to the layer, smectic A phases at thermal equilibrium are optically uniaxial [32]. The planar layers give rise to stepped drops, homogeneous (areas of homogeneously aligned molecules), or pseudoisotropic (optical axis oriented perpendicular to the plane of the slide) textures [4-10] (see Fig. 4b). Since curved layers are formed even under the influence of rather weak forces, focal conic textures (fan shaped or polygonal in thicker layers) are frequently observed in smectic A phases [4-10]. In thin layers, the molecules may spontaneously align normal to the surface, i.e., may adopt pseudoisotropic alignment, giving a dark, essentially textureless field between crossed polarizers.

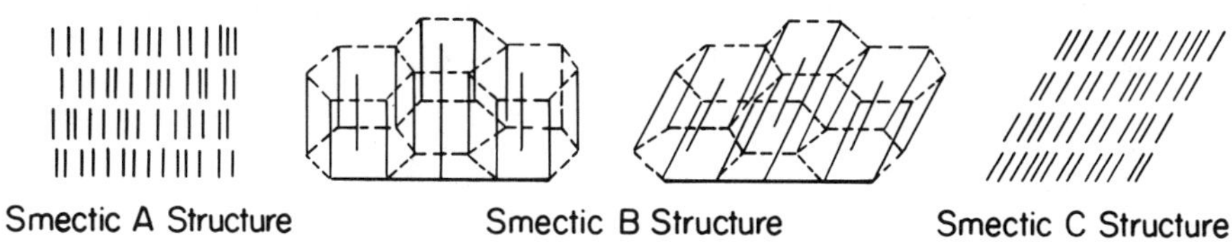

FIG. 8. Schematic representation of molecular arrangements in certain smectic phases.

Some well-known examples of compounds that show smectic A phases are diethyl 4,4'-azoxydibenzoate [1,3]

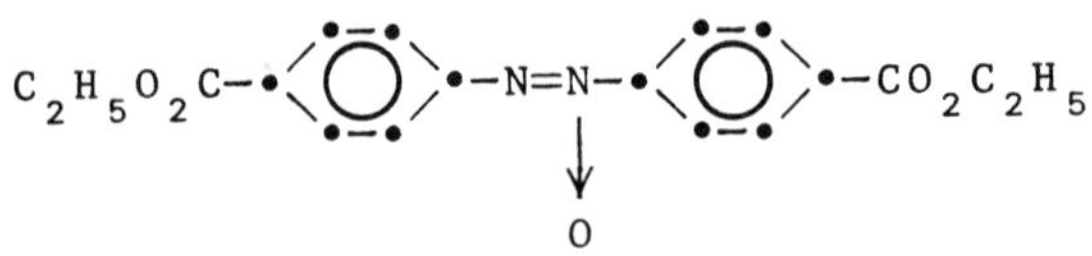

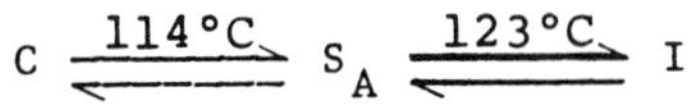

and the homologous series of di-n-alkyl p-azoxy-α-methylcinnamates [33]

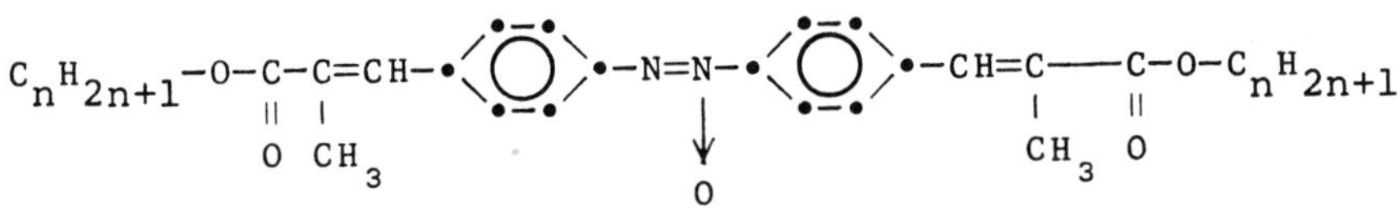

with n = 2 to 12. The ethyl and propyl derivatives exhibit nematic phases as well as smectic A phases, while the higher members, octyl through dodecyl, give smectic C phases in addition to the smectic A phases.

B. Smectic B Phase

The molecules in the smectic B phase are also arranged in layers with parallelism of the long axes and with no long-range order in the direction of the molecular axes. x-Ray scattering data suggest that the layers, however, have the periodicity and rigidity of a two-dimensional solid. Within the layers, the centers of gravity of the molecules are close-packed hexagonal [15,32,34,35]. There is rotational motion about the long axes of the molecules, but the rotational processes are probably correlated [15].

Two different smectic B modifications have been identified [35]:

1. The long axes of the molecules in a hexagonal layer structure are orthogonal to the plane of the structured layers, affording positive optical uniaxial character. An example of this smectic B modification is ethyl 4-(4'-ethoxybenzylideneamino)cinnamate [15].

$$C \underset{}{\overset{81°C}{\rightleftharpoons}} S_B \overset{119°C}{\rightleftharpoons} S_A \overset{157°C}{\rightleftharpoons} N \overset{159°C}{\rightleftharpoons} I$$

2. The long axes of the molecules in a monoclinic (pseudohexagonal) layer structure are tilted relative to the plane of the structured layers giving rise to positive optical biaxial character. An early example of this smectic modification was N,N'-terephthalylidene-bis-(4-n-butylaniline) (TBBA) [23,36].

$$C \overset{112.5°C}{\rightleftharpoons} S_B \overset{144.0°C}{\rightleftharpoons} S_C \overset{172.5°C}{\rightleftharpoons} S_A \overset{198.5°C}{\rightleftharpoons} N \overset{235.5°C}{\rightleftharpoons} I$$

These structures are illustrated schematically in Fig. 8.

Since the primary difference between the two smectic B modifications is the orientation of the principal axis relative to the smectic layer, a difference that is analogous to that observed between smectic A and smectic C phases (see below), there is some support for classifying the tilted smectic B modification as smectic H [14], as discussed further below. In the temperature-composition phase diagrams of binary systems studied to date, the two smectic B modifications are connected by an uninterrupted series of mixed liquid crystals, however, and are classified by Sackmann and Demus [8] as one type. The optical character of the mesophase changes continuously from uniaxial to biaxial [8].

The smectic B layers are not very flexible, and the typical texture is the mosaic texture exhibiting domains inside which the layers are flat. Because of the highly ordered character of the layers, the enthalpy changes associated with the loss of the mesomorphic structure are large compared with those of smectic A and C phases, and the smectic B phase occurs at lower temperatures. The DSC curve for TBBA is given in Fig. 9, and the thermodynamic data are summarized in Table 3 [23].

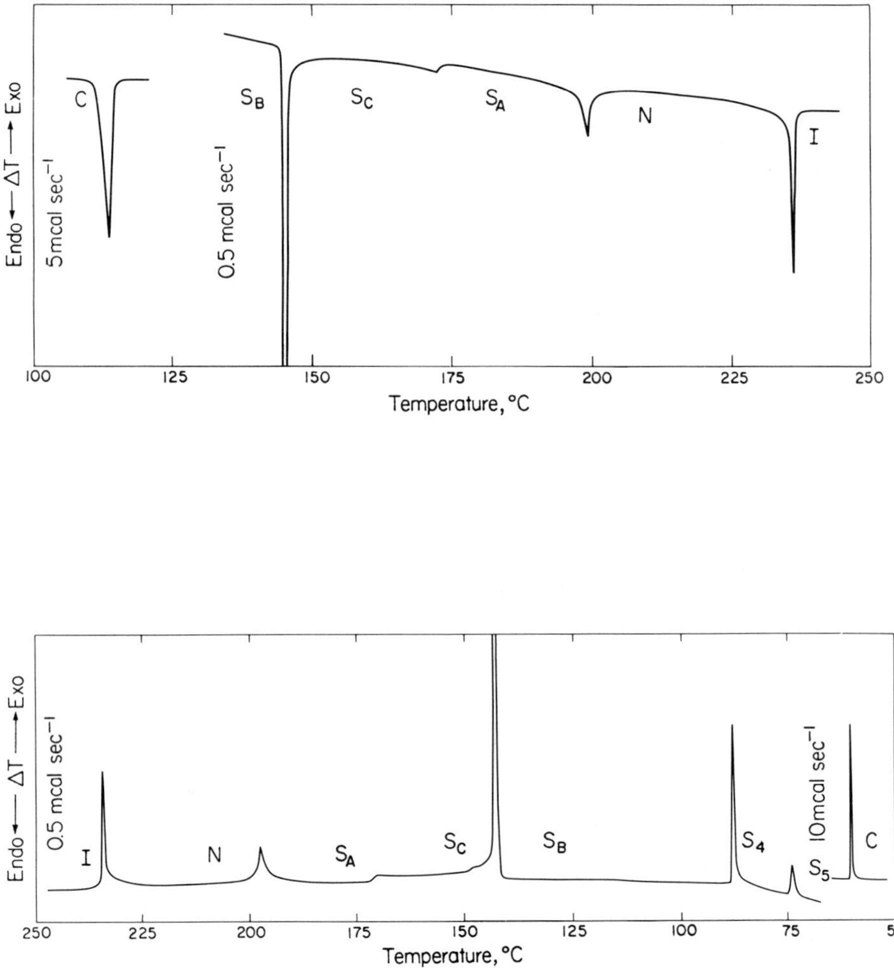

FIG. 9. DSC thermograms for TBBA obtained at a programming rate of 5 °C/min: (a) heating, (b) cooling.

TABLE 3. Transition Enthalpies and Entropies for TBBA[a]

Transition from preceding phase to	Temperature (°C)	Enthalpy (kcal/mol)	Entropy (cal/mol/K)
Crystal - S_B	112.5	4.2	10.9
S_5			
S_G	80.0[b]	0.075	0.21
S_B	89.5	0.26	0.72
S_C	144.0	0.88	2.1
S_A	172.5[c]		
N	198.5	0.16	0.35
I	235.5	0.21	0.41

[a]From Ref. 23.
[b]The S_4-S_5 phase transition in TBBA on cooling occurred at 75.5°C.
[c]Transition heat is small and diffuse.

As discussed above, the texture typical of smectic B phases is the mosaic texture (Fig. 4e). This texture consists of plane layers with each mosaic area having a uniform birefringent color, the color being determined by the orientation of the major molecular axis. The smectic B phase, however, frequently displays the texture of the precursor phase. The focal conic fan-shaped or polygonal texture, or pseudoisotropic alignment, can be imposed on the smectic B modification by a precursor smectic A phase on cooling. There are fewer striations along the lengths of the fans, but the fanlike areas are well defined. Since these textures are not natural for the smectic B phase, they are transient and transform in time to the characteristic mosaic texture. It is probable that the more structured, less flexible smectic B layers cannot accommodate bend deformations so readily as the smectic A phases. This conforming to a typical texture has been termed paramorphosis by Sackmann and coworkers [4].

Because of the similarities in the focal conic fan texture of the smectic A phase and the paramorphotic focal conic fan texture of the smectic B phase, the $S_B \rightleftharpoons S_A$ may be difficult to detect by optical microscopy. At the moment of transition, however, transient transition bars appear in the fanlike areas.

The typical texture for a tilted smectic B phase is also mosaic, but it can exhibit the nematiclike schlieren texture, also observed in smectic C textures. In the latter case, mosaic structures eventually form on prolonged heating. The paramorphotic schlieren texture of the tilted smectic B phase has been observed for 4-n-heptanoyl-4''-bromo-p-terphenyl, which is one of the very few materials reported to give a direct S_B-N transition [37,38].

$$C \underset{}{\overset{178°C}{\rightleftharpoons}} S_E \overset{203.5°C}{\rightleftharpoons} S_B \overset{212°C}{\rightleftharpoons} N \overset{239°C}{\rightleftharpoons} I$$

C. Smectic C Phase

The structure of the smectic C phase is similar to that of the unstructured smectic A phase except that the long axes of the molecules are tilted with respect to the planes of the layers [32,36,39-41] (see Fig. 7). Thus, the optical character is positive biaxial [39], although the biaxial property is not very pronounced. As a consequence of the tilt angle, the layer spacing of the smectic C phase is less than that of the corresponding smectic A phase.

Like smectic A phases, the smectic C phase occurs fairly frequently [1-10,28]. This mesophase may be: (1) the only one observed, (2) followed by a nematic phase, or (3) one of a sequence of phases. The smectic C phase always occurs at lower temperatures than the smectic A phase in a sequence. If it is associated in a sequence with a smectic B phase, the latter is the tilted smectic B modification. The sequence is, as observed for the enantiotropic transitions for TBBA,

$$C \rightleftharpoons S_B \rightleftharpoons S_C \rightleftharpoons S_A \rightleftharpoons N \rightleftharpoons I$$

It is evident that smectic A and C phases are not miscible. The structured analogs of these mesophases, the normal smectic B and the tilted smectic B phases, however, appear to be miscible, and a transition between normal smectic B and tilted smectic B phases has not been reported. It has been suggested by De Vries that the reason for this behavior could be that the smectic layers are free to rotate on each other in the smectic A phases but not in the smectic B and C phases [42].

The tilt angle of the smectic C phase may be constant or temperature dependent [36], although for most smectic C phases the tilt angle is not strongly temperature dependent. Attempts have been made to subdivide the smectic C phases further [43]. Detailed measurements of smectic C_1 phases are: (1) the tilt angle is large, generally 40°, (2) the tilt angle has little temperature dependence, (3) the smectic C_1 phase on heating forms a nematic or isotropic phase directly, and (4) the corresponding nematic phases have pronounced smectic-C-like structure, i.e., skewed cybotactic nematic phases [44]. The 4-n-alkoxybenzoic acids [34], bis-(4'-n-alkoxybenzal)-2-chloro-1,4-phenylenediamines [44], and 4,4'-di-n-alkoxyazoxybenzenes [41] are examples of this modification.

The most characteristic feature of the smectic C_2 phase is the strong temperature dependence of the tilt angle, with the tilt angle approaching zero at the upper temperature limit. Also, the tilt angle is relatively small compared with those of the smectic C_1 phases. Examples of this modification are TBBA, discussed above, and 4-n-octyloxyphenyl 4-(4-n-octyloxybenzoyloxy)benzoate (PBOB) [23].

$$C \underset{}{\overset{87.0°C}{\rightleftharpoons}} S_C \overset{145.0°C}{\rightleftharpoons} S_A \overset{163.0°C}{\rightleftharpoons} N \overset{189.5°C}{\rightleftharpoons} I$$

The structural temperature dependence of the smectic C phase of TBBA is evident in both the textural observations (Figs. 4c and 4d) and the thermal data (Figs. 9a and 9b).

Smectic C structures, like nematic structures, can be twisted by boundary effects or by the addition of optically active compounds. These twisted smectic C modifications behave optically like cholesteric liquids [9,45-47]. Some optically active compounds have been reported to form a twisted smectic phase in addition to a cholesteric [45-47], e.g.,

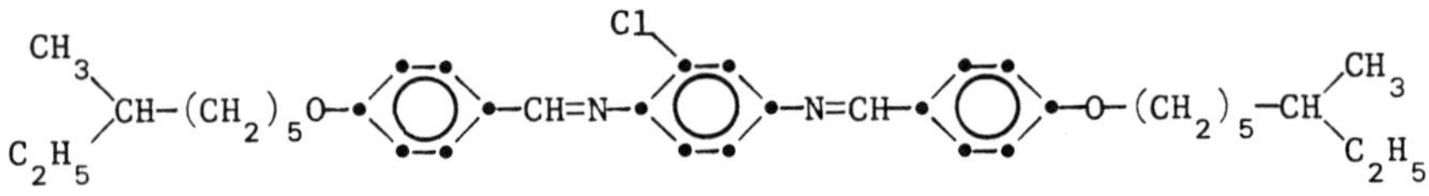

$$C \underset{}{\overset{29°C}{\rightleftharpoons}} S_C \overset{94.4°C}{\rightleftharpoons} Ch \overset{146.5°C}{\rightleftharpoons} I$$

Although the cholesteric phase showed a strong variation of optical rotation with temperature, as is characteristically observed for this mesophase [1-3], the optical activity of the smectic phase showed no measurable temperature dependence. This insensitivity appears plausible if the tilt angle is not temperature dependent and if the pitch is controlled by the interaction between smectic layers rather than between individual molecules.

The changes in enthalpy observed in the DSC curves for TBBA (Fig. 9) and PBOB (Fig. 10) are consistent with closely related structures for the smectic A and C phases and with a temperature-dependent, small, structural change throughout the temperature range of the smectic C phase. The structural temperature dependence of the smectic C phase of TBBA is also evident in the textural observations, as illustrated in Figs. 4c and 4d.

In the smectic A phase of 4-n-heptyloxybenzylidene-4'-n-heptylaniline, the molecular axis has been observed to tilt inside the layers above a critical compressive stress [48]. This mechanical instability can be visualized as a smectic A-to-smectic C phase change induced by compression. The threshold strain for this instability goes rapidly to zero as the S_A-S_C transition temperature is approached. The effect is transient due to rapid relaxation processes.

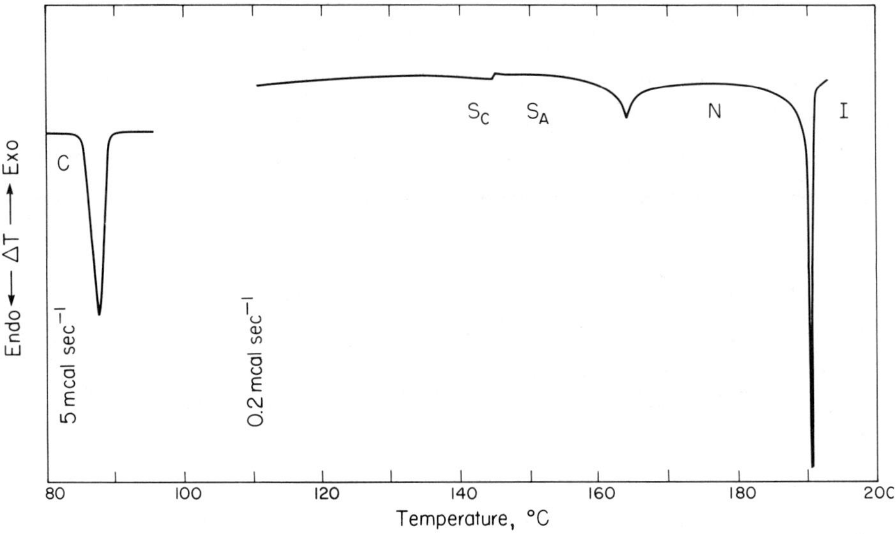

FIG. 10. DSC thermogram for PBOB obtained at a heating rate of 5 °C/min.

The textures characteristic of smectic C phases [4-8] are a function of their phase history. If the mesophase forms from a pseudoisotropic (homeotropic) nematic phase or similarly aligned smectic A phase on cooling, it frequently adopts a texture which is similar to a nematic schlieren texture. If the precursor nematic phase has a schlieren texture, the smectic C phase also usually adopts the schlieren texture. Similarly, if the smectic C phase follows a nematic phase of homogeneous texture, a texture consisting of large areas of homogeneous alignment, the mesophase tends to adopt a homogeneous smectic C texture, which has a mottled, sandy appearance.

Smectic C phases often exhibit focal conic textures when formed on cooling from smectic A phases having focal conic textures. Thin preparations show broken, fan-shaped textures; thicker preparations show broken polygonal textures. It would appear that the tilting of the molecules occurring in the smectic A layers at the S_A-S_C transition leads to packing problems and distortions that result in a less perfect continuum of focal conic domains.

D. Smectic D Phase

The mesophase classified as smectic D possesses a cubic structure and is optically isotropic [49-51]. This mesophase has been observed for two compounds of the homologous series of the 4'-n-alkoxy-3'-nitrobiphenyl-4-carboxylic acids, the hexadecyl and octadecyl derivatives.

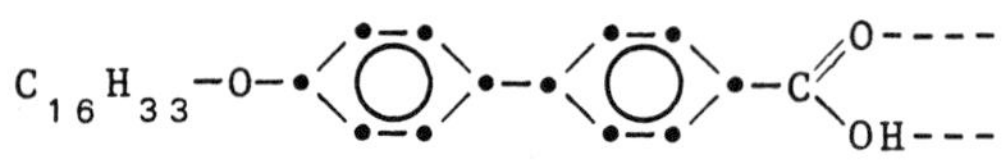

Dimer Formation

$$C \overset{127^\circ C}{\rightleftharpoons} S_C \overset{171^\circ C}{\rightleftharpoons} S_D \overset{197^\circ C}{\rightleftharpoons} S_A \overset{202^\circ C}{\rightleftharpoons} I$$

In the octadecyl derivative, no smectic A phase is observed; the molecular arrangement of the compound on heating passes directly from that of the smectic D phase to that of the isotropic liquid.

$$C_{II} \overset{85^\circ C}{\rightleftharpoons} C_I \overset{125^\circ C}{\rightleftharpoons} S_C \overset{159^\circ C}{\rightleftharpoons} S_D \overset{195^\circ C}{\rightleftharpoons} I$$

x-Ray studies show no evidence of a layered structure, but indicate a cubic lattice with a spacing that corresponds to the length of the dimer. Because of some similarities with lyotropic liquid crystals, where it is not uncommon for two successive anisotropic phases to be separated in the composition phase diagram by a cubic, optically isotropic phase, a model for the molecular arrangement of this mesophase has been proposed that is based on micelle formation and subsequent rearrangement into cubic symmetry [10].

The evidence to date suggests that the basic structural characteristics of the smectic D modification are distinctly different from those of all the other identified smectic phases of thermotropic liquid crystals. The mesophase appears to maintain well-developed long-range order in three dimensions, while remaining highly disordered at the atomic level. (The x-ray diagrams have a diffuse reflection at high angles [50], indicating that the hydrocarbon chains have characteristics typical of the liquid state.) The correspondence between the molecular arrangement of the smectic D phase with that of a cubic phase observed for lyotropic liquid crystals is probably the consequence of the similarity in chemical structure between compounds in which this mesophase has been identified and lipids, i.e., a head group linked to a long hydrocarbon chain.

E. Smectic E Phase

The smectic E modification is a layered structure with a level of order more closely approaching that of the crystalline state [50,52,53] than those observed for the smectic A, B, and C modifications. x-Ray diffraction patterns show three sharp rings at larger angles, compared with only one sharp outer ring for the relatively highly structured smectic B phase. This higher order is consistent with the observation that smectic E phases occur at lower temperatures than types A, B, and C in smectic polymorphism. For example, the sequence of mesophase formation for n-hexyl 4-(4'-phenylbenzylideneamino)cinnamate [53,54] is

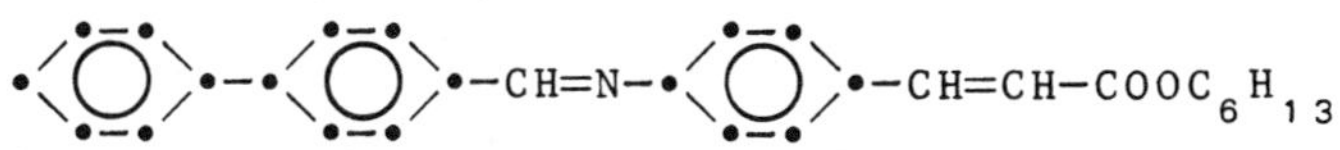

$$C \underset{}{\overset{80°C}{\rightleftharpoons}} S_E \overset{99°C}{\rightleftharpoons} S_B \overset{165.5°C}{\rightleftharpoons} S_A \overset{199°C}{\rightleftharpoons} I$$

Several homologs of this series exhibit similar smectic polymorphism.

The energy changes associated with the various transitions for this liquid crystal are tabulated in Table 4. The strong influence of molecular structure on the transition heats is quite evident from a comparison of the thermodynamic data in Table 4 with those for TBBA given in Table 3. For example, the total enthalpy change associated with the formation of the smectic A phase structure for the n-alkyl 4-(4'-phenylbenzylidene-amino)cinnamates from the isotropic melt is in excess of 1.0 kcal/mol for alkyl chains longer than propyl, which is more than double the corresponding change for smectic A phase formation in TBBA. Thus, correlations between transition enthalpy changes [17,31] and mesophase structure are limited.

On the basis of x-ray diffraction studies, Doucet et al. [52] concluded that the long axes of the molecules in the smectic E modification are normal to the layers. It was concluded that within the layers the molecules have a chevron structure with orthorhombic symmetry. Correlations of the

TABLE 4. Transition Enthalpies and Entropies for n-Hexyl 4-(4'-Phenylbenzylideneamino)cinnamate[a]

Transition from preceding phase to	Temperature (°C)	Enthalpy (kcal/mol)	Entropy (cal/mol/K)
S_E	80	5.07	14.4
S_B	99	0.12	0.32
S_A	165.5	0.85	1.93
I	199	1.15	2.44

[a]From Ref. 52.

orthorhombic array were also observed to exist over several layers. Although there is considerable three-dimensional ordering in the smectic E phase, the absence of high-index Bragg reflections indicates that there is also considerable disorder. The evidence is consistent with a model involving restricted reorientational motion of the molecules about their long axes.

Optically, the smectic E phase has been observed to be uniaxially positive, consistent with the normal orientation of the long molecular axes with respect to the layer planes [8,50].

Other examples of smectic E phases and variants of smectic polymorphism are

4"-n-octyl-4-cyano-p-terphenyl [55]

$$C \xrightleftharpoons{127°C} S_E \xrightleftharpoons{128°C} S_B \xrightleftharpoons{133°C} S_A \xrightleftharpoons{197°C} N \xrightleftharpoons{216°C} I$$

diethyl p-terphenyl-4,4"-carboxylate [50]

$$C \xrightleftharpoons{173.0°C} S_E \xrightleftharpoons{188.5°C} S_A \xrightleftharpoons{259.0°C} I$$

and several disubstituted biphenyls [56].

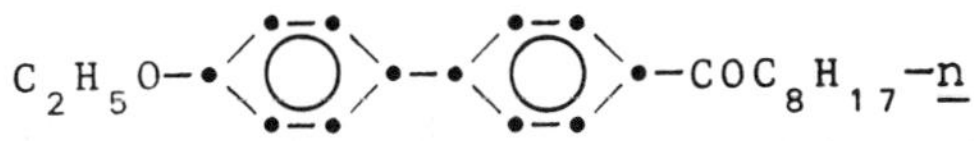

$$C \xrightleftharpoons{108.0°C} S_E \xrightleftharpoons{120.2°C} S_A \xrightleftharpoons{144.8°C} I$$

$$C \xrightleftharpoons{9°C} S_E \xrightleftharpoons{68°C} S_B \xrightleftharpoons{83.9°C} I$$

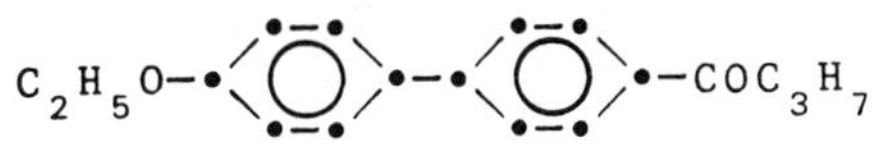

$$C \xrightleftharpoons{123.0°C} S_E \xrightleftharpoons{156.2°C} I$$

Thus, the S_E-Is transition as well as the S_B-Is transition exists.

Smectic E phases exhibit the mosaic textures characteristic of highly structured mesophases; the domains, however, are distinctively striated. As observed for smectic B phases, the smectic E modifications developed from smectic A phases on cooling typically show paramorphotic focal conic fan-shaped textures. Nevertheless, these textures are readily characterized by nontransient concentric bands which cross the fans.

The characteristic texture of the smectic E phases for some compounds, e.g., the n-alkyl 4-(4'-phenylbenzylideneamino)cinnamates, has been reported to be a platelet texture formed from a precursor smectic B phase in the pseudoisotropic condition [53]. In this case, the interference figure was observed to be biaxial positive, suggesting an inclination of the long axes of the molecules with respect to the layer planes. The molecules in the smectic E modification, however, are orthogonal to the layer planes [52]. Both the optical and x-ray data are accommodated by a model involving a chevronlike arrangement of the planes of the molecules viewed in the direction of the long molecular axes, which gives an orthorhombic symmetry [52].

F. Smectic F Phase

The smectic phase modification designated as F has been identified and characterized for 2-(4-n-pentylphenyl)-5-(4-n-pentyloxyphenyl)pyrimidine [57].

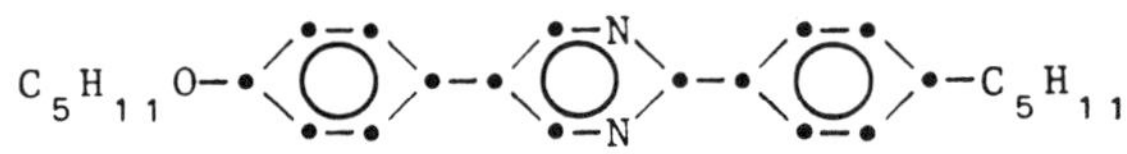

$$C \underset{}{\overset{79°C}{\rightleftharpoons}} S_G \overset{102.7°C}{\rightleftharpoons} S_F \overset{113.8°C}{\rightleftharpoons} S_C \overset{144°C}{\rightleftharpoons} S_A \overset{210°C}{\rightleftharpoons} I$$

The x-ray diffraction pattern for the smectic F phase is similar to that observed for smectic C modifications, indicating that the F modification also is composed of unstructured layers. Optically, however, the smectic F phase is positive uniaxial, while the smectic C phase is positive biaxial [58].

The texture phenomena for the smectic F phase are similar to those for smectic C phases. If the precursor smectic C phase has a focal conic broken fan-shaped texture or a schlieren texture, the smectic F phase shows accordingly a striped fan or schlieren texture. Although the textural differences between the smectic C and smectic F phases are subtle, the transition is readily detectable thermally and is reversible [57]. The thermodynamic data are given in Table 5 to emphasize the distinctiveness of the $S_F \rightleftharpoons S_C$ transition.

TABLE 5. Transition Enthalpies and Entropies for 2-(4-n-Pentylphenyl)-5-(4-n-pentyloxyphenyl)pyrimidine[a]

Transition from preceding phase to	Temperature (°C)	Enthalpy (kcal/mol)	Entropy (cal/mol/K)
S_G	79	8.37	23.8
S_F	102.7	0.14	0.37
S_C	113.8	0.12	0.31
S_A	144	0.05	0.012
I	210	2.74	5.76

[a]From Ref. 57.

Certain homologs of the series of benzylidene anilines investigated by Smith and Gardlund [59] are reported to exhibit smectic polymorphism involving smectic F phases. Identification was made on the basis of miscibility studies with the pyrimidine compound described above [58]. The three Schiff-base liquid crystals are the n-butyl, n-pentyl, and n-heptyl derivatives of N-(4-n-heptyloxybenzylidene)-4'-n-alkylanilines.

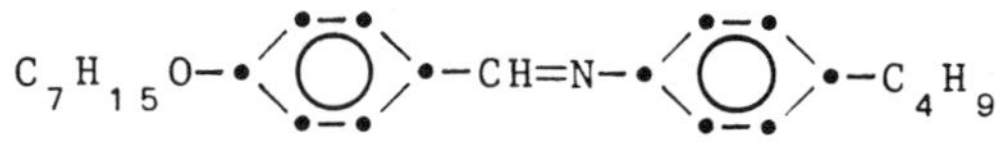

$$C \xrightleftharpoons{32°C} S_F \xrightleftharpoons{63°C} S_C \xrightleftharpoons{65°C} S_A \xrightleftharpoons{74°C} N \xrightleftharpoons{76°C} I$$

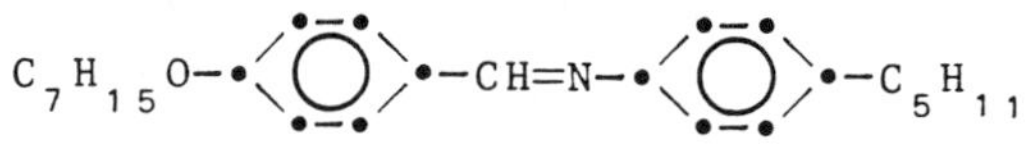

$$C \xrightleftharpoons{23°C} S_G \xrightleftharpoons{58°C} S_F \xrightleftharpoons{64°C} S_C \xrightleftharpoons{68°C} S_A \xrightleftharpoons{80°C} N \xrightleftharpoons{83°C} I$$

Smectic polymorphism reported for the n-heptyl derivative of N-(4-n-heptyloxybenzylidene)-4'-alkylanilines is comparable to that of the n-pentyl derivative given above.

Even on the basis of the limited evidence derived from x-ray, thermodynamic, and microscopic studies, the identification of the smectic F modification as a new mesophase appears well founded. From the x-ray data, it is clear that the mesophase has unstructured layers [57]. The observation thermally of a well-defined, reversible transition between the new smectic modification and the lowest temperature, unstructured smectic phase in smectic polymorphism is evidence of an additional distinctive smectic phase variation. The results of miscibility studies are consistent with this conclusion [57,58].

G. Smectic G Phase

The x-ray diffraction pattern for smectic G phases [57] indicates a molecular arrangement consisting of highly structured layers. The pattern consists of several outer rings, indicating a nonrandom arrangement, in addition to a sharp inner ring corresponding to the layer thickness. Further, the x-ray data indicate a monoclinic lattice with inclined long axes of the molecules in a herringbone array [14,60]. Despite the highly ordered structure, the mesophase shows a liquidlike behavior, and the transition to the mesophase in cooling cycles does not exhibit supercooling.

Consistent with the high level of structural order for mesophase modifications, the smectic G phase occurs at lower temperatures than the smectic A, B, C, and D phases and exhibits the typical mosaic textures as the natural or stable texture [57-59]. A paramorphotic focal conic fan-shaped texture is also observed depending upon the characteristics of the precursor phase and the cooling details. Optically, the smectic G phase is positive biaxial [58].

Compounds exhibiting smectic polymorphism involving smectic G phases have been given above. For example, TBBA gives a monotropic smectic G phase, as well as the enantiotropic smectic A, B, and C phases (see Table 3) [60,61]. The compound for which the smectic G modification was originally identified was 2-(4-n-pentylphenyl)-5-(4-n-pentyloxyphenyl)-pyrimidine [57], whose transition behavior is given in the discussion for the smectic F phase. The smectic sequence is: $C \rightleftharpoons S_G \rightleftharpoons S_F \rightleftharpoons S_C \rightleftharpoons S_A \rightleftharpoons Is$. The benzylidene anilines described in conjunction with the pyrimidine-based compound above show similar sequence behavior. Another homolog of the pyrimidine-based series [62]

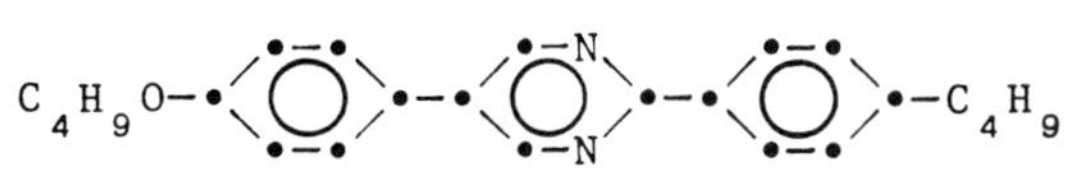

$$C \underset{}{\overset{90.5°C}{\rightleftharpoons}} S_G \overset{113.7°C}{\rightleftharpoons} S_C \overset{120.8°C}{\rightleftharpoons} S_A \overset{214.0°C}{\rightleftharpoons} I$$

gives the variant $S_G \rightleftharpoons S_C \rightleftharpoons S_A \rightleftharpoons Is$.

H. Smectic H Phase

The controversy concerning whether the smectic H phase, identified by de Vries and Fishel [14], is a new modification has apparently been resolved by recent miscibility studies by Richter et al. [61] that contradict some earlier miscibility data [63]. From these studies, it has been deduced that the smectic phase given by N-(4-n-butyloxybenzylidene)-4'-ethylaniline (BBEA) is the smectic G modification.

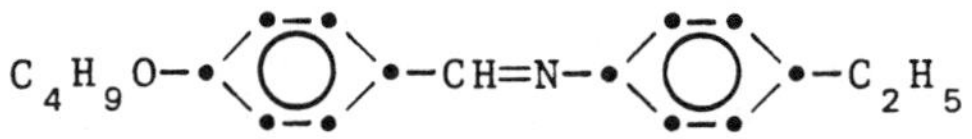

$$C \underset{\longleftarrow}{\xrightarrow{38.7\,°C}} S_G \underset{\longleftarrow}{\xrightarrow{50.5\,°C}} N \underset{\longleftarrow}{\xrightarrow{64.8\,°C}} I$$

The smectic phase given by BBEA forms a region of uninterrupted miscibility with the smectic G phases given by TBBA and 2-(4-n-pentylphenyl)-5-(4-n-pentyloxyphenyl)pyrimidine.

Thus, the $S_G \rightleftharpoons N$ variant exists. With homologs of the N-(4-n-alkoxybenzylidene)-4'-ethylaniline series, the $S_G \rightleftharpoons S_A \rightleftharpoons N$ variant has been observed [64].

The highly ordered nature of the smectic G phase is reflected in the atypically large change in enthalpy associated with the $S_G \rightleftharpoons N$ transition, as illustrated by the DSC curve reproduced in Fig. 11 [65]. The enthalpy change associated with the $S_G \rightleftharpoons N$ transition is 2.8 kcal/mol, compared with a rather low value of 1.4 kcal/mol for the $C \rightleftharpoons S_G$ transition. Also, the total energy loss associated with the development of smectic G order in BBEA on cooling from the isotropic melt is 2.94 kcal/mol, compared with 1.51 kcal/mol for TBBA (see Table 3) and 2.1 kcal/mol for 4-(4-n-propylmercaptobenzalamino)azobenzal [38].

A summary of the various thermotropic liquid crystals identified and their characteristics is given in Table 6.

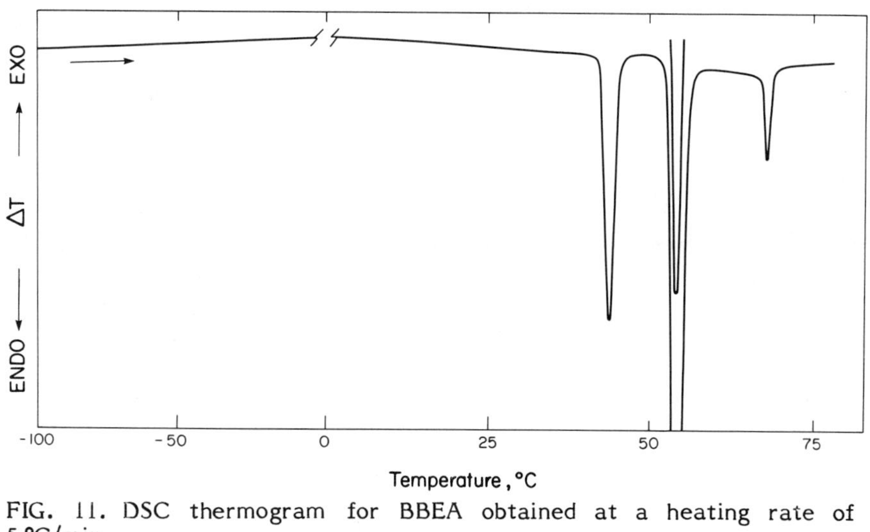

FIG. 11. DSC thermogram for BBEA obtained at a heating rate of 5 °C/min.

TABLE 6. Summary of Characteristics of Thermotropic Liquid Crystals

Modification	Optical Properties	Textures	Structures	Examples	Reference
Nematic	Uniaxially positive	Schlieren; threaded; marbled; homogeneous; pseudoisotropic	Parallelism of long molecular axes	p-azoxyanisole	1, 3
Cholesteric (spontaneously twisted nematic)	Uniaxially negative; or isotropic, optically active	Focal conic; Grandjean steps; homogeneous; isotropic	Twisted nematic structure	Cholesteryl cinnamate	1, 3
Cholesteric (spontaneously twisted smectic)	Biaxial	Focal conic; Grandjean steps	Layer structure	Bis-(p-6-methyloctyl-oxybenzylidene)-2-chloro-1,4-phenylene-diamine	47
Smectic A	Uniaxially positive	Focal conic; (fan shaped, polygonal); stepped drops; homogeneous; pseudoisotropic	Layer structure; molecular axes orthogonal to the layers; random arrangement within layers	Diethyl 4,4'-azoxydi-benzoate	1, 3
Smectic B	Uniaxially or biaxially positive	Mosaic; stepped drops; homogeneous; pseudoisotropic, schlieren	Layer structure; molecular axes orthogonal or inclined to the layers; hexagonal arrangement within the layers	N,N'-Terephthalylidene-bis-(4-n-butylaniline) (TBBA)	36

TABLE 6 continued.

Modification	Optical Properties	Textures	Structures	Examples	Reference
Smectic C	Biaxially positive	Broken focal conic; schlieren; homogeneous	Layer structure; molecular axes inclined to the layers; random arrangement within the layers	4,4'-di-n-dodecyloxy-azoxybenzene (TBBA)	41
Smectic D	Isotropic	Isotropic; mosaic	Cubic structure	4'-n-Octadecyloxy-3'-nitrobiphenyl-4-carboxylic acid	49
Smectic E	Uniaxially positive (biaxially positive)	Mosaic; pseudoisotropic (platelet)	Layer structure; molecular axes orthogonal to the layers; orthorhombic lattice within the layers	Diethyl-p-terphenyl-4,4''-carboxylate	50
Smectic F	Uniaxially positive	Broken focal conic with concentric axes; schlieren	Layer structure; random arrangement within the layers	2-(4-n-Pentylphenyl)-5-(4-n-pentyloxyphenyl) pyrimidine	57
Smectic G	Uniaxially positive	Mosaic	Layer structure; monoclinic lattice with tilted long axes of molecules in herringbone array	2-(4-n-Pentylphenyl)-5-(4-n-pentyloxyphenyl) pyrimidine	57

III. VARIANTS OF POLYMORPHISM OF LIQUID CRYSTALLINE STATES

The variants of polymorphism discussed above are summarized in Table 7. Since twisted structures, such as cholesteric and twisted smectic C phases, are always miscible with normal nematic and smectic C phases, the twisted and nontwisted structures are considered to be of the same phase type. Thus, the twisted phases are not included as additional variants.

TABLE 7. Variants of Polymorphism of Liquid Crystalline Phases

Monomorphism	N	Trimorphism	S_BS_AN
	S_A		S_CS_AN
	S_B		S_GS_AN
	S_C		S_ES_BN
	S_E		S_BS_CN
			$S_ES_BS_A$
			$S_GS_BS_A$
			$S_BS_CS_A$
Dimorphism	S_AN		$S_CS_DS_A$
	S_BN		$S_GS_CS_A$
	S_CN		
	S_GN		
	S_BS_A	Tetramorphism	$S_ES_BS_AN$
	S_CS_A		$S_GS_BS_AN$
	S_ES_A		$S_BS_CS_AN$
	S_ES_B		$S_FS_CS_AN$
	S_BS_C		$S_GS_FS_CS_A$
	S_CS_D		
		Pentamorphism	$S_GS_FS_CS_AN$
			$S_GS_BS_CS_AN$

The sequence of liquid crystalline states as indicated by the observed variants is

$$\begin{matrix} S_G \\ \\ S_E \end{matrix} \; S_B S_F S_C S_D S_A \; \begin{matrix} N \\ \\ Ch \end{matrix}$$

As yet, smectic E and G modifications have not been observed within the same sequence, and their respective order is unknown. The position given the smectic F modification in the sequence is based partly on the knowledge that highly ordered mesophases occur at lower temperatures than the less ordered mesophases. S_B and S_F modifications have not been observed in the same sequence, but on the basis of the level of structural order of the layers, the S_F phase would be expected to be a higher temperature modification relative to the S_B phase. The position of the S_E modification relative to S_G cannot be deduced by similar arguments, however, because both mesophases are characterized by highly structured layers.

As smectic polymorphism for more compounds is studied, additional variants are expected to be observed. For TBBA, for example, a low-temperature smectic phase has been observed in addition to the sequence $S_G S_B S_C S_A N$ [23,60,65]. It is frequently designated S_5. On the basis of the smectic modifications identified and the observed variant sequences, it would appear that S_5 must be either the smectic E phase or a new modification.

IV. MOLECULAR STRUCTURE VERSUS SMECTIC PHASE FORMATION

Certain guidelines exist regarding the features of molecular structure that favor liquid crystal formation [1,3,5,6,30]. Many molecular features are common to both nematic and smectic phase formation (see Chap. 2):

1. The molecule is long relative to its diameter. Flat segments, such as aromatic rings, favor liquid crystal formation.
2. The molecule has some rigidity along its long axis; double bonds and ring structures enhance rigidity.
3. The molecule may have strong dipoles and/or easily polarizable

groups. Liquid crystal formation is favored if both are present and the strong dipole is on the molecular axis.

4. The molecule may have weak polar groups at its extremities, although this characteristic is less influential than those listed above.
5. Molecular interactions may be involved that favor liquid crystal formation, such as: dipole-dipole interactions, ion-dipole interactions, induced dipole interactions, and possibly, hydrogen bonding.
6. Anisotropy of the cohesive forces between elongated molecules is essential. If intermolecular forces are very weak, the cohesive forces may be inadequate to maintain order in the fluid state. If the intermolecular forces are too strong, however, the melting point may be elevated to such a high temperature that thermal motion will prevent the existence of the ordered mesophase after melting occurs. Therefore, for liquid crystal formation, it is essential that the cohesive forces operating between elongated molecules be both anisotropic and of appropriate magnitude.

Common molecular architecture for liquid crystalline compounds is two or more planar ring systems, e.g., aromatic nuclei, joined by a central linkage and having appropriate terminal groups,

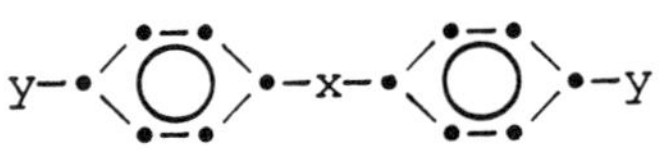

where x represents the central linkage and y the terminal group [1,3,5,6,30,66,67]. Aromatic nuclei are rigid, polarizable, and planar. Mesophase formation is favored if the central group that joins the rings is itself rigid in nature, e.g., involves a multiple bond, a system of conjugated double bonds, or a ring formed by dimerization of carboxyl groups, and if the rings are para substituted. The more extended the conjugated, rigid, central part of the molecule, the more thermally stable the liquid crystal will be.

Replacement of a terminal hydrogen in a molecule by another substituent enhances the potential of the system to form liquid crystals. Highly polarizable substituents increase intermolecular forces of attraction and increase mesophase stability; $-C_6H_5$ and $-NHCOCH_3$ are effective in promoting both nematic and smectic phase stability. A common pattern of behavior in various homologous series is that the lower homologs are nematic phases, the middle members exhibit smectic phase formation followed by a nematic phase, and the long-chain members are purely smectic phases. These trends of transition temperature in homologous series are illustrated in Fig. 12. For smectic phase formation, the intermolecular lateral attractions are essential, but they must be considerably greater than the terminal attractions. With the long-chain members of a homologous series of mesomorphic compounds, the parts of the molecules containing the dipolar (e.g., aromatic ring) and polarizable units, e.g.,

$$\text{-COO-} \qquad \text{-CH=N-} \qquad \text{-}\underset{\downarrow \atop 0}{\text{N}}\text{=N-}$$

are separated further, resulting in a weakening of the terminal attraction.

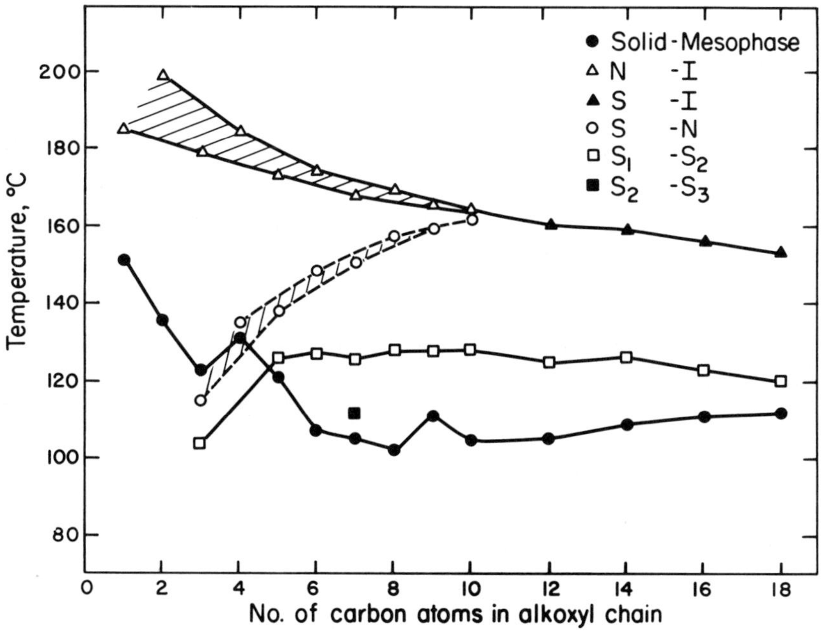

FIG. 12. Effect of chain length on transition temperatures for a homologous series of 4-n-alkoxylbenzylidene-4'-aminoazobenzenes [50].

With lateral substituents, opposing forces are operating. (1) The long molecular axes may be forced apart by the substituent, thereby reducing intermolecular forces of attraction, which is especially detrimental to smectic phase stability. (2) Substitution of a ring hydrogen with a group that increases the molecular polarizability, and possibly also the molecular dipolarity, should increase lateral intermolecular attractions, enhancing liquid crystal thermal stability, especially that of the smectic phase. This effect is apparent in the studies by Goldmacher and Barton [68].

If a compound has a suitable molecular structure for liquid crystal formation, certain conditions must be satisfied if the smectic phase is to form:

1. The molecules must have a tendency to form a layered crystal lattice.
2. The terminal intermolecular attractive forces in the crystal must be relatively weak, so that the fusion point of the material will not be too high.
3. As the crystal fuses, the lateral intermolecular attractive forces must be adequate to retain the layered structure of the smectic phase.

In general, a long, narrow, rodlike molecule with dipoles acting across the long molecular axes ought to enhance smectic phase formation. More specific guidelines that can be accepted with a reasonable degree of certainty regarding molecular structures that favor smectic phase structuring are [30]:

1. Carboxylic acid salts and amine salts are predominantly smectogenic compounds.
2. Terminal groups that favor smectic phase formation are ring-COOR, ring-CH=CHCOOR, ring-$CONH_2$, etc.
3. Smectic phase formation is favored by increasing the length of terminal n-alkyl chains (see Fig. 12).
4. Terminal substituents such as ring-C_6H_5, ring-$NHCOCH_3$, and ring-$OCOCH_3$ favor smectic more than nematic phase formation.
5. Terminal substituents such as ring-CN, ring-NO_2, and ring-OCH_3 strongly enhance thermal stability of nematic phases and often decrease that of smectic phases.

6. Lateral substituents that broaden a molecule diminish markedly the tendency of a compound to form mesophase structures. Substituents that are weakly dipolar, e.g., methyl, or large and only moderately dipolar, e.g., iodine, are more detrimental to smectic than to nematic phase thermal stability.
7. Lateral substituents that do not broaden the molecule enhance mesophase thermal stability; the smectic phase thermal stability is especially enhanced if the substituent is dipolar.
8. Adverse steric effects within the molecule markedly reduce the potential of a compound to form mesophases, in particular, smectic phases.

REFERENCES

1. G.H. Brown and W.G. Shaw, Chem. Rev., 57:1049 (1957).
2. I.G. Chistyakov, Sov. Phys. Cryst., 5:916 (1961).
3. G.W. Gray, Molecular Structures and the Properties of Liquid Crystals, Academic, New York, 1962.
4. H. Sackmann and D. Demus, Mol. Cryst. Liq. Cryst., 2:81 (1966).
5. H. Sackmann and D. Demus, Fortschr. Chem. Forsch., 12:349 (1969).
6. G.H. Brown, J.W. Doane, and V.D. Neff, A Review of the Structure and Physical Properties of Liquid Crystals, CRC Press, Cleveland, Ohio, 1971.
7. P.G. de Gennes, Mol. Cryst. Liq. Cryst., 21:49 (1973).
8. H. Sackmann and D. Demus, Mol. Cryst. Liq. Cryst., 21:239 (1973).
9. H. Sackmann, Pure Appl. Chem., 38:505 (1974).
10. A. Saupe, in Liquid Crystals and Plastic Crystals, Vol. 1 (G.W. Gray and P.A. Winsor, eds.), Halstead, New York, 1974, p. 8.
11. E. Friedel, C.R. Acad. Sci., Paris, Ser. A.B., 180:269 (1925).
12. K. Herrmann and A.H. Krummacher, Z. Kristallogr. Mineral. Petrogr., A, 81:317 (1932).
13. J. Falgueirettes and P. Delord, in Liquid Crystals and Plastic Crystals, Vol. 2 (G.W. Gray and P.A. Winsor, eds.), Halstead, New York, 1974, p. 62.
14. A. de Vries and D.L. Fishel, Mol. Cryst. Liq. Cryst., 16:311 (1972).

15. M. Lambert and A.M. Levelut, in Anharmonic Lattices, Structural Transitions, and Melting (T. Riste, ed.), Noordhoff, Leiden, 1974, p. 375.
16. D. Demus, K.H. Kölz, and H. Sackmann, Z. Phys. Chem. (Leipzig), 249:217 (1972).
17. R.S. Porter, E.M. Barrall, and J.F. Johnson, Acct. Chem. Res., 2:53 (1969).
18. A.V. Galanti and R.S. Porter, J. Phys. Chem., 76:3089 (1972).
19. C.W. Griffen and R.S. Porter, Mol. Cryst. Liq. Cryst., 21:77 (1973).
20. S.E.B. Petrie, H.K. Bücher, R.T. Klingbiel, and P.I. Rose, Eastman Org. Chem. Bull., 45, No.2 (1973).
21. E.C.H. Hsu and J.F. Johnson, Mol. Cryst. Liq. Cryst., 20:177 (1973).
22. A.S. Marshall, J.P. Van Meter, and S.E.B. Petrie, paper presented at the IV International Liquid Crystal Conference, Kent State University, Kent, Ohio, USA, 1972, Abstract No. 115.
23. J.R. Flick, A.S. Marshall, and S.E.B. Petrie, in Liquid Crystals and Ordered Fluids, Vol. 2 (J.F. Johnson and R.S. Porter, eds.), Plenum, New York, 1974, p. 97.
24. R.D. Ennulat, Mol. Cryst. Liq. Cryst., 3:405 (1967).
25. S.L. Arora, T.R. Taylor, and J.L. Fergason, in Liquid Crystals and Ordered Fluids, Vol. 1 (J.F. Johnson and R.S. Porter, eds.), Plenum, New York, 1970, p. 321.
26. A. de Vries, Mol. Cryst. Liq. Cryst., 24:337 (1974).
27. G.C. Fryberg, E. Gelerinter, and D.L. Fishel, Mol. Cryst. Liq. Cryst., 16:39 (1972).
28. S. Diele, P. Brand, and H. Sackmann, Mol. Cryst. Liq. Cryst., 16:105 (1972).
29. G.R. Luckhurst and A. Sanson, Mol. Cryst. Liq. Cryst., 16:179 (1972).
30. G.W. Gray, in Liquid Crystals and Plastic Crystals, Vol. 1 (G.W. Gray and P.A. Winsor, eds.), Halstead, New York, 1974, p. 103.
31. E.M. Barrall and J.F. Johnson, in Liquid Crystals and Plastic Crystals, Vol. 2 (G.W. Gray and P.A. Winsor, eds.), Halstead, New York, 1974, p. 254.
32. P.G. de Gennes, The Physics of Liquid Crystals, Clarendon, Oxford, 1974.

33. H. Arnold, D. Demus, H.J. Koch, A. Nelles, and H. Sackmann, Z. Phys. Chem., 240:185 (1969).

34. I.G. Chistyakov, L.S. Schabischev, R.I. Jarenov, and L.A. Gusakova, Mol. Cryst. Liq. Cryst., 7:279 (1969).

35. A. Levelut and M. Lambert, C.R. Acad. Sci., Paris, Ser. A.B., 272:1018 (1971).

36. T.R. Taylor, S.L. Arora, and J.L. Fergason, Phys. Rev. Lett., 25:722 (1970).

37. G.W. Gray, K.J. Harrison, and J.A. Nash, J. Chem. Soc. Chem. Commun., 431 (1974).

38. D. Demus, M. Klapperstück, R. Rurainski, and D. Marzotko, Z. Phys. Chem. (Leipzig), 246:385 (1971).

39. T.R. Taylor, J.L. Fergason, and S.L. Arora, Phys. Rev. Lett., 24:359 (1970).

40. A. de Vries, Acta Crystallogr., A25:135 (1969).

41. I.G. Chistyakov and V.M. Chaikovskij, Mol. Cryst. Liq. Cryst., 7:269 (1969).

42. A. de Vries, J. Phys. (Paris), Lett., 35:139 (1974).

43. A. de Vries, J. Phys. (Paris), Colloq., 36:Cl-1 (1975).

44. A. de Vries, Mol. Cryst. Liq. Cryst., 10:219 (1970).

45. M. Leclercq, J. Billard and J. Jacques, Mol. Cryst. Liq. Cryst., 8:367 (1969).

46. W.Z. Urbach and J. Billard, C.R. Acad. Sci., Paris, Ser. A.B., B274:1287 (1972).

47. W. Helfrich and C.S. Oh, Mol. Cryst. Liq. Cryst., 14:289 (1971).

48. R. Ribotta, R.B. Meyer, and G. Durand, J. Phys. (Paris), Lett., 35:161 (1974).

49. D. Demus, G. Kunicke, J. Neelsen, and H. Sackmann, Z. Naturforsch., 23a:84 (1968).

50. S. Diele, P. Brand, and H. Sackmann, Mol. Cryst. Liq. Cryst., 17:163 (1972).

51. G.W. Gray and P.A. Winsor, Mol. Cryst. Liq. Cryst., 26:305 (1974).

52. J. Doucet, A.M. Levelut, M. Lambert, L. Liebert, and L. Strzelecki, J. Phys. (Paris), Colloq., 36:Cl-13 (1975).

53. D. Coates, G.W. Gray, and K.J. Harrison, Mol. Cryst. Liq. Cryst., 22:99 (1973).
54. D. Coates and G.W. Gray, J. Phys. (Paris), Colloq., 36:Cl-365 (1975).
55. G.W. Gray, J. Phys. (Paris), Colloq., 36:Cl-337 (1975).
56. D. Demus, L. Richter, C.E. Rürup, H. Sackmann, and H. Schubert, J. Phys. (Paris), Colloq., 36:Cl-349 (1975).
57. D. Demus, S. Diele, M. Klapperstuck, V. Link, and H. Zaschke, Mol. Cryst. Liq. Cryst., 15:161 (1971).
58. M.J. Billard, C.R. Acad. Sci., Paris, Ser.A.B., B280:573 (1975).
59. G.W. Smith and Z.G. Gardlund, J. Chem. Phys., 59:3214 (1973).
60. J. Doucet, A.M. Levelut, and M. Lambert, Phys. Rev. Lett., 32:301 (1974).
61. L. Richter, D. Demus, and H. Sackmann, J. Phys. (Paris), Colloq., 37:C3-41 (1976).
62. S. Diele, private communication, 1975.
63. W.Z. Urbach and J. Billard, paper presented at IV International Liquid Crystal Conference, Kent State University, Kent, Ohio, 1972.
64. A. de Vries, J. Phys. (Paris), Lett., 35:157 (1974).
65. J.P. Van Meter and S.E.B. Petrie, paper presented at the 162nd National Meeting of the American Chemical Society, Washington, D.C., 1971, Abstract No. Phys. 177.
66. G.H. Brown, Amer. Scientist, 60:64 (1972).
67. J.P. Van Meter, in Supplement A: The Chemistry of Double-Bonded Functional Groups, Part 1 (S. Patai, ed.), John Wiley and Sons, London, 1977, p. 93.
68. J. Goldmacher and L.A. Barton, J. Org. Chem., 32:476 (1967).

Chapter 5

POLYPEPTIDE LIQUID CRYSTALS

Donald B. DuPré

Department of Chemistry
University of Louisville
Louisville, Kentucky

and

Edward T. Samulski

Department of Chemistry and Institute of Materials Science
University of Connecticut
Storrs, Connecticut

I. MACROMOLECULAR LIQUID CRYSTALS

A. Introduction

It is not widely appreciated that a number of high molecular weight polymeric substances exist in or can form a liquid crystal or mesomorphic state. Over the last 25 years, the terminology used to characterize molecular organization in thermotropic liquid crystals has frequently recurred in qualitative descriptions of macromolecular order in the solid state of amorphous polymers [1]. Moreover, several distinctly mesomorphic polymeric systems have been elucidated. Certain block copolymers, for example, separate into different phases in the solid state or the swollen gel state generated by solvating one block preferentially. In these phases the aggregated block sequences of polymer are organized into a regular macrostructure. The macrostructures conform to the constraints imposed by a covalently linked chain of blocks or sequences of monomer units and are related to structural arrangements encountered in lyotropic liquid crystals consisting of small amphiphilic molecules (see Sec. I.B) [2]. Liquid crystallinity has also been demonstrated in solutions of rigid polyarimides, a comparatively new class of high-temperature and ultrahigh-strength polymers [3]. The conjunction in these polymers of a mesomorphic state with macromolecular order may, during fiber production, facilitate the formation of the highly oriented solid state prerequisite for good strength. In general, however, the phenomenon of liquid crystallinity in macromolecular systems is of particular interest due to the implications for biological systems, real or model, which are typically composed of large molecules. After some brief remarks about the biological significance of mesomorphism, we shall focus attention on recent advances in a particular class of biopolymer liquid crystals, synthetic polypeptide solutions.

B. Liquid Crystals in Biological Systems

Almost half a century ago Friedel noted mesomorphic characteristics in myelin. Since then it has been recognized that many biological structures exhibit the properties of liquid crystals [4,5]. Considerable concentrations of mesomorphic compounds have been found in many parts of the body, often as sterol derivatives. Living tissue, such as muscle, tendon, ovary, adrenal cortex, and nerve, exhibits many of the optical birefringent properties that are characteristic of liquid crystals. Massive deposits of mesomorphic cholesterol derivatives have been found in the kidneys, liver, brain, spleen, marrow, and aorta walls. Certain living sperms possess a mesomorphic state. Also, solutions of viruses, collagen, hemoglobin of sickle cell anemia, native protein, nucleic acids, and fibrinogen show resemblances to the state.

Now that structures with the characteristics of liquid crystals have been observed in the extracts of animal tissue [5], increasing attention is being focused on their role in structure and biochemical function. In particular, the dynamic-plastic quality of the cellular structure (i.e., capacity for change in shape and material interchange) has been attributed to liquid crystallinity. It may be that catalytic processes at the cellular level find a favorable environment in the mesomorphic structure. For example, this highly organized phase which can dissolve other components without qualitatively changing its own nature would provide a matrix for orientation and juxtaposition of precursor molecules, facilitating highly stereospecific chemical reactions required for the synthesis of the complex optically active molecules found in nature. The unusual combination of lability and lateral cohesion makes the liquid crystal state "biologically useful."

The class of mesomorphic compounds that has received the most attention in this context is the phospholipids. This attention is well deserved since phospholipids comprise some 30% of the human brain, are a major component of the myelin sheath (the fatty sheath wrapped around nerve fibers), and are found in blood corpuscles and in the chloroplasts of plants. Phospholipids are amphiphilic molecules which possess two contrasting parts: one lipophilic (hydrophobic) in character, the other hydrophilic. The former part consists of two hydrocarbon chains, 14 to 18 carbon atoms in length, and the latter is a polar group containing phosphorus and nitrogen atoms. An example is α-cephalin (phosphatidylethanolamine) with its structural formula shown below.

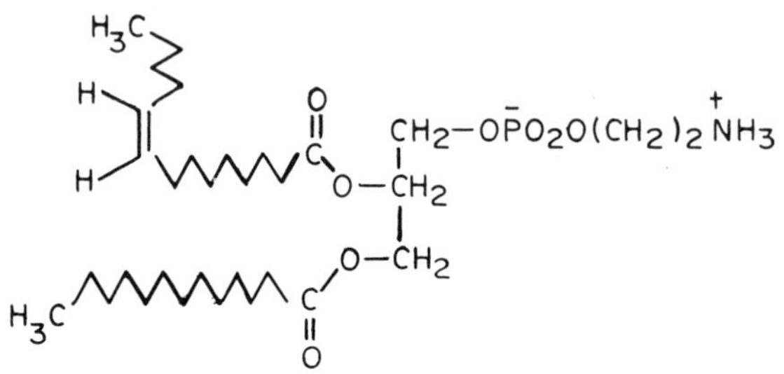

The degree of unsaturation and the hydrocarbon chain lengths found in naturally occurring phospholipids depend on the type of cell membrane from which they are extracted.

The nature of the supramolecular organization found in the liquid crystal phases of phospholipids has been extensively investigated. Brain lipids have been identified in several different types of liquid crystal structures [4]. Although phospholipids are relatively insoluble in water, the amphiphilic nature of the molecules nevertheless causes them to assume specific arrangements in water. The two most prevalent of these are lamellar and hexagonal. The lamellar is composed of alternating polar and hydrocarbon layers, similar to the smectic structure in thermotropic mesophase; the hexagonal structure consists of long cylinders which contain water molecules and the polar groups of the phospholipid molecules, while the hydrocarbon chains extend outward radially from the surface [5].

A particularly striking discovery made from infrared studies of phospholipids is the high degree of mobility in the hydrocarbon section of these molecules [6]. The infrared spectra of the hydrocarbon chains resemble those of a liquid at room temperature although the physical appearance of phospholipids is that of a solid. Since the melting point of the anhydrous phospholipids is above 200° C, these observations suggest that phospholipids exist as liquid crystals at room temperature.

It has also been shown that the mesomorphic transition temperature of liquid crystal phospholipids is dependent on the amount of water in the lipid phase [7]. The degree of unsaturation and its perturbing effect on the conformation of the hydrocarbon chain also influence the mesomorphic transition temperature. As the chains of the phospholipid become more unsaturated, the transition temperature is lowered. Similar-order transitions have been reported among hydrocarbon chains of intact membranes of microorganisms [8]. Furthermore, pure phospholipids extracted from the microorganisms exhibit the same temperature dependence on the degree of unsaturation as that observed in intact membranes. The degree of unsaturation in the latter was varied by controlling the growth medium of the microorganism. These findings are strong evidence for a liquid crystalline organization of the phospholipids in living cells.

In addition to the bilayer lipid structure of cell membranes, mesomorphic character has been ascribed to other cellular components in vivo and in vitro. Prominent in this category are the nucleic acids. In certain cellular organelles such as chromosomes, DNA occurs in a concentrated form. Bouligand has pointed out striking similarities between the optical properties derived from the underlying supramacromolecular organization of the concentrated DNA phases and those observed in other well-known cholesteric textures [9]. He proposes a very intriguing scheme for the structural arrangement of DNA in dinoflagellate chromosome to account for these observations. The double helical filament of DNA alternates direction in a uniplanar arrangement with neighboring sequences of the DNA molecule parallel to each other. These planes of DNA in turn are twisted relative to one another generating a cholestericlike texture. [The cholesteric twist is a consequence of the inherent asymmetry of the DNA double helix and the resulting asymmetric <u>intermolecular</u> potential (see Sec. II.C.2b).] Concentrated aqueous solutions of nuclei acids were shown earlier to exhibit a cholesteric texture in vitro [10,11].

Mesomorphic behavior has also been described in a more remote segment of the in vivo spectrum, the exocuticle of certain classes of beetles. Neville and Caveny have demonstrated that the bright iridescent colors reflected from the surface of <u>Scarabaeid</u> beetles originates from a petrified cholesteric structural arrangement of chitin crystallites in the exocuticle [12]. It is suggested that this cholesteric texture forms spontaneously in a mobile liquid crystal phase present during the initial stages of the exocuticle growth cycle.

Viruses, complex assemblages of protein and nucleic acid, also form liquid crystal phases in solution. It was discovered 35 years ago that the rod-shaped tobacco mosaic virus (TMV) readily forms a birefringent phase in aqueous solution [13]. As an aqueous solution of TMV is concentrated, birefringent liquid crystalline tactoids appear suspended in the isotropic phase. In this thixotropic gel the tactoids coalesce into a homogeneous liquid crystalline phase on increasing the TMV concentration and/or changing the solution pH. More recently it has been shown that fd-type filamentous bacterial viruses form a birefringent phase in concentrated solution [14]. The gellike solution adopts a cholesteric structure in the liquid crystal phase. Iridescent colors are reflected from highly concentrated gels of the fd virus; in dry gels the cholesteric pitch is comparable to the wavelength of light. The cholesteric structure derives from the asymmetry in the intermolecular forces between chiral solute particles. The fd virus tertiary structure is chiral; the viruses are rodlike with a chiral arrangement of essentially α-helical proteins on the exterior [15].

The liquid crystal phase has been directly implicated in at least two major degenerative molecular diseases: atheroschlerosis and sickle cell anemia. A complex cholesterol-phospholipid-lipoprotein mesophase has been identified in the initiation and maintenance of atheromatous deposits on the aortic intima in dissected human and rabbit arteries [16]. The paracrystalline nature of this precursor to plaque build-up with its resultant loss of arterial elasticity and atherosclerotic stenosis is dramatically evident when viewed under the polarization microscope. The familiar liquid crystal spherulitic structures are seen to exhibit the birefrigent and optical rotation properties characteristic of the cholesteric mesomorphic texture.

The presence in red blood cells of hemoglobin S (Hb-S), the mutant hemoglobin of sickle cell anemia, has been shown to lead to the formation of paracrystalline (read: liquid crystalline) aggregates inside the cell under conditions of low oxygen tension. The morbid aggregates ultimately arrange themselves into a gellike material composed of long fibers which extend the entire length of the cell and distort its usual shape [17]. The low solubility of Hb-S causes a lowered osmotic pressure which tends to diffuse surplus water out of the cells, collapsing the cell membrane. The cells thus become deformed or visibly "sickled." This bizarre shape creates flow abnormalities which impede proper capillary circulation. Stagnation of the capillary system and further decrease of the oxygen tension in the blood stream results. Physiological consequences of this molecular disease include anemia, vascular stasis, hemorrhage, thrombosis, and exhaustion of bone marrow.

Much more work needs to be done before the true significance of the liquid crystal state in biological systems is understood. However, it is striking that mesomorphic modifications are singularly well suited to provide the delicate balance of organization and lability so characteristic of life processes. It seems probable that liquid crystallinity has important biological consequences, first because of its presence in living material and second because of its unusual dependence on slight changes in composition or in the physical and chemical environment for its formation, continuation, or cessation.

II. LIQUID CRYSTAL POLY-γ-BENZYL-L-GLUTAMATE

The most extensively studied class of macromolecular liquid crystals is that of the synthetic polypeptides, which in their rigid, rodlike helical conformations have been shown to exhibit the mesomorphic behavior of all three major liquid crystal structures: nematic, cholesteric, and smectic. In the remainder of this chapter we will focus attention on the properties of a single such polypeptide, poly-γ-benzyl-L-glutamate, as it is currently the best characterized macromolecular liquid crystal. The mesomorphic phase of PBLG was qualitatively reviewed in 1970 [18]. Herein we emphasize the more recent findings in this liquid crystal.

A. The Poly-γ-Benzyl-L-Glutamate Macromolecule

Poly-γ-benzyl-L-glutamate (PBLG) is a homopolymer of the L enantiomorph of a single amino acid with the basic repeat unit

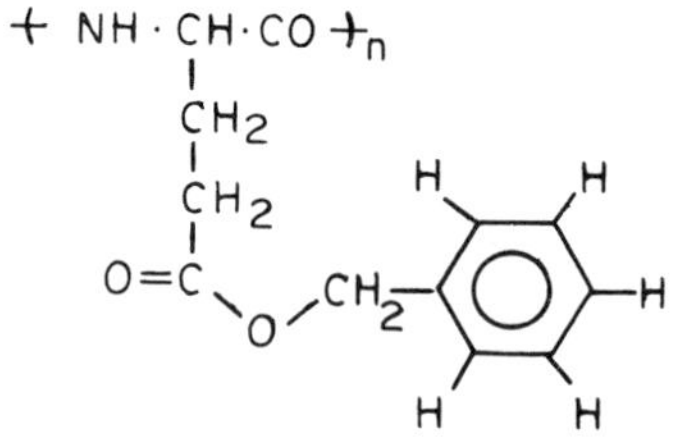

The polymer may be prepared in high degrees of polymerization ($n > 1{,}000$). Due to its ready availability and good solubility characteristics, PBLG has undoubtedly been the most extensively studied synthetic polypeptide. The fact that the monomer is not a naturally occurring amino acid has not diminished its usefulness as a model for protein structure.

In 1951, Perutz [19] used oriented fibers of PBLG in the first x-ray crystallographic proof of Pauling and coworkers' then-new proposal [20] of the α-helical conformation. Since that time PBLG has on numerous occasions been used as the classic model of a rigid, rodlike macromolecule in solution. Research on PBLG, usually involving some aspect of the helix-coil transition, continues to be published weekly.

The α-helical conformation itself has the following characteristics: the backbone chain of the polypeptide repeats exactly every 18 amino acid residues; there are 3.6 residues per turn of the helix. Each successive residue translates 1.5 Å along the helix axis. There is a hydrogen bond between the N-H of one residue and the C=O of the amino acid four residues down the chain.

It has been known for some time that PBLG will maintain its α-helical conformation in a number of helicogenic solvents: solvents of low enough dielectric constant not to interfere with the primary organization of the rod-shaped polymeric backbone, its hydrogen-bonded network, and the delicate secondary structure of the pendant side chains. As a result of this unique geometry, PBLG in helicogenic solvents is characterized by dielectric and diamagnetic susceptibilities which are anisotropic. Some of the fundamental physical parameters of the isolated macromolecule are presented in Table 1. The dimensions of the macromolecule are determined with the assumption that the PBLG backbone assumes the extended, rigid

TABLE 1. The Poly-γ-benzyl-L-glutamate (PBLG) Macromolecule

α-Helix; n Amino Acid Residues
Length = n x 1.5 Å
Diameter <30 Å (the diameter of the α-helical core out to βC is 6.4 Å)
Side-chain volume ≈2/3 of molecule
Residue molecular weight = 219
Density = 1.27
Dielectric anisotropy, $\Delta\varepsilon = \varepsilon_{\parallel} - \varepsilon_{\perp} > 0$
Diamagnetic anisotropy, $\Delta\chi = \chi_{\parallel} - \chi_{\perp} > 0$
Birefringence, $\Delta n = n_{\parallel} - n_{\perp} = 0.025 > 0$

α-helix conformation. The internal degrees of freedom in the side chain yield a variety of side-chain secondary structures and lead to considerable variation in the PBLG diameter. In Fig. 1 two extreme possibilities for side-chain arrangements are shown.

B. PBLG as a Lyotropic Liquid Crystal

At sufficiently high concentrations of polymer, PBLG and a number of other polypeptides listed in Table 2 form lyotropic liquid crystals in appropriate solvents or solvent mixtures. PBLG exhibits textures and properties similar to all three major liquid crystal classes; i.e., the nematic, smectic, and cholesteric mesophases (Table 3). It is curious that a mesophase develops even in some solvents that do not, in more dilute solution, support the necessary helical configuration of the macromolecule (e.g., dichloroacetic acid). (Recent theoretical treatments of the helix-coil transition in a system of macromolecules with intermacromolecular interactions have considered this apparent anomaly [21,22].)

1. Formation

The formation of the liquid crystal proceeds in stages as the concentration of polymer is increased. When the polypeptide exceeds a certain limiting concentration in a solvent, the solution separates into two

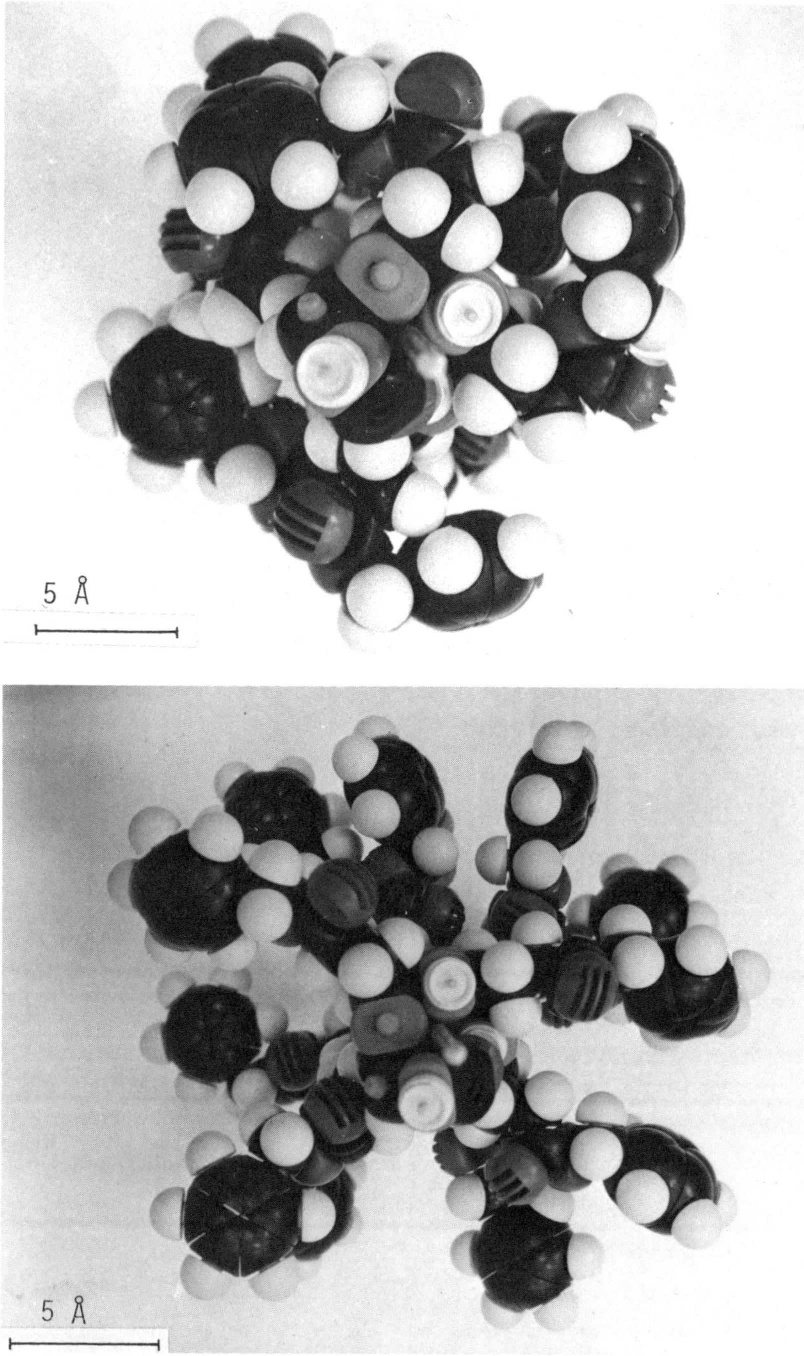

FIG. 1. Projections of PBLG diameter with extreme side-chain conformations.

TABLE 2. Polypeptides Forming Liquid Crystals in Suitable Organic Solvents

Polypeptide $-(NH.CHR.CO)_n-$	R
Poly-L-glutamic acid	$-CH_2CH_2COOH$
PLGA esters:	
γ-methyl	$-CH_2CH_2COOCH_3$
γ-ethyl	$-CH_2CH_2COOC_2H_5$
γ-benzyl	$-CH_2$-$CH_2COOCH_2C_6H_5$
(L and D isomers)	
Poly-β-benzyl-L-aspartate	$-CH_2COOCH_2C_6H_5$
Polylysine	$-(CH_2)_4-NH_2$
Poly-ε-carbobenzyloxy-L-lysine	$-(CH_2)_4NHCOOCH_2C_6H_5$
Copolymer of equal parts of PBLA and PDGA	

TABLE 3. Lyotropic Liquid Crystal Solutions of Poly-γ-benzyl-L-glutamate

Texture	Solvent
Cholesteric	$CHCl_3$, CH_2Cl_2, dioxane, pyridine, m-cresol, DCA, DMF, quinoline, thiophene
Nematic	CH_2Cl_2-dioxane (≈ 4:1 vol.)[a]
Smectic	benzene

[a]Racemic mixture of D and L isomers also forms a nematic liquid crystal.

liquid phases. The polypeptide-rich phase is birefringent and departs from the more dilute medium initially in the form of spherical liquid droplets or "spherulites." If the two-phase solution is cooled or the polypeptide concentration further increased, the droplets grow in size and coalesce, forming a continuous birefringent fluid. The birefringent solution is a lyotropic liquid crystal.

The PBLG liquid crystal forms in a number of solvents and mixed solvent systems, including benzene, nitrobenzene, chloroform, N,N-dimethylformamide (DMF), pyridine, dioxane, dichloromethane, isomers of dichloroethene, dichloroacetic acid (DCA), and m-cresol. It has been shown that in all circumstances in which PBLG forms a liquid crystal phase in solution it is in the extended helical molecular conformation.

Robinson and coworkers [11,23-26] pioneered the detailed optical investigations of liquid crystal PBLG solutions and designated the polymer concentration at which the birefringent phase initially appears the A point and that at which the entire solution becomes continuously birefringent the B point. The A and B points depend on the particular solvent, but no correlation with solvent properties is apparent. In any particular solvent, however, the A and B points are related to the degree of polymerization n. The higher the value of n, the lower the polymer concentration at the occurrence of the A and B points.

2. Rod-Ordering Theory

These qualitative observations alone suggest that the rodlike shape of the polypeptide molecule, and not specific polymer-solvent interactions, are responsible for the formation of the liquid crystal. Theoretical investigations of the possibility that phase separation or self-ordering could occur in concentrated solutions of rodlike solutes were initially carried out by Onsager [27], Isihara [28], and Flory [29,30]. Flory's model is readily grasped conceptually and is qualitatively described below. It should be noted, however, that Straley has recently criticized this model, suggesting that the Onsager approach is the proper one for long rods [31].

Flory applied the lattice model for polymer solutions to a system composed of noninteracting, impenetrable, rodlike solute particles and inert, spherical solvent molecules [30]. The former were assumed to occupy a continuous linear array of sites on the lattice, while the latter occupy single lattice sites. This model treatment indicates that in dilute solution the rodlike particles are free to adopt random orientations. However, as the concentration of the rodlike particles is increased, a point is eventually reached beyond which randomness of orientation for these particles is no longer compatible with the number of lattice sites at their disposal. There is a larger number of ways to arrange the rods on the lattice, or a higher entropy of mixing, for solutions in which the solute particles assume orientations locally correlated to one another (a semi-parallel arrangement of the rodlike solute particles) than for isotropic solutions (randomly oriented solute). For solute concentrations above the limiting concentration point (A), theory predicts that the solution will separate into two phases, the dilute phase being isotropic in its arrangement of the solute particles, while the more concentrated phase would be anisotropic. If the solute particle has a large axial ratio, X = length: diameter, the solute volume fraction at the A point ϕ_A is given approximately by

$$\phi_A = \frac{8}{X}\left(1 - \frac{2}{X}\right)$$

This phase separation is unique in that it occurs as a consequence of the rodlike shape of the solute particles; no attractive forces are required.

Experimental values for the A and B points were determined optically by noting the concentrations of polypeptide at which phase separation and coalescence occurred. It was also reported that the viscosity of the solution below the A point was about 2.5 times that of the more concentrated liquid crystal phase [11]. Subsequently, a study of shear viscosity for PBLG solutions with polypeptides of differing degrees of polymerization (axial ratio, X) as a function of polypeptide concentration has been shown to yield independent determinations of the A point [32]. In Fig. 2 the calculated dependence of the A and B points on the axial ratio from both Onsager's and Flory's models are shown. The open symbols are experimental values for the A points, the closed symbols are for the B points, in various PBLG solvent systems. The axial ratio X was varied by choosing different molecular weight PBLG (length = 1.5 Å x n).

Flory suggested that the discrepancy between his theory and experiment may be due partly to the idealization of PBLG as an impenetrable rod [33,34]. The PBLG side chain contains six single bonds about which restricted rotation certainly occurs. Penetration of solvent into the flexible side-chain portion of the macromolecule (roughly two-thirds of the macromolecular volume) is expected, and undoubtedly, mixing of solvent with the flexible side chain accounts for the good solubility

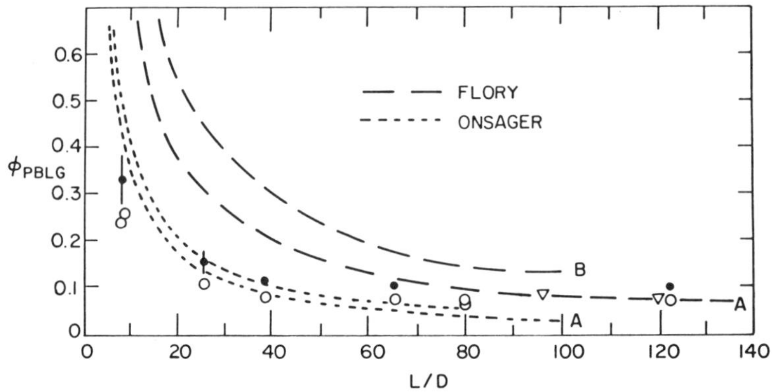

FIG. 2. Volume fraction of PBLG at A and B points. PBLG-dioxane, optical measurements (Ref. 26) o A points, • B points; shear rate determination of A points ∇ (Ref. 32).

characteristics of PBLG. (The solubility of the homolog PBLA is much less than PBLG.)

In a recent comparison of rod-ordering theories Straley implies that the Onsager theory best accounts for the experimental results [31]. However, before a critical comparison of theory and experiment can have any significance, more experiments on better characterized polypeptide samples must be performed. The PBLG samples used in the cited experiments were polydisperse. A distribution of axial ratios could be expected to have a marked effect on the calculated phase boundaries [35]. Since the actual distribution of molecular weight can be readily measured with chromatographic techniques and can be altered by mixing polypeptide samples, rod-ordering theories incorporating this effect could be readily tested. Apart from these considerations, the experimental results in Fig. 2 clearly indicate that the predominant factor in the formation of the lyotropic PBLG liquid crystal is the rodlike shape of the polypeptide.

3. Thermodynamics

The Flory lattice model yields a phase diagram describing the equilibrium between isotropic solution and liquid crystalline solution of impenetrable rods in terms of the axial ratio of the rod, the polymer-solvent interaction parameter, and the degree of rod order in the liquid crystalline phase [30]. Miller and coworkers, using a variety of techniques--nuclear magnetic resonance (NMR) spectroscopy, polarization microscopy, hydrodynamic properties, calorimetry and isopiestic studies--have experimentally determined the phase diagram of PBLG in dimethylformamide (DMF) [36-42]. Their PBLG-DMF phase diagram is shown in Fig. 3. It possesses the gross features predicted by the Flory model [30]. Incorporating side chain-solvent mixing into the model, in the form of a thermodynamic interaction parameter between solvent and side chain, improves the agreement between theory and experiment [42]. Minton has obtained a similar phase diagram using a lattice treatment of rodlike particles in a homogeneous medium in order to explain the gelation (mesomorphism) of sickle cell hemoglobin [17]. Recently, Okamoto et al. have measured the solvent-polymer interaction parameter in PBLG solutions as a function of polypeptide concentration [43]. Noting the disagreement between theory and experiment these workers question the current state of lattice theories of rodlike molecules.

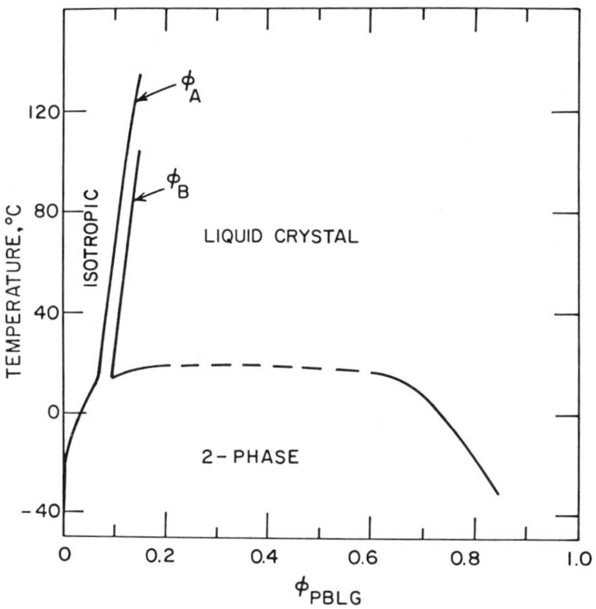

FIG. 3. Phase diagram for PBLG-dimethyl formamide. (Miller et al., Liquid Crystals and Ordered Fluids, Vol. 2, (J.F. Johnson and R.S. Porter, eds.), Plenum, New York, (1973, p. 243).

4. Viscoelasticity

Polypeptide liquid crystals, as all other mesophases, are characterized by five viscoelastic constants that are fundamental to the establishment of the phase and determine its extraordinary sensitivity to external perturbations. In liquid crystals, three elastic moduli, $K_{ii} = (K_{11}, K_{22}, K_{33})$, pertain to the description of bulk properties. These moduli correspond to the forces necessary to induce in the director field $\bar{n}$ of the material ($\bar{n}$ specifies the average local alignment of constituent molecules) deformations of splay, twist, and bend, respectively (Fig. 4). Such deformations may be a result of mechanical stress or natural thermal excitation. Typically, the values of the K_{ii} are some 22 orders of magnitude less than the elastic constants associated with "normal" solids. This accounts for the unusual sensitivity of liquid crystals. In true solids, these moduli are masked by the more significant constants of the stress-strain tensor. In liquids, elastic curvatures are not sustained.

Two viscosity coefficients, γ_1 and γ_2, are also of importance in the response of the liquid crystal. γ_1 is a measure of the coupling of angular motion of $\bar{n}$ with the vorticity of the fluid (i.e., "twist"), whereas γ_2 relates to viscous drag in bend.

Measurements of the K_{ii} and γ_i coefficients have been made in a number of thermotropic liquid crystals. The importance of these material

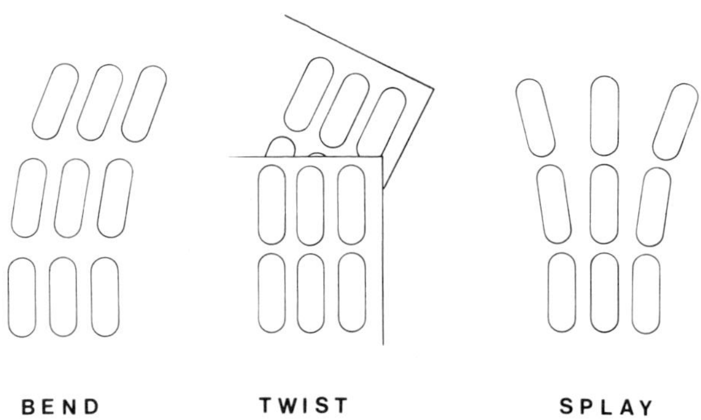

FIG. 4. Schematic illustration of bend, twist, and splay deformations.

parameters for biological structures, which in many cases resemble lyotropic mesomorphic structures, is exemplified in the recent work of Helfrich [44,45] who developed a theory of the elasticity of lipid bilayers. According to theory, vesicles (closed bilayer films) need not be spherical but may be deformed by electric and magnetic fields and by excess pressure. Several red blood cell shapes can be understood on the basis of curvature elasticity [46]. Studies of K_{22} and γ_1 have been conducted recently on the PBLG liquid crystal. The results of these investigations are discussed in Sec. II.D.3.

C. Cholesteric Structures

1. Optical Properties

The optical properties of PBLG solutions have been well studied. Robinson [11,23-26] found that sufficiently concentrated solutions of the polypeptide in appropriate solvents (see Table 3) behave in some aspects as cholesteric liquid crystals. They display, for example, high optical rotatory powers (as high as 10^5 degrees per millimeter) not attributable to the properties of individual PBLG molecules. This suggests a form optical rotation of a twisted superstructure within the medium. When viewed in the polarization microscope the cholesteric phase is characterized by regions of regular striations arranged in a swirllike or fingerprint pattern (Fig. 5).

The optical retardation has been shown to vary in a regular manner between equidistant lines of extinction. The observations are consistent with a regular variation of refractive index along the axis of a helicoidal superstructure that meanders through the medium. The distance between the striations is one-half the pitch P of the structure. For PBLG, P/2 is of the order of 3-100 μm and depends on concentration in an inverse manner ($P/2 \sim 1/C^2$). The magnitude of P/2 is not sensitive to the molecular weight of the polypeptide. Between the A and B points, where the liquid crystal has not totally formed, changes in the overall concentration of polymer do not affect the striation separation since only the relative amounts of isotropic and anisotropic fluid are changed in the two-phase region.

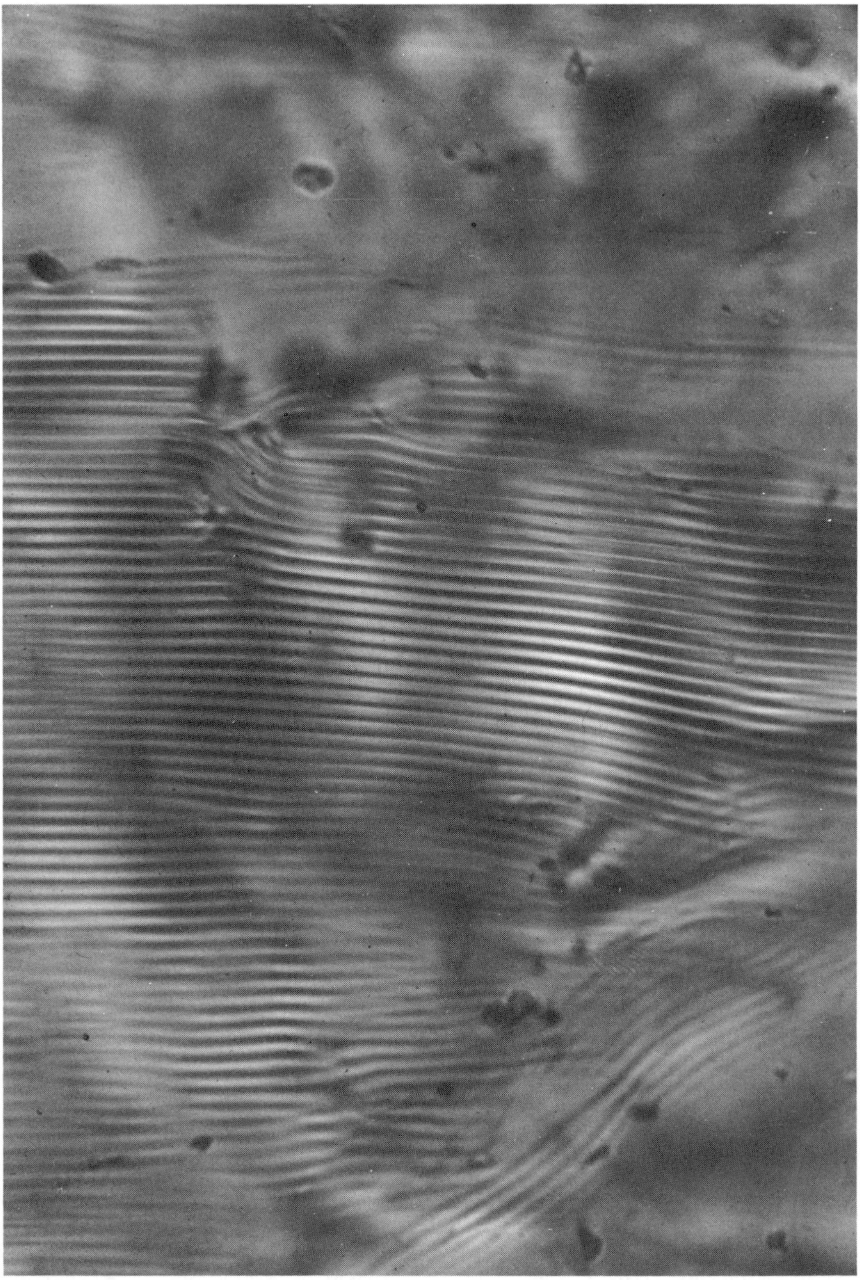

FIG. 5. Photomicrograph of PBLG-dioxane liquid crystal under crossed polars. The distance between equispaced retardation lines is one-half the pitch of the cholesteric texture.

Unlike other thermotropic cholesterics, however, PBLG solutions do not exhibit dramatic iridescent colors when viewed in reflected white light. Their absence is due to the fact that P/2 is well beyond the wavelength of light even for highly concentrated solutions of the polymer. Attempts to get P/2 in the optical region by concentrating the polymer result in opaque gels. (An exception is found in solutions of the ethyl isomer of polyglutamic acid [26].)

If the anisotropic fluid is confined in a thin (< 1 mm) rectangular cell, a well-defined pattern of lines forms. When the direction of viewing is along the thin dimension of the cell, one observes that the equally spaced lines generally align parallel to the walls of the container, and that, further away from the walls, small uniform areas appear; these are surrounded by the parallel lines but within them no lines are observed. Left standing, these uniform polygonal areas grow as two or more neighboring areas unite. This pattern suggests that the parallel lines are layers seen edge on, while the uniform areas are layers lying parallel to the plane of the liquid container interface (normal to the direction of viewing). A simple model which facilitates visualizing the optical pattern consists of close-packed groups of nested transparent polyhedra of progressively increasing size. The parallel lines (Fig. 5) represent a set of equally spaced, parallel sides of a nested set of polyhedra seen edge on. The uniform areas (top and bottom right, Fig. 5) are the sides of the smallest polyhedra in the set as seen through the transparent sides of the larger polyhedra in the set.

Quantitative investigations of the optical properties of the polypeptide liquid crystal are summarized below:

1. When the birefringent liquid crystal is viewed with one polar (either polarizer or analyzer), the parallel lines are invisible when lines are perpendicular to the electric vector of the light.
2. When seen between crossed polars, the alternating light and dark parallel lines are replaced by parallel bands of very bright birefringent colors. The variation in the colors indicates that the retardation oscillates through a regular series of local maxima and minima as the viewer moves along a perpendicular to the parallel bands. The color change that occurs when a compensator is inserted shows that the direction of high refractive index is nearly parallel to the lines.

3. In solutions with more regular patterns, bright colors appear in the uniform polygonal areas when they are observed with crossed polars. These colors cannot be extinguished with a quartz wedge and are reproducible depending on the thickness of the area. The color changes in a continuous manner when one polar is rotated, and with monochromatic light, two extinction positions are found 180° apart. The extinction position varies considerably with the wavelength of light being used.

The last observation (3) signifies the presence of a very high-form optical rotation* with a correspondingly high rotatory dispersion. Using the polarization microscope as a polarimeter, form optical rotations of 20,000 to 150,000 degrees per millimeter have been measured in uniform areas of PBLG solutions.

Other equally remarkable optical properties are associated with this type of liquid crystal. For example, when white light is incident on the surface of the PBLG liquid crystal, selective reflection takes place over a small region of the spectrum, the wavelength of maximum reflection varying with the angle of incidence in accordance with Bragg's Law [26]. At normal incidence, the reflected light is circularly polarized; one circular component is totally reflected over a spectral range of some 200 Å, while the other passes through unchanged. Also, quite contrary to usual experience, the reflected wave has the same sense of circular polarization as that of the incident wave.

---

*The form optical rotation referred to here should not be confused with the behavior of an optically active molecule in the vicinity of an absorption. The latter optical rotation occurs in isotropic solutions which contain optically active molecules and has its origin in the selective absorption of one circularly polarized component of the light. The form optical rotation which we are discussing is caused by the supramolecular arrangement of the molecules in the liquid crystal and originates in the selective reflection of circularly polarized light by this arrangement. It is so large that the contribution from ordinary optical rotation coming from the asymmetric polypeptide helices may be neglected.

As noted earlier, the highest refractive index of the PBLG molecule is parallel to the helix axis [47]. This fact taken together with the sign of the birefringence in the liquid crystal pattern and the other optical features implies that the helical axes lie parallel, or nearly parallel, to the equally spaced retardation lines. Hence, a line of constant retardation is a plane in which the helical axes lie, seen edge on, and in this plane the helices assume a certain preferred orientation. The direction of preferred orientation changes in a continuous manner as the viewer moves along at right angles to the retardation lines (along the cholesteric axis). These optical characteristics are also present in the birefringent "spherulites" which are suspended in the isotropic solutions at concentrations between the A and B points. The "spherulites" contain additional features which are described in detail by Robinson [26]. A prominent feature exhibited by the "spherulites" is a radial line of dislocation. The pattern of retardation lines in the "spherulites" also suggests a cholesteric arrangement of the PBLG rods.

2. Supramolecular Structure

A model for the gross molecular organization of cholesteric liquid crystals has been proposed by de Vries [48] and Robinson [23], wherein it is assumed that cholesteric mesophases are composed of a large number of thin birefringent molecular "layers" or planes with the principal axis of successive planes turning through a small spiral angle (Fig. 6). Within each "identifiable layer," molecules lie with their long axes approximately parallel. Molecules in one plane therefore have a slight angular displacement relative to those in adjacent planes. The displacement is cumulative, occurring about an axis of torsion z which is normal to the individual planes. z is not macroscopically aligned in the absence of external orientational influences (solution-container interface, magnetic or electric fields, shear, etc.).

The properties of thermotropic liquid crystals of the cholesteric structure have been adequately rationalized in terms of this framework. The observations outlined above on the PBLG liquid crystal show that, in the solvents mentioned, the cholesteric structure is an accurate representation of the supramolecular arrangement of the polypeptide. For a more

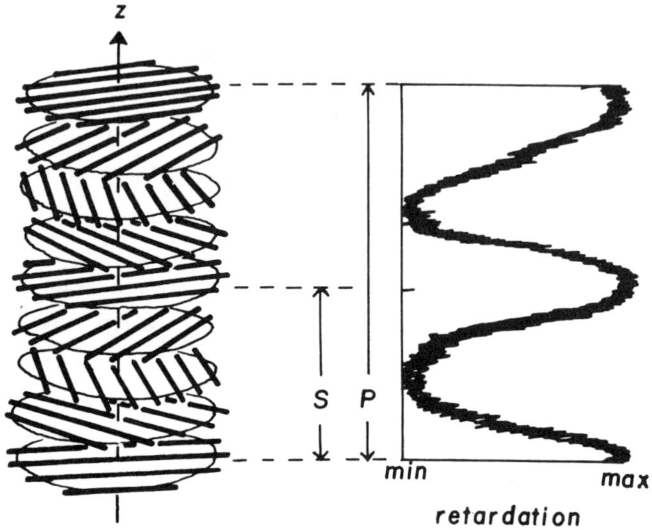

FIG. 6. Model for cholesteric texture; polypeptide rods are represented by the parallel dark lines; z is the axis of torsion or cholesteric optical axis.

detailed description of the relationship of molecular organization and supramolecular topologies to the corresponding optical patterns observed in these in cholesterics, the reader is referred to an elegant series of papers by Bouligand [49-53].

a. Near-Neighbor Polypeptide Association in the Mesomorph: x-Ray Analysis. x-Ray diffraction patterns obtained from polypeptide liquid crystals give information on the nature of the distribution of the helical macromolecules in the cholesteric phase and on the manner in which this distribution changes with polymer concentration. Such studies on cholesteric PBLG have shown that the rodlike molecules pack in something approaching a hexagonal arrangement, as schematically illustrated in Fig. 7. The cholesteric twist minimally distorts this local ordering and does not distort the hexagonal microscopic arrangement; the local structure results in well-defined x-ray reflections. The sharpness of the x-ray pattern was noted to increase with increasing polymer concentration, indicating a more ordered, or in this case, a more tightly packed microstructure. Theoretical

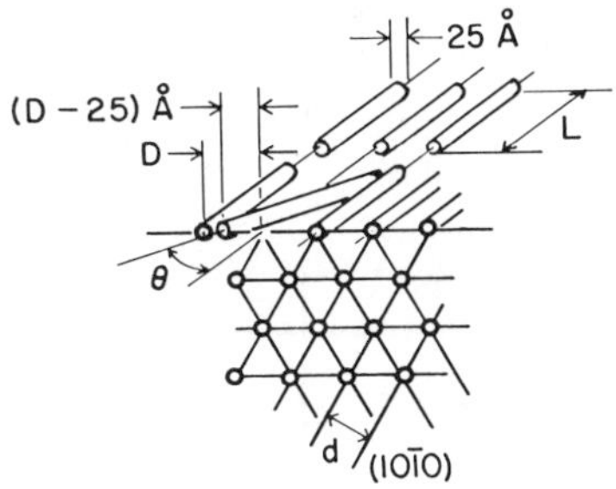

FIG. 7. Model illustrating the packing of PBLG rods on a hexagonal net.

considerations [25] of the effect of concentration of rigid rods on the distance d between the (10$\bar{1}$0) planes of the hexagonal net predict that:

$$d^2 = \frac{W\sqrt{3} \times 10^{24}}{2N_0 h} \cdot \frac{1}{C}$$

where d is in Angstroms; W = 219, the residue weight in PBLG; N_0 = Avogadro's number; h = 1.5 Å, the length of the projection of a single residue along the helix axis; and C is the concentration of polymer in grams per 100 ml. The experimental values of d versus polymer concentration (open circles) are compared with theory in Fig. 8. The absolute values of d confirm that the liquid crystal is, at the molecular level, composed of discrete single molecules and not aggregations of two or more helices.

From the dimensions of the hexagonal net it is possible to calculate the distance of closest approach between the exterior surfaces of two neighboring helices $\underline{D}$ if one assumes a value for the helix diameter D (see Fig. 1). Values of $\underline{D}$ between +15 Å and -4 Å have been measured for varying concentrations of the PBLG liquid crystal using a rather large value of D = 28 Å. The negative sign for $\underline{D}$, which occurs at the higher polymer concentrations, indicates that the side chains of neighboring helices could overlap with one another.

Considering further this hexagonal model, it is possible to show that the molecular arrangement of the polypeptide helices in the liquid crystal imposes considerable constraint on macromolecular rotational motion. On the hexagonal net the long, rodlike molecules are allowed to execute only a very small angular rotation θ (defined in Fig. 7) about an axis through its

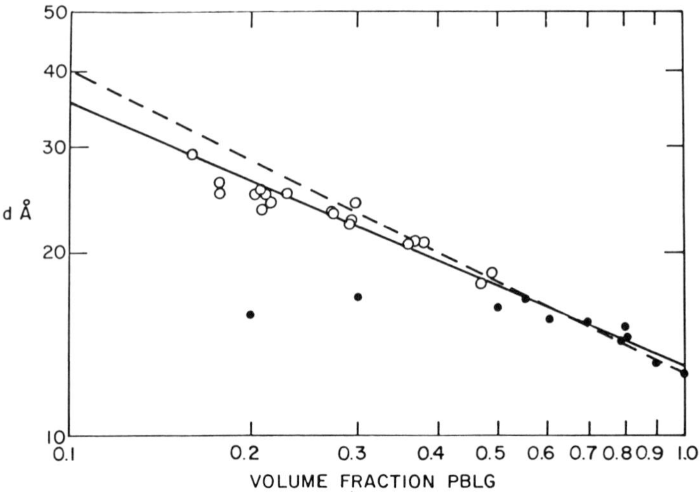

FIG. 8. The intermacromolecular spacing versus PBLG concentration in liquid crystals, o ≈ PBLG-dioxane (Ref. 24) and in solid plasticized films •. (E.T. Samulski, Ph.D. Thesis, Princeton University, 1969).

center before coming into contact with its nearest neighbor. Typically, the helices are restricted to oscillations of no more than 1° of arc.

The twist angle per molecule about the z axis of the cholesteric structure which is required to give the observed pitch is much less than θ. It is easily seen that the z axis passes through P/2d molecules over a distance equal to half the pitch. This number is of the order of 10^4-10^5 at typical PBLG liquid crystal concentrations. The resultant angle of twist per molecule, given by 2d/P x 180 °, is therefore only 10^{-2}-10^{-3} degree of arc.

b. Origin of the Spontaneously Twisted Texture. The origin of the spontaneously twisted macromolecular suprastructure in concentrated polypeptide solutions is a point for much speculation. Robinson et al. [24] suggested that long-range ($1/r^3$) electric dipole forces are responsible for the spontaneous twist of the cholesteric structure. The asymmetric array of permanent dipoles on the helical polypeptide backbone was suggested to impose a unidirectional torsion biased to one side of the nearly parallel helices.

A peculiarity of the polypeptide liquid crystal not readily accounted for by a long-range dipolar origin of cholesteric twist is a change in sense of the cholesteric twist in mixed solvent systems. In Fig. 9, Robinson's data for PBLG in CH_2Cl_2-dioxane mixtures [11] are shown with additional studies [54] in other mixed solvent systems. For a certain ratio of solvents the PBLG mixed solvent liquid crystal is nematic! In all solvents PBLG retains the right-handed α-helical conformation [55]; the chiral sense (right-handedness) of the asymmetric distribution of permanent dipoles on the helix backbone remains unchanged.

A proposed alternate model free of the above disadvantage suggests that the twist is generated by short-range dispersion or van der Waals forces [56]. This model rests on the one feature all cholesteric macromolecular liquid crystals have in common: an asymmetric helical "shape." The theory is that natural thermal oscillations of the closely packed helices in the liquid crystal would be biased to one side of the

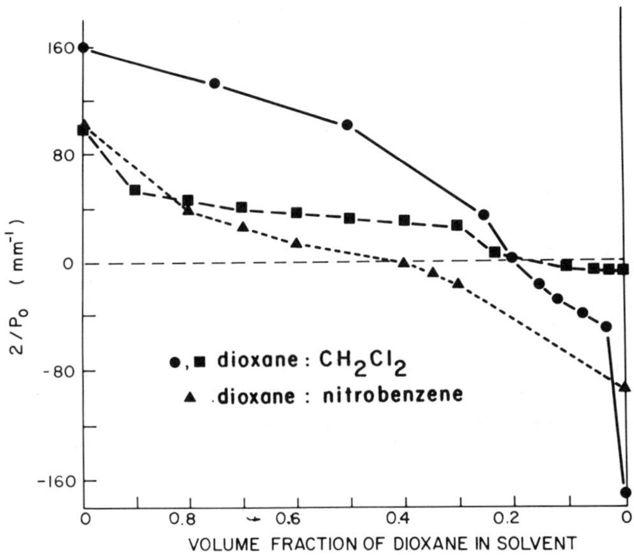

FIG. 9. Optical rotation (2/P) for PBLG liquid crystals in mixed solvent systems. ● PBLG-dioxane: CH_2Cl_2 (Ref. 11); ■ PBLG (MW 310,000; 20% wt/vol) in dioxane - CH_2Cl_2; ▲ PBLG (MW 310,000; 20% wt/vol) in dioxane - nitrobenzene.

parallel position because of the chirality of the van der Waals surface of the polymer molecule. This is analogous to the barrier to rotation on one side of the parallel position evident in two screws of the same sense whose threads are interlocked. The biasing of oscillations would in turn lead to a unidirectional twist over large distances, and the sense of this twist would be dependent on the sense of the chirality of the polymer outer surface. In this case reversal of the chain sense does not affect the chiral sense; a random array of antiparallel helices (found in the solid state [57]) can generate the cholesteric torsion.

The macromolecular surface chirality in a given solvent would be opposite for the D and L enantiomorphs of polybenzylglutamate. This accounts for the nematic phase observed in solutions of equal parts PBLG and PBDG. In order to explain the results shown in Fig. 9, one must propose that the surface chirality of PBLG depends on specific polymer-solvent interactions which preferentially "stabilize" certain side-chain conformations. By appropriate arrangement of the PBLG side chains one can change the screw sense of the surface chirality of the macromolecule from right-handed to left-handed while maintaining the same right-handed α-helical backbone conformation. The space-filling models of PBLG schematically shown in Fig. 10 have side-chain secondary structures which differ in the sense of their surface chirality. The models were constructed using the dihedral angles of the side-chain bonds derived from calculated low-energy side-chain conformations [58].

This essentially short-range dispersion force model for the cholesteric texture in PBLG solutions may give some insight into the relationship between these lyotropic systems and thermotropic cholesterics. Goossens has presented a molecular statistic theory of cholesterics which indicates that the cholesteric twist has its origin in an asymmetric contribution to the intermolecular dispersion energy which is caused by short-range electric dipole-quadrupole interactions [59]. This asymmetric interaction is present only when the component molecules have no center or plane of symmetry, e.g., only chiral molecules exhibit the cholesteric texture.

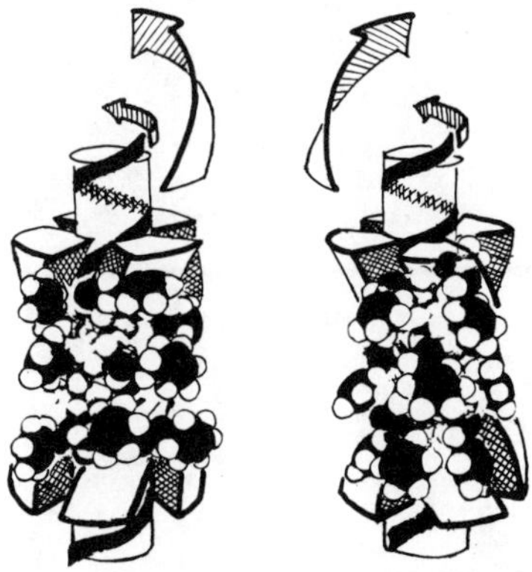

FIG. 10. After space-filling models of PBLG with side-chain secondary structures exhibiting right-handed and left-handed van der Waals surfaces (Ref. 56, 58).

D. Nematic Structure

PBLG also exists in the nematic structure which is either solvent- or field-induced. Figure 11 is a photograph of nematic PBLG which illustrates characteristic optical properties of the nematic texture.

1. Pitch Compensation

It was noted above that the sign of the form optical rotation of cholesteric solutions of PBLG in various solvents may be positive or negative. In an appropriate mixture of the two solvents (Fig. 9), the liquid crystal becomes "compensated," resulting in a cholesteric mesophase with infinite pitch (by definition, a nematic liquid crystal). An equimolar mixture of the D and L isomers of PBLG also results in a "compensated" untwisted nematic texture.

2. Cholesteric-to-Nematic Transition Induced by Fields

An unusual phenomenon occurs when certain types of cholesteric liquid crystals are subjected to an electric or a magnetic field. If the field is sufficiently high the cholesteric structure is transformed into one

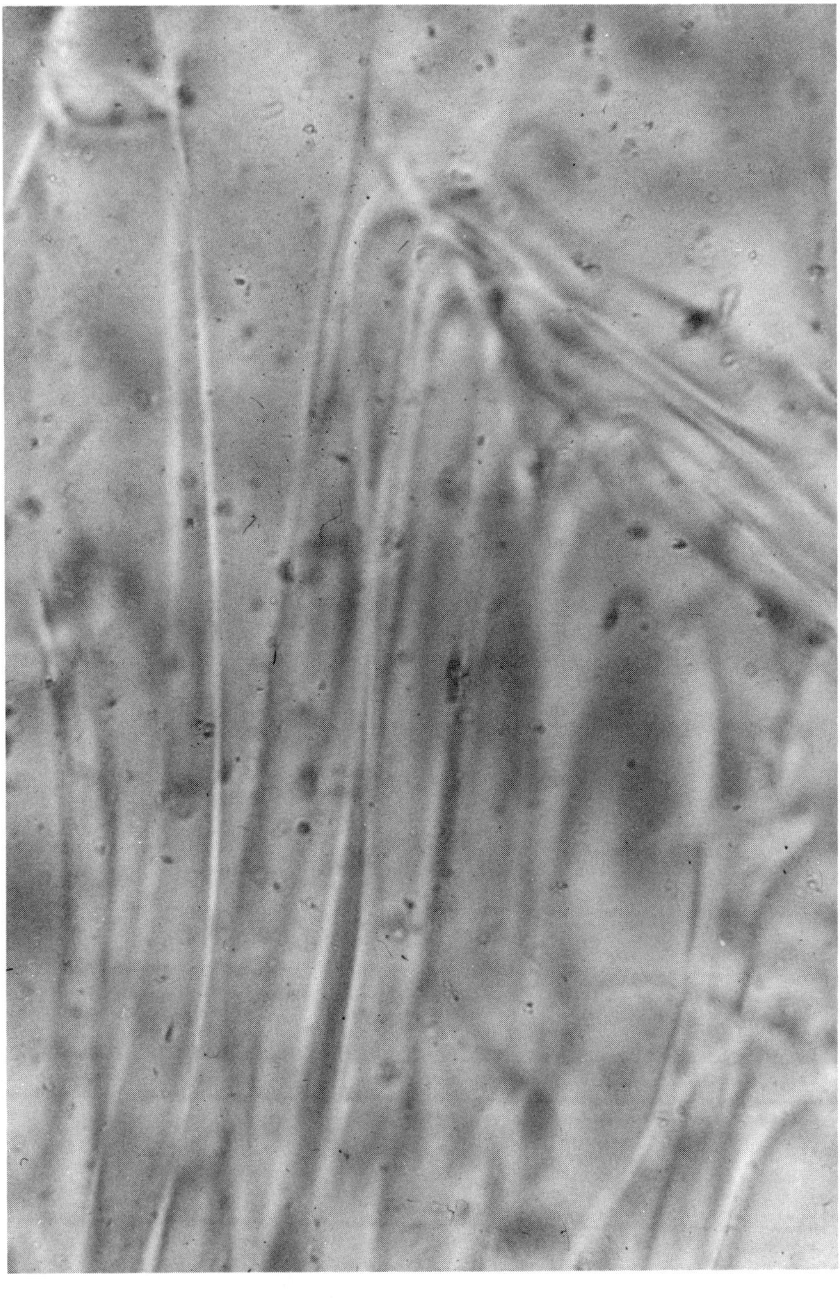

FIG. 11. Photomicrograph of nematic PBLG-dioxane liquid crystal.

resembling an aligned nematic structure. The transition can only occur if the anisotropic dielectric or diamagnetic susceptibilities of the individual molecules are positive quantities, which is seen from Table 1 to be the case for PBLG in the helical conformation. As a result PBLG molecules will tend to align themselves so that their long axes are parallel to the field direction. At some critical magnetic field strength H_c the mesophase would be expected to lose the high optical rotatory power and diffraction properties associated with the cholesteric phase. The relationship [60,61] between the zero field pitch P_0 and H_c involves K_{22}:

$$H_c = \frac{\pi 2}{2}\left(\frac{K_{22}}{\Delta\chi_m}\right)^{1/2}\frac{1}{P_0} \qquad (1)$$

The analogous critical electric field would be of the same form with $\Delta\varepsilon/4\pi$ replacing $\Delta\chi_m$, where $\Delta\varepsilon$ is the anisotropy of electric susceptibility, also a positive number. For most cholesterics of moderate periodicity the critical field can be as high as 100 kG, and the effect is not readily observable. In PBLG solutions, however, the pitch is considerably larger, reducing the field requirements for the transition.

A functional form for the dependence of the cholesteric pitch P on the magnetic field strength has also been derived by deGennes [60,61] which, in reduced form, is given by:

$$P/P_0 = \left(\frac{2}{\pi}\right)^2 K(k)E(k) \qquad (2)$$

where K(k) and E(k) are elliptical integrals of the first and second kind, respectively, and k is defined through the relation, $H/H_c = k/E(k)$, and varies between 0 and 1. As the field increases the pitch of the structure is predicted to increase slowly at first and then diverge logarithmically as the critical field is approached (solid curve of Fig. 12).

The field-induced cholesteric-to-nematic transition of this lyotropic liquid crystal has in course been followed by NMR [62-65], infrared dichroism [66,67], optical methods [68], and most recently by magnetic susceptibility [69] and laser light-beating spectroscopic techniques [70].

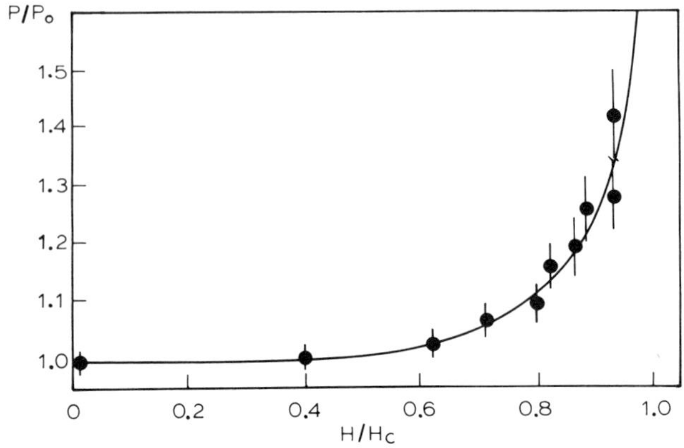

FIG. 12. Dependence of the pitch P on magnetic field strength H for liquid crystal PBLG of molecular weight 310,000 in dioxane (20% wt/vol). Reduced plot where P_o is the pitch in the absence of the field and H_c is the critical field. $P_o = 30 \pm 1 \mu m$, $H_c = 5.06 \pm .02$ kG. The solid curve is after the theory of deGennes (Ref. 60, 61).

3. Twist Elastic Constant

Field-induced helical unwinding has been observed optically in thermotropic liquid crystal mixtures prepared to be of larger than normal pitch [71-73]. Measurements of K_{22} on several of these doped systems have been made [72-75]. The authors and their colleagues have investigated liquid crystal solutions of PBLG, where K_{22} was determined in various solvents as a function of concentration, temperature, and molecular weight [69,76-78]. For the PBLG solutions examined, H_c is less than 20 kG, and the critical electric field for the transition occurs around 350 V/cm. It should be noted that even in strong fields the transition to nematic order in the polypeptide system is slow, taking as long as 100 hr. before equilibrium is attained. Addition of trace amounts of trifluoroacetic acid markedly reduces the equilibration time [54,79].

A typical plot of the dependence of the pitch on magnetic field strength is shown in Fig. 12, which confirms the validity of applicability of the theory developed for thermotropics in these lyotropic systems. In Table 4, values of K_{22}, calculated from Eq. (1), P_0, H_c, and $\Delta\chi_m$ are summarized for different molecular weight PBLG in various solvents.

TABLE 4. Parameters of 20% (wt/v) Solutions of 310,000 MW PBLG in Three Solvents That Support the Cholesteric Structure

Solvent	P_o (μm)	H_c (kG)	$K_{22}:\Delta\chi_m$ (emu.cgs)	$\Delta\chi_m$ (emu/cm^3)	K_{22} (dyne)
Dioxane	30	5.06	9.6	$0.64 \pm 0.07 \times 10^{-8}$	6.22×10^{-8}
Chloroform	42	6.87	33.7	$0.63 \pm 0.05 \times 10^{-8}$	2.1×10^{-7}
Dichloromethane	80	5.23	72.3	$0.74 \pm 0.05 \times 10^{-8}$	5.4×10^{-7}

The value of K_{22} initially found for the 310,000 molecular weight (MW) sample in dioxane is comparable to those obtained in the more common thermotropic liquid crystals (10^{-8} dyne) [77]. K_{22} for a sample of higher degree of polymerization (MW = 550,000) in a different solvent (CH_2Cl_2) turned out to be 582×10^{-8} dyne [69,70]. A wide variability of this elastic modulus, which does not occur in thermotropic systems, is now evident. K_{22} in the PBLG liquid crystal could be made to vary over a wide range by adjustment of solvent and polymer molecular weight. Only a slight concentration dependence of K_{22} throughout the liquid crystal range was, however, noted (Fig. 13). The variations can be at least partially understood in terms of changing intermacromolecular interactions and recent statistical hard rod theories of liquid crystal elasticity [78].

a. Solvent Effects. Table 4 demonstrates that choice of solvent dramatically alters the pitch of the cholesteric texture. There seems to be no straightforward correlation between P_0 and particular solvent properties (e.g., solvent polarity, dielectric constant, solubility parameter, etc.) On the other hand, $\Delta\chi_m$ is the same for PBLG in each of these solvents. The magnetic anisotropy is essentially that of the polypeptide; the solvent molecule is only slightly ordered in the liquid crystal, and principle values of the solvent susceptibility tensor are nearly the same. The observed $\Delta\chi_m$ of PBLG may be divided into two contributions associated with the helical backbone and side chain, $\Delta x_{(helix)}$ and $\Delta x_{(sidechain)}$. They are related to $\Delta\chi_m$ observed in Eq. (3):

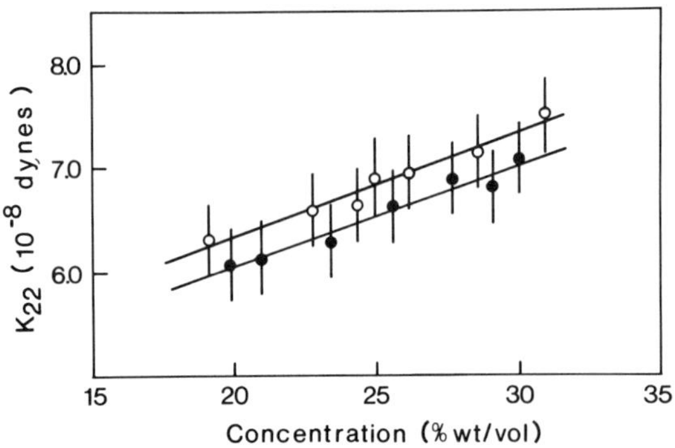

FIG. 13. Plot of the twist elastic constant K_{22} versus PBLG concentration. The open circles are the data points for a 310,000-MW PBLG sample in dioxane at 20° C, the closed circles are the same sample at 25° C.

$$\Delta\chi_m = (\chi_{\|} - \chi_{\perp})S = \Delta\chi_{(helix)}S + \Delta\chi_{(sidechain)}S_\beta S \qquad (3)$$

where $\chi_{\|}$ and $\chi_{\perp}$ are the components of the susceptibility $\|$ and $\perp$ to the helix, the order parameter S describes the average orientation of the helix relative to the nematic director, and S_β is the average orientation of the side chain relative to the helix axis [54,69]. Measurements of $\Delta\chi_m$ for polyglutamic acid, which lacks the highly anisotropic phenyl ring on its sidechain, suggest that the second term in Eq. (3) contributes about 40% of the total value for $\Delta\chi_m$. Since $\Delta\chi_m$ is essentially the same in CH_2Cl_2 and dioxane, the diamagnetic projections of the proposed side-chain conformations exhibiting opposite chiral senses (see Sec. III.B.2) must be similar in both solvents.

The sixfold change in K_{22} in going from dioxane to CH_2Cl_2 suggests that the solvent can attenuate intermacromolecular interactions. Since H_c and $\Delta\chi_m$ are almost insensitive to solvent, the nearly 10-fold change in K_{22} (Table 4) comes from the change in P_0 through Eq. (1).

b. Effect of Temperature: Helix Order Parameter. Heating eventually causes the PBLG liquid crystal to melt into an isotropic solution of randomly dispersed helices [80]. The α-helical conformation itself is stable up to high temperatures (200° C in m-cresol) [81,82]. The randomization of the liquid crystal prior to melting can be followed through changes in the order parameter S.

$\Delta\chi_m$ is proportional to S [Eq. (3)] and decreases as the temperature increases (Fig. 14). P_0 increases linearly with temperature while H_c is inversely proportional to the temperature [78]. The latter result enables the calculation of the temperature dependence of $K_{22}/\Delta\chi_m$. Saupe has shown that in nematic liquid crystals $K_{22} = K_{22}^N S^2$, where K_{22}^N is temperature independent [83]. Using this relationship we infer that the ratio $K_{22}/\Delta\chi_m$ is proportional to S and will reflect the functional dependence of S on temperature. In Fig. 15 the temperature dependence of $K_{22}/\Delta\chi_m$ is shown. Both Figs. 14 and 15 show that in contrast to thermotropic liquid crystals, S changes smoothly with temperature up through the clearing point, suggesting that the transition from an ordered system of rods to an unordered system is not first order. The higher order

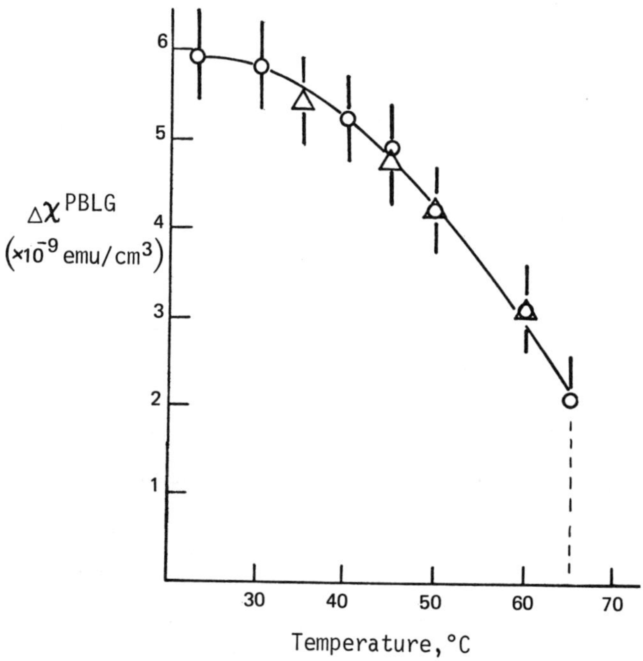

FIG. 14. Plot of diamagnetic anisotropy $\Delta\chi$ for PBLG-dioxane liquid crystal versus temperature. The solution becomes isotropic at ≈ 80° C, ⊙ increasing T, △ decreasing T.

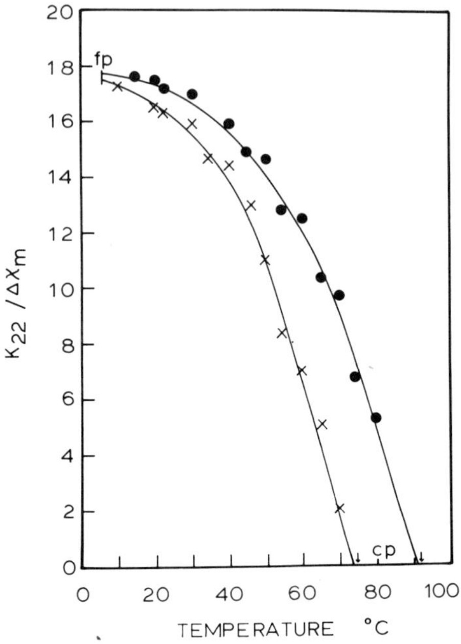

FIG. 15. Plot of the $K_{22}:\Delta\chi_m$ ratio versus temperature. The circles represent the data points for a 30% PBLG sample (310,000 MW) and the crosses are for a 25% PBLG sample (300,000 MW), both in dioxane.

of the transition has analogy in the helix-to-random coil phase transition of single, unassociated polypeptides. In this case, however, we are observing the thermal dissolution of the supramolecular organization of the cholesteric structure rather than the unraveling of individual PBLG macromolecules.

The temperature variation of the twist elastic constant can be derived from the data in Figs. 14 and 15. K_{22} decreases with increasing temperature (Fig. 16) as do elastic constants in more common materials [84].

c. Molecular Weight Dependence. In thermotropic liquid crystals changes in molecular structure from one liquid crystal to another are rather subtle, and K_{22} varies only slightly from compound to compound. As noted above, the value of K_{22} of liquid crystal PBLG of molecular weight

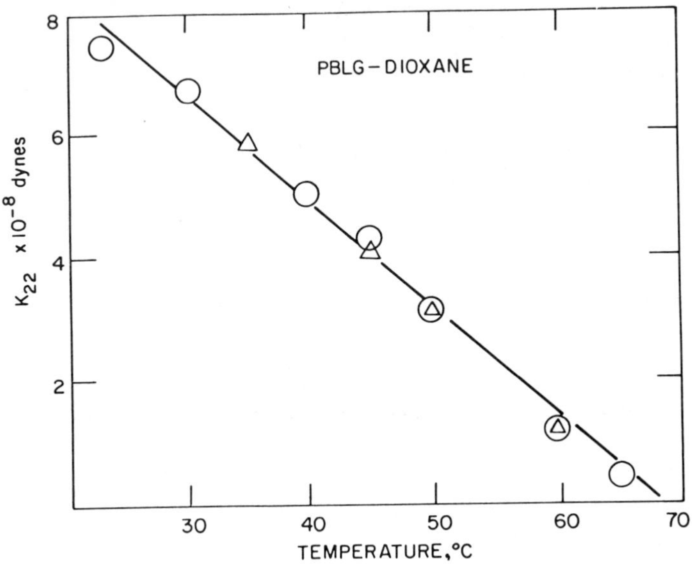

FIG. 16. Plot of K_{22} versus temperature derived from data in Figs. 14 and 15; ⊙ increasing T, △ decreasing T.

550,000 in dichloromethane is a factor of 90 larger than the value found on a 310,000 molecular weight sample in dioxane at the same temperature.

The unusual variability of K_{22} in polypeptide liquid crystals can be partially explained by changing interparticle interactions induced by solvent and recent theoretical findings on the elasticity of a suspension of long, hard rods [85,86]. Straley [86] has calculated the elastic constants K_{ii} of a hard rod liquid crystal using the Onsager [27] approach and finds that:

$$K_{22}/\rho^2 L^4 DkT \approx 0.02 \tag{4}$$

where ρ is the number density, L is the helix length, and D is the rod diameter. This relationship accounts for most of the difference in K_{22} for these two particular systems. The difference in molecular weight of the solute is dominant in its influence. The axial ratios L:D for the 550,000- and 310,000-MW PBLG are 150 and 85, respectively. According to Eq. (4), this alone would lead to a factor of 10 larger K_{22} for the higher molecular weight polypeptide. The additional factor of 9 may easily be the result of the different solvents involved (a ninefold increase was noted above in changing the solvent from dioxane to dichloromethane) and polymer concentration through the density term. Another factor that could influence the value of K_{22} is the effective diameter of the helical rod; D may be differentially modified by solvation of the macromolecule.

The theory cited contains no potential energy term other than hard rod repulsions. The observed clearing point is, therefore, not in the physics of the model. However, at a given temperature in the mesophase regime it is valid to compare system to system once empirical solvent effects are introduced, since they apparently considerably influence the potential function.

4. Magnetic Resonance in Polypeptide Liquid Crystals

NMR techniques have been used to obtain information about the molecular dynamics in the phase itself and, in more conventional experiments, about small guest molecules oriented in the liquid crystal host. For example, high-resolution NMR can be used to monitor the transition from a dilute isotropic solution of randomly dispersed PBLG helices to the concentrated liquid crystalline solutions. The sharp proton NMR spectrum of PBLG with clearly resolved resonances for the various chemically shifted protons of the peptide residue collapses into a broad structureless absorption several kilohertz wide when the PBLG concentration is increased through the isotropic-liquid crystal transition [87] (A point, Sec. II.B). The broadening occurs because the rotation of the PBLG molecules is gradually reduced and static dipolar interactions between pairs of protons in PBLG are not averaged to zero. These same dipolar interactions in guest molecules (molecules having a smaller number of protons) in a liquid crystal host or solvent appear as hyperfine splittings in their NMR spectra and lead to useful information about guest molecular structure. The isotropic-to-liquid crystal phase transition is also apparent in the solvent molecular dynamics as reflected in abrupt changes in the solvent proton spin-lattice relaxation time, T_1 [88]. With deuterated solvents, similar discontinuities in the deuteron T_1 at the phase boundary were not apparent [89].

The polypeptide T_1 is frequency dependent. This is often the case in macromolecules when the correlation time for rotational reorientation approaches the time scale of the NMR experiment (the inverse of the Larmor frequency) and may occur in isotropic solutions [90]. Frequency-dependent T_1s are also characteristic of the liquid crystal phase; they result from modulation of dipolar interactions by long-range hydrodynamic fluctuations. The frequency dependence of PBLG proton T_1s in the liquid crystal phase has been described in terms of a distribution of correlation times arising from the range of molecular flexibility in the peptide residue (backbone and side chain) [89] and in terms of the dynamic constraints imposed by the close-packed helices in the liquid crystal [91]. The latter description has been augmented by electron spin resonance studies of spin-labeled PBLG [92].

The dipolar splittings in the proton NMR spectra of the solvent were used in the original observation of the field-induced cholesteric-to-nematic transition in PBLG liquid crystals [62-64]. Since that time NMR spectra of small guest molecules dissolved in this phase have elucidated the inherent order of this liquid crystal [65], structural parameters of the guest molecules (in some cases the solvent) [93-95], and the factors influencing guest orientation [79,96,97]. In general, the guest orientation is considerably smaller than that in thermotropic hosts (guest order parameters in polypeptide liquid crystals are approximately 0.001). Also, specific guest-polypeptide interactions (e.g., hydrogen bonding) and not geometric constaints (in contrast with thermotropic systems) are primary in determining guest orientation in the liquid crystal [97].

E. Smectic Structure

There is no direct optical or structural characterization that conclusively demonstrates that PBLG or any other polypeptide forms a lyotropic liquid crystal with the smectic organization. The high specific viscosity and low effective dipole moment of PBLG in concentrated benzene solution, however, are suggestive of a two- to three-dimensional rod ordering in smecticlike arrangements.

Electro-optic and intrinsic viscosity studies on concentrated benzene solutions indicate that the macromolecular organization of PBLG is such as to produce a large birefringence but an effective molecular dipole moment significantly smaller than that expected for the free polymer rod [98]. For example, specific intrinsic Kerr constants of PBLG (MW = 37,000) in DMF and benzene yield estimates of the permanent dipole moment of the polymer of 1,450 and 805 Debye, respectively. Both DMF and benzene are solvents that are not expected to disrupt the primary structure of the α-helix of the individual polypeptide molecules. The low value of the effective dipole in benzene is suggestive of a side-by-side, antiparallel association of the macromolecules, an organization reminiscent of the smectic liquid crystal structure. The value in DMF is assumed to represent the best estimate for the free rod of this degree of polymerization [99]. Furthermore, the intrinsic viscosity of PBLG in benzene is unusually high, indicating substantial molecular aggregation. Upon the addition of small amounts of a polar solvent to the benzene-PBLG liquid crystal, the phase undergoes a change to a more isotropic system in which the major form of association appears to be antiparallel but involves relatively few molecules. Further addition of the polar component causes the structure to break up into essentially individual solute molecules [98]. The presumed smectic-to-free rod transition is not observed in solvents similar to benzene, such as dioxane, but aggregation is apparently present [100].

It has been shown that solid films of PBLG cast from solvents such as CH_2Cl_2 or $CHCl_3$ retain the cholesteric structure found in the prenascent liquid crystal [56,63,101]. Diffusion studies of benzene through films cast from benzene-PBLG and CH_2Cl_2-PBLG solutions indicate that the morphological structure of the membranes is substantially different [102]. Those cast from benzene exhibit a greater permeability than those cast from methylene chloride. The diffusion coefficient at infinite dilution of the former exceeds that of the latter by some 550%, although the solubility of benzene in the films differs by only 18% and the densities of the membranes are identical. Again, a different molecular organization of liquid crystal PBLG in benzene may be inferred.

F. Miscellaneous Studies

1. Rotational Viscosity Coefficient

The rate of reorientation of the nematic director in a magnetic field allows the determination of γ_i [69,103]. γ_i is dependent on polypeptide axial ratio, concentration, and solvent composition. Addition of 1-2% trifluoroacetic acid to the PBLG-dioxane causes a marked decrease in γ_i [54], perhaps through disruption of polypeptide aggregation [100]. The values of γ_i are much larger than those found in thermotropic systems [104].

2. Light Scattering

Iizuka et al. proposes that clusters of PBLG molecules existing in the liquid crystal phase give rise to light-scattering patterns and changes in these patterns when the liquid crystal is subjected to orienting electric fields [105]. Similar studies on solid PBLG films cast from oriented liquid crystalline solutions give evidence of several types of morphological structure [105-107].

3. Oriented Solid State

When the solvent is evaporated from an oriented nematic PBLG solution, a highly oriented film forms. x-Ray diffraction patterns from such films [108] are similar to those obtained from mechanically drawn polypeptide fibers [57]. The patterns are amenable to interpretation; details of the PBLG helical conformation can be obtained. They suggest that subtle differences in polypeptide conformation may be correlated with the solvent from which the film has been cast [108].

4. Induced Circular Dichroism

Achiral molecules dissolved in thermotropic cholesteric liquid crystals exhibit circular dichroism (CD). The sign of the CD is dependent on the twist sense of the cholesteric structure [109]. This observation has been extended to lyotropic PBLG liquid crystals [110]. Liquid crystal-induced CD could be valuable as a probe of microscopic structure in compensated polypeptide liquid crystals, i.e., critical solvent mixtures (Fig. 9).

RECENT DEVELOPMENTS

This chapter was devoted to liquid crystalline properties of polypeptide solutions as until recently these were the only fully characterized polymer liquid crystals. Since the preparation of this manuscript in 1976, a substantial body of data on other nonpeptide polymer liquid crystals has been published or released in patent, abstract, or conference form. There now exists a wide range of thermotropic and lyotropic macromolecular liquid crystals, some of which have immediate practical applications. For example, data on the synthesis and physical properties of a series of polyarylamides that exhibit lyotropic liquid crystallinity in suitable solvents has been published [111]. A representative polymer of this class is polybenzamide of repeat structure

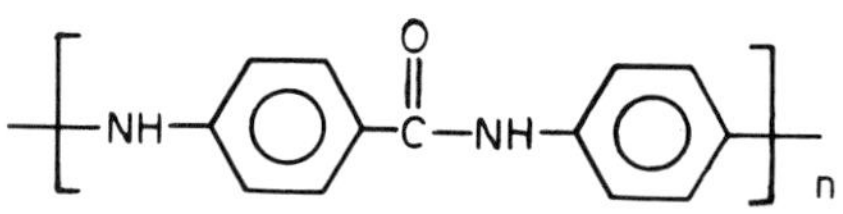

a highly linear macromolecule due not to intramolecular hydrogen bonding but as a result of the planarity and rigidity of the amide linkage and adjacent aromatic groups. These polymers form anisotropic solutions above a certain critical concentration in solvents such as strong acids, ureas, and dialkylamide-salt solutions (wicked conditions in comparison to those discussed above). When spun, the anisotropic arrangement of macromolecules provides an effective precursor for the formation of ultrahigh-strength fibers and films. Additional reviews of polypeptide liquid crystals are also now available [112,113].

Acknowledgment

This work was supported in part by a NIH research grant (PHS Grant AM 17497-02) from the National Institute of Arthritis, Metabolism, and Digestive Diseases, by Research Corporation, and by the Petroleum Research Fund administered by the American Chemical Society.

REFERENCES

1. H. Mark and A.V. Tobolsky, Physical Chemistry of High Polymeric Systems, High Polymers, Vol. 2, Interscience, New York, 1950.
2. B.R. Gallot, Pure Appl. Chem., 38:1 (1974).
3. S. Frenkel, J. Polym. Sci., Polym. Symp., 44:49 (1974).
4. V. Luzzati and F. Husson, J. Cell. Biol., 12:207 (1962).
5. D. Chapman and G.T. Stewart, in Plastic Crystals and Liquid Crystals, Vol. 1, (G.W. Gray and P.A. Winsor, ed.) Halsted, Chichester, 1974, Chap. 6.
6. D. Chapman, The Structure of Lipids, Wiley, New York, 1965.
7. D. Chapman, Sci. J., 1:32 (1965).
8. J.M. Stein, in Molecular Associations in Biological and Related Systems, Advances in Chemistry Series, No. 84, American Chemical Society, Washington, D.C., 1969.
9. Y. Bouligand, M.O. Soyer, and S. Puiseux-Dao, Chromosoma, 24:251 (1969).
10. M. Spencer, W. Fuller, M.H.F. Wilkins, and G.L. Brown, Nature, 194:1014 (1962).
11. C. Robinson, Tetrahedron, 13:219 (1961).
12. A.C. Neville and S. Caveney, Biol. Rev., 44:531 (1969).
13. J.D. Bernal and I. Fankuchen, J. Gen. Physiol., 25:111 (1941).
14. J. Lapoint and D.A. Marvin, Mol. Cryst. Liq. Cryst., 19:269 (1973).
15. D.A. Marvin and E.J. Wachtel, Nature, 253:19 (1975).
16. D.M. Small and G.G. Shipley, Science, 185:222 (1974).
17. A.P. Minton, J. Mol. Biol., 82:483 (1974).
18. A.V. Tobolsky and E.T. Samulski, in Plastic Crystals and Liquid Crystals, Vol. 1, (G.W. Gray and P.A. Winsor, eds.), Halsted, Chicester, 1974, Chap. 4, p. 175 of Ref. 5.
19. M.F. Perutz, Nature, 167:1053 (1951).
20. L. Pauling, R.B. Corey, and H.R. Branson, Proc. Natl. Acad. Sci., USA, 37:205 (1951).
21. M.K. Grover and R. Zwanzig, Biopolymers, 13:2103 (1974).
22. E.A. DiMarzio and M. Bishop, Biopolymers, 13:2331 (1974).
23. C. Robinson, Trans. Faraday Soc., 52:571 (1956).

24. C. Robinson, J.C. Ward, and R.B. Beevers, Disc. Faraday Soc., 25:29 (1958).
25. C. Robinson and J.C. Ward, Nature, 180:1183 (1957).
26. C. Robinson, Mol. Cryst., 1:467 (1966).
27. L. Onsager, Ann. N.Y. Acad. Sci., 51:627 (1949).
28. A. Isihara, J. Chem. Phys., 19:1142 (1951).
29. P.J. Flory, Proc. Roy. Soc., Ser. A, 243:70 (1956).
30. P.J. Flory, Proc. Roy. Soc., Ser. A, 243:73 (1956).
31. J.P. Straley, Mol. Cryst. Liq. Cryst., 22:333 (1973).
32. J.J. Hermans, Jr., J. Colloid Sci., 17:638 (1962).
33. P.J. Flory, J. Polym. Sci., 49:105 (1961).
34. P.J. Flory and W.J. Leonard, J. Amer. Chem. Soc., 87:2102 (1965).
35. H.T. Peterson, D.E. Martire, and M.A. Cotter, J. Chem. Phys. 61:3547 (1974).
36. K.D. Goebel and W.G. Miller, Macromolecules, 3:64 (1970).
37. E.L. Wee and W.G. Miller, J. Phys. Chem., 75:1446 (1971).
38. J.H. Rai and W.G. Miller, Macromolecules, 5:45 (1972).
39. J.H. Rai and W.G. Miller, J. Phys. Chem., 76:1081 (1972).
40. J.H. Rai and W.G. Miller, Macromolecules, 6:257 (1973).
41. J.H. Rai, W.G. Miller, and R.G. Bryant, Macromolecules, 6:262 (1973).
42. W.G. Miller, C.C. Wu, E.L. Wee, G.L. Santee, J.R. Rai, and K.G. Goebel, Pure Appl. Chem., 38:37 (1974).
43. A. Akamoto, K. Kubo, and K. Ogino, Bull. Chem. Soc. Jap., 47:1054 (1974).
44. W. Helfrich, Phys. Lett., 43A:409 (1973).
45. W. Helfrich, Z. Naturforsch., 28C:693 (1973).
46. W. Helfrich and H. Deuling, Vth International Liquid Crystal Conference, Stockholm, Sweden, 1974; J. de Phys., C-1:327 (1975).
47. E.J. Ambrose and A. Elliot, Proc. Roy. Soc., Ser. A, 205:47 (1951).
48. H. deVries, Acta Crystallogr., 4:219 (1951).
49. Y. Bouligand and M. Kleman, J. Phys. (Paris), 31:1041 (1970).
50. Y. Bouligand, J. Phys. (Paris), 33:715 (1972).
51. Y. Bouligand, J. Phys. (Paris), 34:603 (1973).
52. Y. Bouligand, J. Phys. (Paris), 34:1011 (1973).
53. Y. Bouligand, J. Phys. (Paris), 35:215 (1974).

54. R.W. Duke, D.B. DuPré, W.A. Hines, and E.T. Samulski, J. Amer. Chem. Soc., 98:3094 (1976).
55. A.R. Downie, A. Elliott, W.E. Hanby, and B.R. Malcolm, Proc. Roy. Soc., Ser. A, 242:325 (1975).
56. E.T. Samulski, Ph.D. Thesis, Dept. Chemistry, Princeton University, 1969.
57. A. Elliott, Poly-α-amino Acids, (G.D. Fasman, ed.), Dekker, New York, 1969, Chap. 1.
58. H.A. Scheraga, J.F. Yan, and G. Vanderkooi, J. Chem. Phys., 49:2713 (1968).
59. W.J.A. Goossens, Mol. Cryst. Liq. Cryst., 12:237 (1971).
60. P.G. deGennes, Solid State Commun., 6:163 (1968).
61. P.G. deGennes, Mol. Cryst. Liq. Cryst., 7:325 (1969).
62. S. Sobjima, J. Phys. Soc. Jap., 23:1070 (1967).
63. E.T. Samulski and A.V. Tobolsky, Macromolecules, 1:555 (1968).
64. M. Panar and W.D. Phillips, J. Amer. Chem. Soc., 90:3880 (1968).
65. R.D. Orwoll and R.L. Vold, J. Amer. Chem. Soc., 93:5335 (1971).
66. E. Iizuka, Biochim. Biophys. Acta, 243:1 (1971).
67. E. Iizuka and Y. Go, J. Phys. Soc. Jap., 31:1205 (1971).
68. W.J. Toth and A.V. Tobolsky, Polym. Lett., 8:531 (1970).
69. C. Guha-Sridhar, W.A. Hines, and E.T. Samulski, J. Chem. Phys. 61:947 (1974).
70. R.W. Duke and D.B. DuPré, Macromolecules, 7:374 (1974).
71. R.B. Meyer, Appl. Phys. Lett., 14:208 (1969).
72. G. Durand, L. Leger, F. Rondelez, and M. Veyssie, Phys. Rev. Lett., 22:227 (1969).
73. R. Regaya, H. Gasparoux, and J. Prost, Rev. Phys. Appl., 7:85 (1972).
74. L. Leger; Solid State Commun., 10:697 (1972).
75. C. Williams and P.E. Cladis, Solid State Commun., 10:357 (1972).
76. C. Guha-Sridhar, W.A. Hines, and E.T. Samulski, J. Phys. (Paris), 36:270 (1975).
77. R.W. Duke and D.B. DuPré, J. Chem. Phys., 60:2759 (1974).
78. D.B. DuPré and R.W. Duke, J. Chem. Phys., 63:143 (1975).
79. P. Laszlo, A. Paris, and E. Marchal, J. Phys. Chem., 77:2925 (1973).
80. V.G. Baranov, S. Y. Frenkel, Bi Zhu-Chan, and T.I. Volkov, Vysokomol. Soedin., 8:957 (1966).

81. J.S. Franzen, J.B. Harry, and C. Bobik, Biopolymers, 5:193 (1967).
82. K. Tsuji, H. Ohe, and H. Watanabe, Polym. J., 4:553 (1973).
83. A. Saupe, Z. Naturforsch., 15a:810 (1960).
84. C. Kittel, Introduction to Solid State Physics, 3rd ed., J. Wiley and Sons, New York, 1967, p. 123.
85. R.G. Priest, Phys. Rev., A7:720 (1973).
86. J.P. Straley, Phys. Rev., A8:2181 (1973).
87. D. Doskocilova and B. Schneider, Macromolecules, 6:76 (1973).
88. W.A. Hines and E.T. Samulski, Macromolecules, 6:794 (1973).
89. B.M. Fung and T.H. Martin, J. Chem. Phys., 61:1697 (1974).
90. R. Kimmich and F. Noack, Z. Elektrochem., 75:269 (1971).
91. W.A. Hines and E.T. Samulski, J. Polym. Sci., C44:11 (1974).
92. E.L. Wee and W.G. Miller, J. Phys. Chem., 77:182 (1973).
93. M.P. Klein, D. Gill, and G. Kotowycy, Chem. Phys. Lett., 2:677 (1968); J. Amer. Chem. Soc., 90:6870 (1968).
94. B.M. Fung, M.J. Gerace, and L.S. Gerace, J. Phys. Chem., 74:83 (1970).
95. M.J. Gerace and B.M. Fung, J. Chem. Phys., 53:2984 (1970).
96. J.M. Anderson and A.C.F. Lee, J. Mag. Res., 4:160 (1971).
97. E.T. Samulski and H.J.C. Berendsen, J. Chem. Phys., 56:3920 (1972).
98. J.C. Powers, Jr., and W.L. Peticolas, Biopolymers, 9:195 (1970).
99. A. Wada, J. Polym. Sci., 45:145 (1960).
100. A.K. Gupta, C. Dufour, and E. Marchal, Biopolymers, 13:1293 (1974).
101. E.T. Samulski and A.V. Tobolsky, Nature (London), 216:997 (1967).
102. Y. Mohadger, T.K. Kwei, and A.V. Tobolsky, Macromolecules, 4:755 (1971).
103. R.W. Filas, L.E. Hajdo, and A.C. Eringen, J. Chem. Phys., 61:3037 (1974).
104. H. Gruler, J. Chem. Phys., 61:5408 (1974).
105. E. Iizuka, T. Keira, and A. Wada, Mol. Cryst. Liq. Cryst., 23:13 (1973).
106. G.L. Wilkes, Mol. Cryst. and Liq. Cryst., 18:165 (1972).
107. G.L. Wilkes, Polym. Lett., 10:935 (1972).
108. E.T. Samulski and A.V. Tobolsky, Biopolymers, 10:1013 (1971).

109. F.D. Saeva and J.J. Wysocki, J. Amer. Chem. Soc., 93:5928 (1971).

110. F.D. Saeva, Pure Appl. Chem., 38:25 (1974).

111. P.W. Morgan, Polymer Preprints, 17:47 (1976).

112. E.T. Samulski, in Liquid Crystalline Order in Polymers (A. Blumstein, ed.), Academic Press, New York (1978).

113. E. Iizuka, Adv. Polym. Sci., 20:79 (1976).

Chapter 6

CHOLESTERIC LIQUID CRYSTAL-INDUCED CIRCULAR DICHROISM

Franklin D. Saeva*

Xerox Corporation
Webster Research Center
Webster, New York

I. INTRODUCTION

Cholesteric liquid crystals display many extraordinary properties due to the existence of a macroscopic helical structure. They are best known for their exceptionally large optical rotation, i.e., several hundred thousand degrees per millimeter, and for their iridescent colors due to the selective reflection of circular polarized light in the region of the pitch band. It is in

*Present Affiliation: Research Laboratories, Eastman Kodak Company, Rochester, New York, 14650

this region of the pitch band that most of the previous optical investigations of cholesterics have been concentrated. For a more detailed description of the nature of cholesteric mesophases see Chap. 3.

In this chapter the extrinsic absorptive circular dichroism exhibited by molecules within the cholesteric mesophase will be treated. This optical property allows one to probe the macroscopic internal cholesteric structure as well as the alignment of the probe molecule on a molecular level.

We present here some mechanistic aspects, properties, and uses of the cholesteric liquid crystal-induced circular dichroism (LCICD) phenomenon. We define LCICD as the extrinsic circular dichroism observed in liquid crystals as a consequence of the dissymmetric array of aggregated molecules, be they the guest or the host. To introduce and place this topic into perspective, a number of definitions and background information will be presented prior to the discussion of the LCICD phenomenon.

A. Definitions

Molecular structures may exist as nonsuperimposable mirror images (enantiomers), the screw sense of which may be either left- or right-handed. This screw sense is termed the chirality [1]. Since there is no energy difference between enantiomers, a diastereomeric relationship must exist to distinguish between them, i.e., a relationship that is not enantiomeric. Figure 1 presents the enantiomeric relationship between left- and right-handed helices.

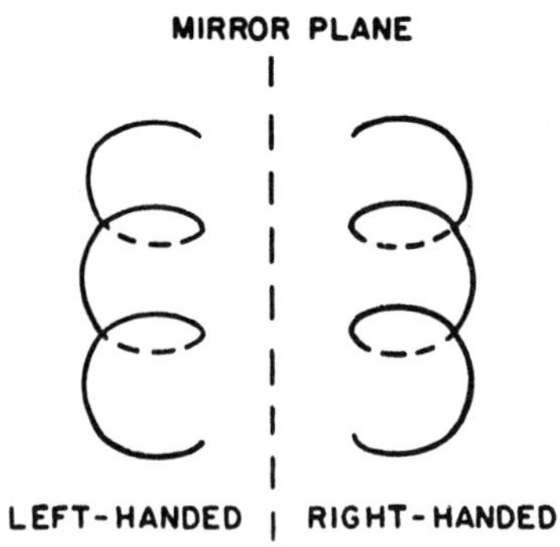

FIG. 1. Enantiomeric helices of opposite chirality.

A chiral substance exhibits optical rotatory power in that it has the ability to distinguish between left or right circular polarized light by refraction ($n_L \neq n_R$), absorption ($\varepsilon_L \neq \varepsilon_R$), or reflection. In these expressions, where n and ε are the refractive index and molar extinction coefficient, respectively, the subscript defines the sense of circularly polarized light.

For optical rotation to be displayed the speeds (refractive indices) of left- and right-handed circular polarized light must be unequal. Plane-polarized light (as shown in Fig. 2) vibrating in a single plane, e.g., x-y plane, may be thought to be the result of two circularly polarized beams rotating in opposite directions which are in phase, of the same frequency and one-half the amplitude. As plane-polarized light travels through a

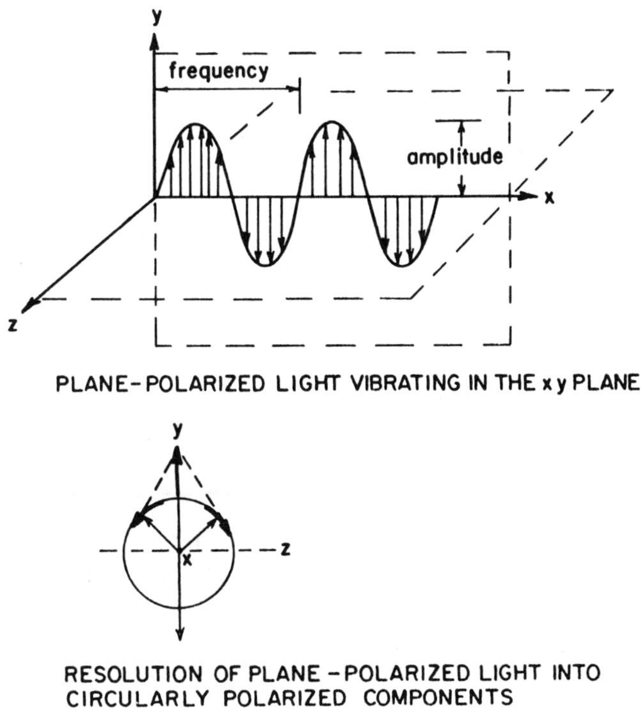

FIG. 2. Schematic representation of linear and circular polarized light.

chiral medium the resultant vector of the two circularly polarized components will be rotated either clockwise or counterclockwise depending on which component travels faster through the medium. This effect is called circular birefringence. An apparent optical rotation also results if linear polarized beams, i.e., $n_{\parallel} \neq n_{\perp}$, travel at different speeds through a medium resulting in a phase shift between the two beams. This effect, which is clearly distinguishable from circular birefringence, is called linear birefringence. Rotation of the sample between crossed polarizers will change linear but not circular birefringence.

The difference in the absorption of left- and right-handed circularly polarized light is called circular dichroism (CD), i.e., $\varepsilon_R \neq \varepsilon_L$, where ε_R and ε_L are the molar extinction coefficients for right- and left-handed circular polarized light, respectively. CD is a signed quantity, as is optical rotation, dependent on which circular component is absorbed to a greater extent. By convention, $\varepsilon_L > \varepsilon_R$ is termed positive CD and $n_L > n_R$ positive rotation [1]. Optical rotation occurs at all wavelengths, while CD, on the other hand, occurs only in the region of absorption. The wavelength dependence of CD and optical rotation (ORD) in the region of absorption is called a Cotton effect. Linear dichroism, in contrast, results from the preferential absorption of linear polarized light such that $\varepsilon_{\parallel} \neq \varepsilon_{\perp}$. Again, one can distinguish between linear and circular dichroism by simply rotating the sample normal to the incoming light beam. Linear dichroism effects will be altered while CD will not.

B. Natural Optical Activity

Naturally optically active compounds capable of rotating the plane of polarized light can be divided into two main classes. In the first class, the compounds are chiral on the molecular level and exhibit optical activity as a solid, liquid, or gas. Compounds in this class are intrinsically chiral, e.g., d-camphor, and l-menthol.

Intrinsically chiral compounds can be further subdivided into the following classes:

1. Inherently dissymmetric chromophores
2. Coupled oscillators
3. Dissymmetrically perturbed symmetric chromophores

Typical examples of inherently dissymmetric chromophores are systems such as biaryls and disulfides. Coupled oscillators include compounds with nonconjugated chromophores such as homoconjugated olefins. Symmetrical chromophores such as carbonyl, aromatic rings, and double bonds may be among those dissymmetrically perturbed. Rotatory power is normally significantly larger for the inherently dissymmetric than dissymmetrically perturbed symmetric chromophores.

The rotatory strength (R_K), which is the product of electric and magnetic dipole moments u_e and u_m, respectively, is a measure of the interaction of a chromophore with its dissymmetric environment. For intrinsically chiral molecules optical rotation and CD as a function of wavelength provide valuable structural information.

The second class of compounds we classify as extrinsically chiral since optical rotatory power occurs only in the solid state, e.g., quartz. Extrinsic CD has also been observed in a small class of compounds (e.g., urea [2] and benzil [3] which are not optically active in solution but gain their optical activity from a chiral molecular organization in the crystalline solid state. Molecules in this unusual class of compounds are arranged in a helical form in the crystal and can be translated from one to the other by a trigonal screw axis [3].

Certain achiral molecules also display extrinsic CD while complexed to polypeptides in an α-helical conformation [4-7], when dissolved in chiral solvents [8,9] or as aggregated dyes [10].

C. Faraday Effect

Achiral molecules also display optical rotatory power in the presence of magnetic fields. This magnetically induced optical activity, called the Faraday effect, is quite different from natural optical activity. In the Faraday effect a magnetic field is applied parallel to the direction of light propagation. All compounds, irrespective of their chemical structure, under these conditions exhibit optical rotation and CD which are proportional to the field strength and cos θ, which is the angle the magnetic field makes with the direction of the incoming light beam. The application of magnetically induced optical activity in structural chemistry is concerned mainly with the nature of optical transitions [11].

The previous discussion pertains to the isotropic (random) distribution of isolated molecules or complexes with dimensions small compared to the wavelength of light and where linear birefringence and dichroism is absent. exist.

D. Optical Properties of Cholesteric Mesophases

The nature of and the chemical structures providing cholesteric mesophases were treated in Chap. 3. Along with exhibiting linear and circular birefringence and linear and circular dichroism, cholesterics possess intrinsic as well as extrinsic optical rotatory power. It is the extrinsic rotatory power, especially the CD, which will be treated in this chapter.

Molecules in cholesteric mesophases are normally intrinsically chiral. Nematic mesophases (see Chap. 2) which possess achiral molecules may be readily transformed into a cholesteric system by the addition of chiral molecules or any other dissymmetric perturbation [12], e.g., chiral surface treatment.

The helical structure described schematically in Fig. 3 of the cholesteric mesophase can be disrupted by heating above the cholesteric-isotropic temperature or by the application of electric [13] and magnetic fields [14]. The "nematiclike" layers within the cholesteric mesophase are slightly twisted from one another forming a macroscopic helical structure. The optical properties of cholesteric systems have been treated theoretically by Mauguin [15] and de Vries [16]; they considered the cholesteric mesophase as being composed of chirally stacked birefringent plates. In their theoretical treatment they did not take into consideration the linear dichroism of the nematiclike layers, however. Out of this early work the relationship between the pitch and wavelength of maximum reflection (λ_0) was determined, i.e., $\lambda_0 = nP$, where n is the mean refractive index of the cholesteric medium, and P is the pitch of the helical structure. In the region of λ_0, circularly polarized light which does not match the helix sense of the cholesteric mesophase is selectively reflected while the circularly polarized component that matches the helix sense of the mesophase is selectively transmitted [15-18] (see Fig. 4). For light propagating along the helix axis incident plane-polarized light outside the region of λ_0 is completely transmitted although the left- and right-handed

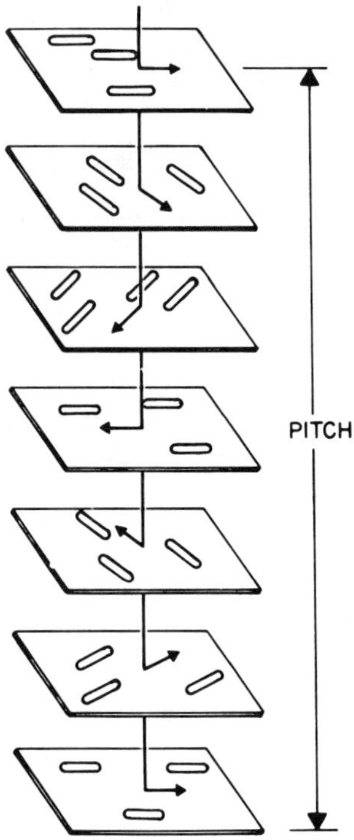

FIG. 3. Schematic representation of a cholesteric mesophase.

circularly polarized components travel at somewhat different speeds through the medium. The enormous CD signal observed in the region of λ_0 is due to this selective reflection of circular polarized light. Positive CD in the region of λ_0 indicates a right-handed helical structure and negative CD a left-handed structure.

The shapes of the optical rotation and CD curve for a typical cholesteric mesophase composed of sterol derivatives are represented in Fig. 5. The optical rotatory power of cholesteric mesophases has already been discussed in detail in Chap. 3 and will not be treated here. The ORD (x-x line) and CD (solid line) of a thin film ($\approx 10 \mu m$) of a cholesteric mesophase composed of sterol derivatives, e.g., cholesteryl nonanoate (CN)-cholesteryl chloride (CC) mixture, with λ_0 at 500 nm is consistent with the de Vries-Mauguin equation [15,16].

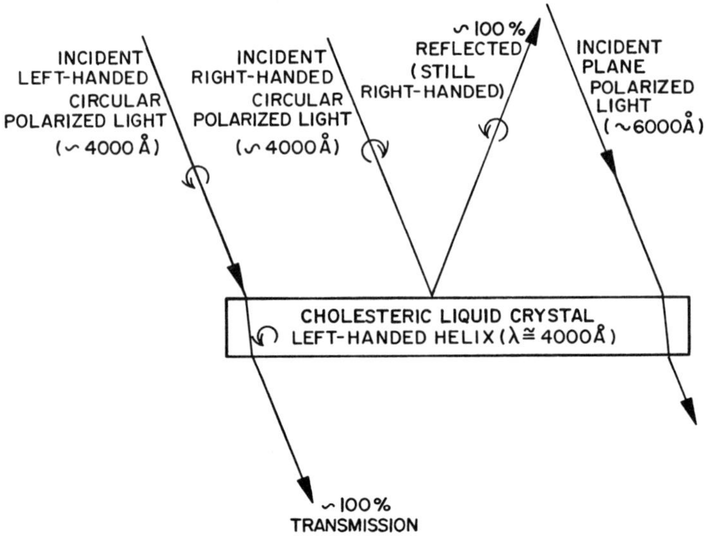

FIG. 4. Selective reflection of circular polarized light by cholesteric liquid crystals.

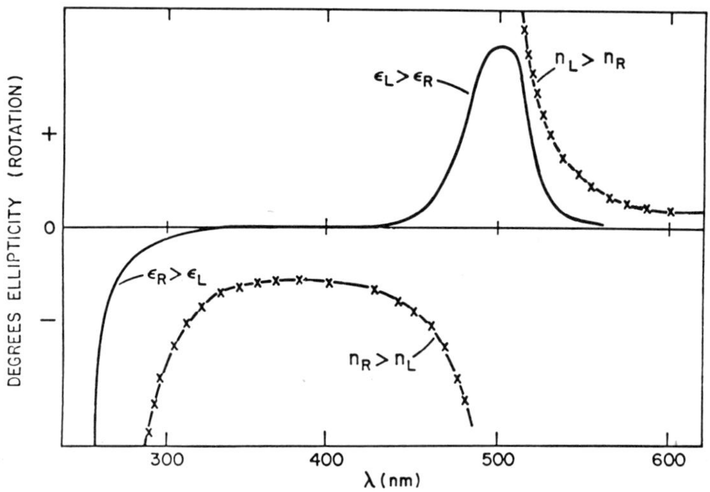

FIG. 5. Schematic representation of the optical rotation and circular dichroism, as a function of wavelength, for a typical cholesteric mesophase composed of cholesteryl derivatives with λ_0 at 500 nm.

II. EXTRINSIC CIRCULAR DICHROISM IN CHOLESTERIC MESOPHASES

A. Liquid Crystal-Induced Circular Dichroism in Achiral Solutes

When one ventures into the region of absorption of the cholesteric matrix molecules or an achiral solute such as N-(p-methoxybenzylidene)-p-butyl aniline (1) in a cholesteric mesophase, CD is observed within the optical transitions. Initial observations of this effect showed the CD to be of an extremely high intensity, i.e., 10^2-10^3 higher, than that due to natural CD [19].

Figure 6 presents the CD and absorption spectra of left- and right-handed cholesteric mesophases in which 1 has been dissolved. Selective reflection of circularly polarized light from the cholesteric pitch band [17] occurs between 500 and 600 nm while anil absorption lies between 200

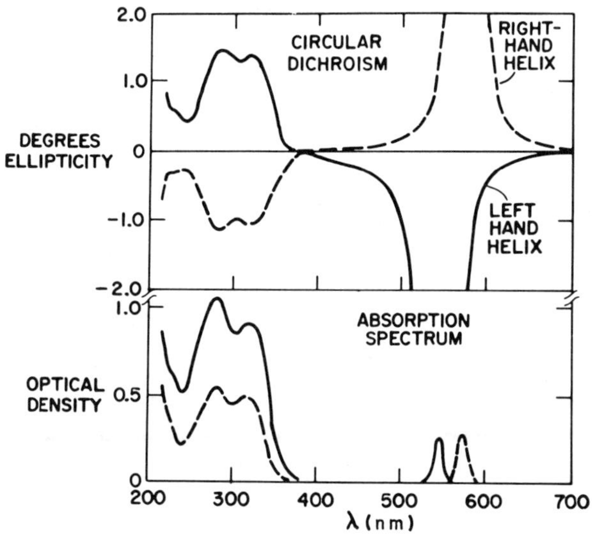

FIG. 6. Circular dichroism (upper) and absorption spectrum (lower): ---, 6.3-μm film of N-(p-methoxybenzylidene)-p-butylaniline (52.5 mg) in 10.0 g of 27.7:72.3 wt% cholesteryl chloride -cholesteryl nonanoate (right-hand helix); -, 11.7-μm film of 1 (63.19 mg) in 10.0 g of 90.6:9.4 wt% 2:3 (left-hand helix). (Reprinted from Ref. 12, p. 5929, by courtesy of the American Chemical Society.)

and 390 nm. The sign of the anil LCICD is opposite to that of the pitch-band CD of the cholesteric mesophase. A variety of chemical structures (i.e., azo compounds, aromatic ketones, anils) all show this effect, even though the solutes are not liquid crystalline themselves. The sign of the LCICD bands of 1 is dependent upon the chirality of the cholesteric helix, being positive for a left-handed helix and negative for a right-handed helix for λ_0 at a wavelength greater than the absorption by the solute. When the cholesteric environment is converted to an inactive nematic mesophase either by use of a compensated mixture of cholesteryl compounds [20] or by conversion to a nematic mesophase with an electric field [13], LCICD no longer exists. This indicates that a macroscopic helical structure, not just a chiral isotropic solvent, must be present for the extrinsic CD effect.

B. Infrared Circular Dichroism

Very shortly after the initial observations of LCICD [19], Stegemeyer and Mainusch report similar observations [21,22] of induced Cotton effects of achiral solutes in cholesteric mesophases by ORD. Schrader and Korte [23] subsequently investigated the infrared rotatory dispersion (IRD) of cholesteric matrices themselves. The cholesteric compositions were produced from the addition of 2 mol% d-carvone to a nematic [24] (Licristal®) which produced rotatory dispersion curves of about 7000° per millimeter for the host material vibrational bands. l-Carvone in Licristal® produced ORD curves for the host of opposite sign to the d-carvone mixtures. Chabay, in a similar study, reported the infrared circular dichroism (IRCD) [25] due to absorption and scattering of cholesteric liquid crystals in planar Grandjean texture. IRCD spectra were obtained on an instrument previously described [26], and anisotropy factors ($\Delta\varepsilon/\varepsilon$ max) up to 0.39 were observed for the cholesteric matrix vibrational transitions.

The rotational strength and sign of the interaction of light with the cholesteric matrix molecules were found to be a function of the inverse difference between the wavelength of absorption λ_{ab} and the λ_0 of the cholesteric pitch band: $R_K = f(\lambda_{ab} - \lambda_0)^{-1}$ [25]. Similar IRCD studies of 1 chirally perturbed by the addition of a chiral solute were reported by Mason et al. [27].

C. Origin of Liquid Crystal-Induced Circular Dichroism

LCICD of ordered achiral solutes possessing transition moments which are polarized perpendicular to one another have been found to display CD signals of opposite sign [28]. LCICD spectra of N-ethylcarbazole dibenzothiophene, and dibenzofuran show the 0-0 band for the singlet $1L_b \leftarrow 1A$ (short axis in-plane) transition to be of positive CD sign (see Fig. 7) and the $1L_a \leftarrow 1A$ transition (long axis in-plane) to be of negative CD sign in a right-handed helical structure ($\lambda_0 > \lambda_{ab}$).

The understanding of LCICD sign changes with the polarization of optical transitions as well as the position of λ_0 relative to λ_{ab} was first provided by Sackmann and Voss [29]. These authors and later Holzwarth and Holzwarth [30] considered the origin LCICD of solute and cholesteric matrix molecules, respectively. They extended the theory of electromagnetic radiation from the nonabsorbing to the absorbing case by adding a frequency-dependent complex distribution to the spiraling dielectric tensor of the liquid crystal itself.

As previously mentioned, cholesteric liquid crystals can be thought as being composed of nematiclike layers which behave as linearly birefringence and linear dichroic plates where the local optical axes of subsequent layers are twisted by a small angle which is a function of the pitch of the mesophases. Solute molecules may orient themselves within the nematic layers such that their long molecular axes align parallel to the local optic axes and matrix molecules. Cholesteric liquid crystals behave as optically negative ($n_0 > n_e$) unaxial crystals with the optic axis being also the helix axis (h).

In the three-dimensional coordinate system (x, y, z) presented in Fig. 8, the y axis coincides with the helix axis, and the x axis is assigned parallel to the local optic axis ℓ. The x axis then rotates with ℓ counterclockwise toward the viewer as in a left-handed helix. In cholesterics the refractive index along the local optic axis of the nematiclike layer (n_x), which is equal to n_0, is larger than the refractive index parallel to the z axes (n_z), which is identical to the refractive index in the direction of the helix axis (n_y), and which may be referred to as n_e.

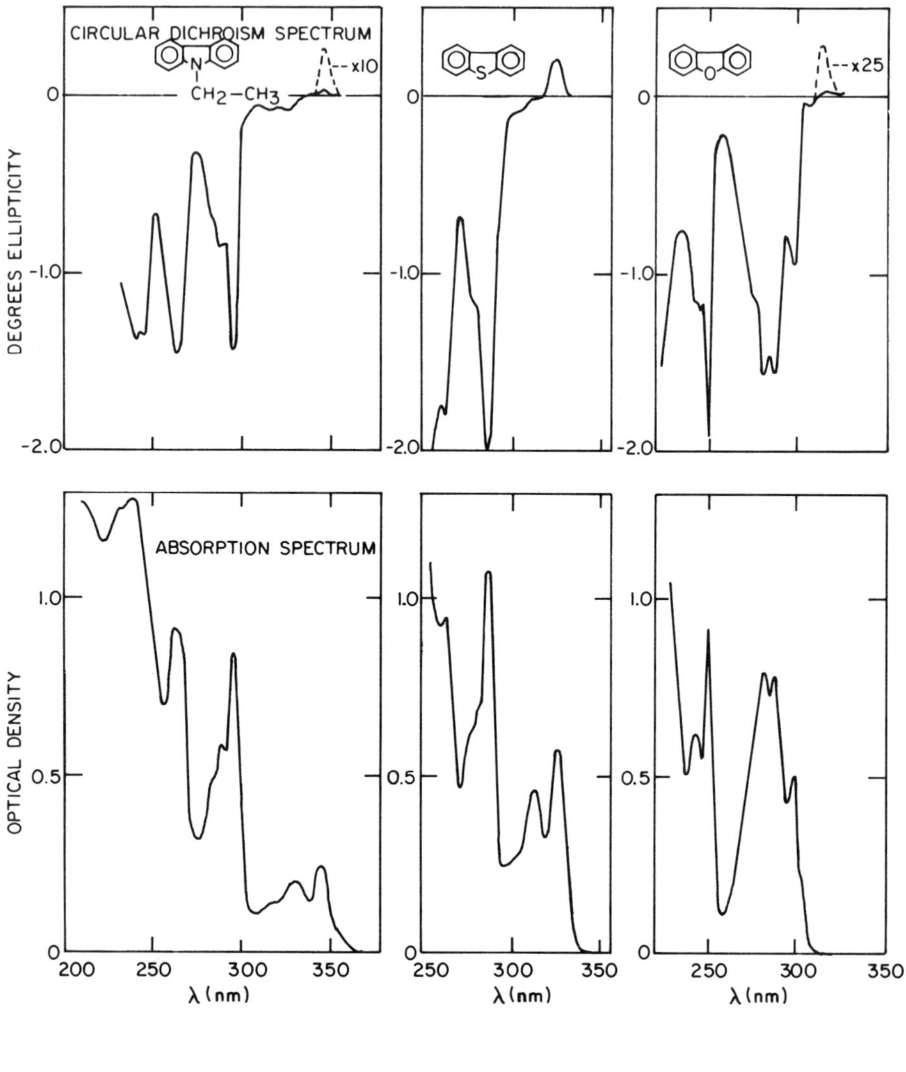

FIG. 7. Circular dichroism (upper) and absorption (lower) spectra of N-ethylcarbazole, dibenzofuran, and dibenzothiophene (1.0 wt%) in 60:40 (wt%) cholesteryl chloride-cholesteryl nonanoate (right-handed helix) as ≈ 5-μm film. (Reprinted from Ref. 28, p. 5135, by courtesy of the American Chemical Society.)

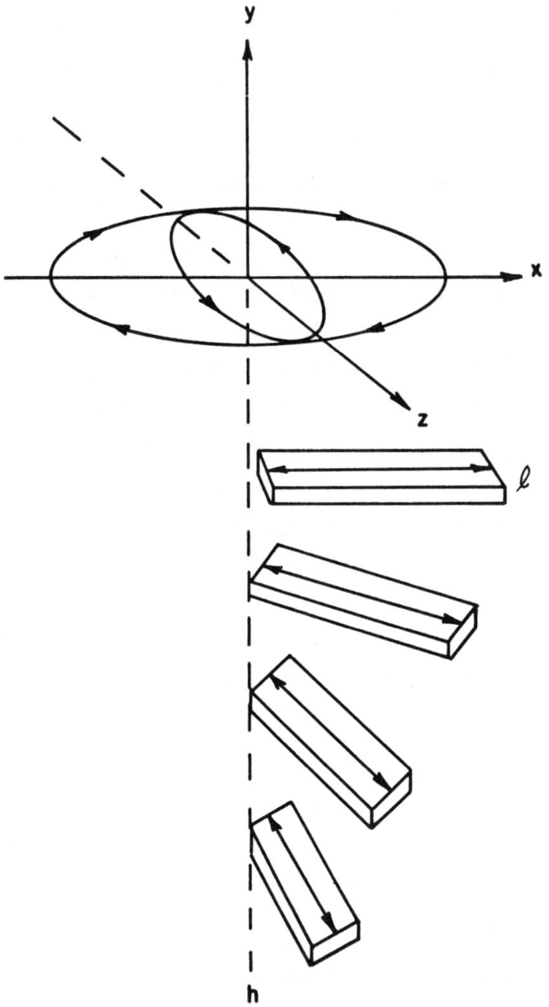

FIG. 8. Three-dimensional coordiante system (x, y, z) where y axis coincides with the helix axis h and the x axis is assigned parallel to the local optic axis ℓ of the nematic layer.

Sackmann and Voss [29] and Holzwarth and Holzwarth [30] started with the theory of de Vries [16] and Mauguin [15] and considered that a plane-polarized wave is split into two elliptically polarized components which are rotating in opposite directions and traveling at different velocities. The axes of the ellipses are parallel to the x and z axes, and the major axis of one wave coincides with the minor axis of the other. The axes of the ellipses rotate with the local optical axis of each nematic layer. Optical transition moments polarized parallel to ℓ will absorb more energy from the elliptically polarized wave with the maximum amplitude parallel to ℓ than from the wave with the minor axis parallel to ℓ. The elliptically polarized component with its major axis parallel to ℓ is rotating with the same chirality as the cholesteric helix while the other elliptically polarized wave is rotating with a chirality opposite to the chirality of the cholesteric helix [16]. A linear polarized wave of a wavelength that is absorbed by a liquid crystal or solute molecules will become elliptically polarized, and CD is a measure of the ratio of the major to minor axes of the ellipse. LCICD is then predicted to be based on the linear dichroism of the solute or liquid crystal molecule [29]. For the case where λ_{ab}/λ_0 is large compared to the relative birefringence of the medium and the orientation of the optical transition moment with the local optic axes is other than 45° [29] the following relationship was theoretically derived:

$$\varepsilon_L - \varepsilon_R = (n_x - n_{z(y)})(\varepsilon_x - \varepsilon_{z(y)}) \frac{\lambda_0}{2n\lambda_{ab}} \tag{1}$$

For very large values of P the CD is inversely proportional to the relative birefringence of the medium and P, while also being directly proportional to the average birefringence and λ_{ab} [28]. Holzwarth and Holzwarth [30] in a more detailed treatment take into consideration the birefringence introduced by absorbing species. Ramaseshan et al. [31] have also developed a rigorous theory of the optical properties of cholesteric liquid crystals taking into account the effect of absorption.

From Eq. (1) it can be seen that the LCICD sign for ordered molecules within a cholesteric mesophase is dependent on the sign of the birefringence of the cholesteric matrix, wavelength of absorption relative to the pitch of helical structure [32], and polarization of the optical transition, as well as chirality of the cholesteric helix. The LCICD sign variation with the mesophase chirality [19,29], polarization of the optical transition, and pitch band relative to the wavelength of absorption have been verified experimentally [28,29,32]. The concentration independence and pitch dependence of the molecular ellipticity $[\theta]$, of an achiral solute within the cholesteric mesophase [32] is entirely consistent with Eq. (1).

The CD has been shown experimentally to be dependent on and can be calculated from the linear dichroism of the optical transition [29,33-36].

The origin of LCICD for ordered chromophores in cholesterics has been adequately demonstrated. It has been proposed that at least a second mechanism would contribute to LCICD [28]. As a result of this second mechanism, an unordered chromophore exposed to the helical mesophase would exhibit LCICD [28]. Recently, Mason and Peacock [37] have reported the LCICD spectrum of a spherical solute (molybdenum hexacarbonyl). Due to the symmetry of the solute, $\varepsilon_x - \varepsilon_z = 0$, and the CD contribution from the previously discussed mechanism for ordered chromophores should be zero. Mason has suggested that the spherical solute gives rise to extrinsic CD due to the differential effective Lorentz radiation field for the left and right circularly polarized light arising from the large circular birefringence of the medium [37]. For a left-handed cholesteric solvent the CD absorption of an achiral solute at the wavelength λ_{ab} is given by Eq. (2) [37].

$$\frac{\varepsilon_L - \varepsilon_R}{\varepsilon_o} = (n_x - n_{z(y)})^2 \frac{P/\lambda_0^3}{2\left(\frac{1}{\lambda_{ab}^2} - \frac{1}{\lambda_0^2}\right)\left(n^2 + 2\right)} \tag{2}$$

The sign of the LCICD effect, as suggested by Mason [37], should change as the pitch-band position is altered from a longer to a shorter wavelength than the solute absorption. The validity of Eq. (2) has yet to be verified.

D. Liquid Crystal-Induced Circular Dichroism in Polypeptide Cholesterics

LCICD has also been observed in lyotropic cholesteric systems. Certain synthetic polypeptides, e.g., poly(γ-benzyl-L-glutamate) (PBLG), are known to exist in an α-helical conformation in a variety of solvents [38]. Concentrated solutions of these polypeptides in helix-supporting solvents readily form birefringent lyotropic cholesteric liquid crystalline mesophases. In this case the helical polypeptide is analogous to the cholesteryl derivative in thermotropic cholesteric systems. See Chap. 5, "Polypeptide Liquid Crystals" [39,40], for further details on these mesophases.

Achiral molecules, such as anthracene and pyrene, which do not exhibit detectable induced circular dichroism in dilute or concentrated isotropic solutions of PBLG have been found to exhibit LCICD in the anisotropic birefringent lyotropic cholesteric mesophase formed by PBLG in helix-supporting solvents such as chloroform, methylene chloride, and dioxane [41]. The extrinsic CD of the achiral solute is associated with the formation of the birefringent cholesteric mesophase and disappears when the α-helical molecules of PBLG become randomly orientated by means of small changes in PBLG concentration.

The observed LCICD for anthracene in the lyotropic cholesteric mesophase formed by PBLG in dioxane is different than that observed in thermotropic cholesteryl mesophases composed of cholesteryl derivatives [32]. Figure 9 presents the LCICD and electronic spectra of anthracene dissolved in thermotropic and lyotropic cholesteric mesophases. The LCICD spectrum of anthracene between 300 and 400 nm in PBLG-dioxane (18:82 wt%) shows CD bands of a single sign which follow the absorption spectrum quite closely. The CD for the pitch band and anthracene absorption bands are both of negative sign indicating a left-handed helical structure. The chirality of the cholesteric helix in dioxane is of opposite chirality to the helicity of the polypeptide [43]. The chirality of the cholesteric helix in chloroform, however, is opposite to that in dioxane and methylene chloride, i.e., right-handed. The LCICD spectrum of anthracene in a right-handed thermotropic cholesteric mesophase composed of cholesteryl chloride-cholesteryl nonanoate (60:40 wt%), on the other hand, shows CD bands of both positive and negative signs between 300 and 400 nm whose relative intensities do not follow its absorption spectrum [41].

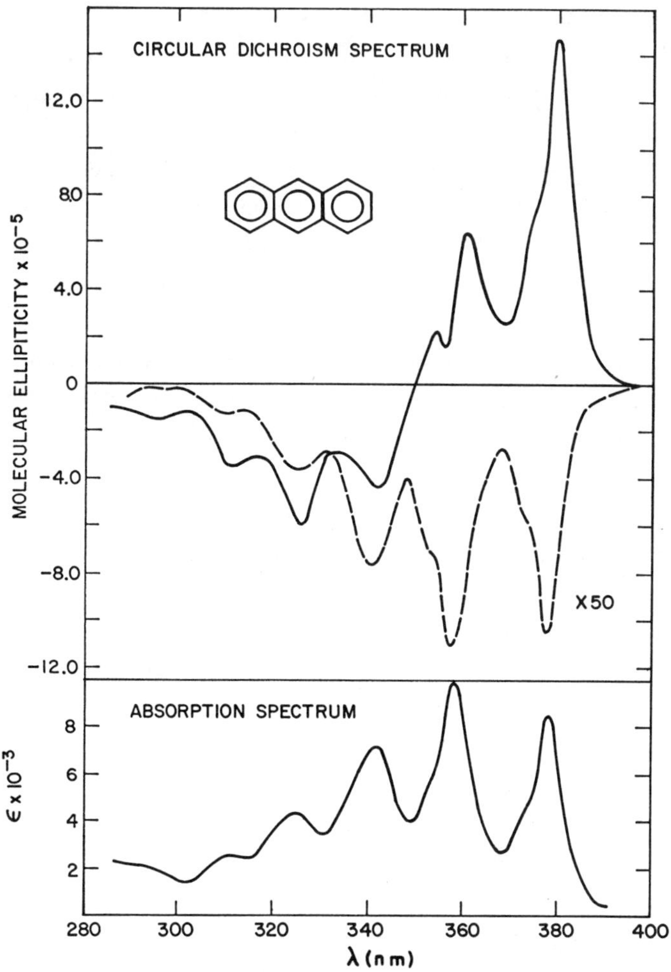

FIG. 9. Circular dichroism (upper) and absorption (lower) spectra of anthracene in (——) a thermotropic cholesteric mesophase composed of 60:40 (wt%) cholesteryl chloride-cholesteryl nonanoate (pitch ≈ 9μm); (---) a lyotropic cholesteric mesophase composed of PBLG-dioxane (18:82 wt%) (MW PBLG = 125,000, pitch ≈ 18μm). (Reprinted from Ref. 32, p. 7659, by courtesy of the American Chemical Society.)

The variation in anthracene LCICD between the two cholesteric systems has been tentatively attributed to the difference in ability of the mesophases to order the solute physically [41]. In the thermotropic system the preferred conformation of the anthracene is such that it aligns its long axis parallel to the long axis of the liquid crystal molecules [33]. In the lyotropic system, however, alignment of the long and short axes of anthracene perpendicular to the long axis of the PBLG α helix appear to be equally preferred. This conclusion is consistent with the relative sizes of the solute anthracene (≈ 10 Å) and PBLG molecules (≈ 850 Å for MW 125,000) [43]. The ordering of solute molecules by the liquid crystal matrix would be most effective for solutes comparable in size and shape to the liquid crystal molecules.

Enantiomeric lyotropic cholesteric mesophases are produced by PBLG and PBDG, respectively, in dioxane solvent as indicated by the LCICD spectra for anthracene in Fig. 10 [41]. The chirality of the cholesteric helix in the lyotropic mesophase can be established by determining the sign of the cholesteric pitch band CD. Once a correlation between the sign of the CD for the pitch and an LCICD band has been made the chirality of the cholesteric helix may be determined simply from the sign of the LCICD [50] providing the position of λ_0 is known. The chirality of the cholesteric helices formed by PBLG and PBDG in dioxane (shown in Fig. 10) are left- and right-handed, respectively.

Molecular ellipticity values for anthracene in PBLG of two different molecular weights, i.e., 46,000 and 125,000, were identical within experimental error for samples of virtually identical pitch. The lower molecular weight samples were found to relax from the Grandjean to the focal conic texture in 1-2 hr while the higher molecular weight sample took 24-30 hr [41].

LCICD was also observed within the electronic transitions of the polypeptide itself [42]. Rotational strengths for the polypeptide absorption bands are approximately 10^2-10^3 larger in the cholesteric mesophase than in isotropic solutions. The $1L_b \leftarrow 1A$ electronic transition of the phenyl group in PBLG shows large positive CD when the polypeptide forms a cholesteric mesophase [41].

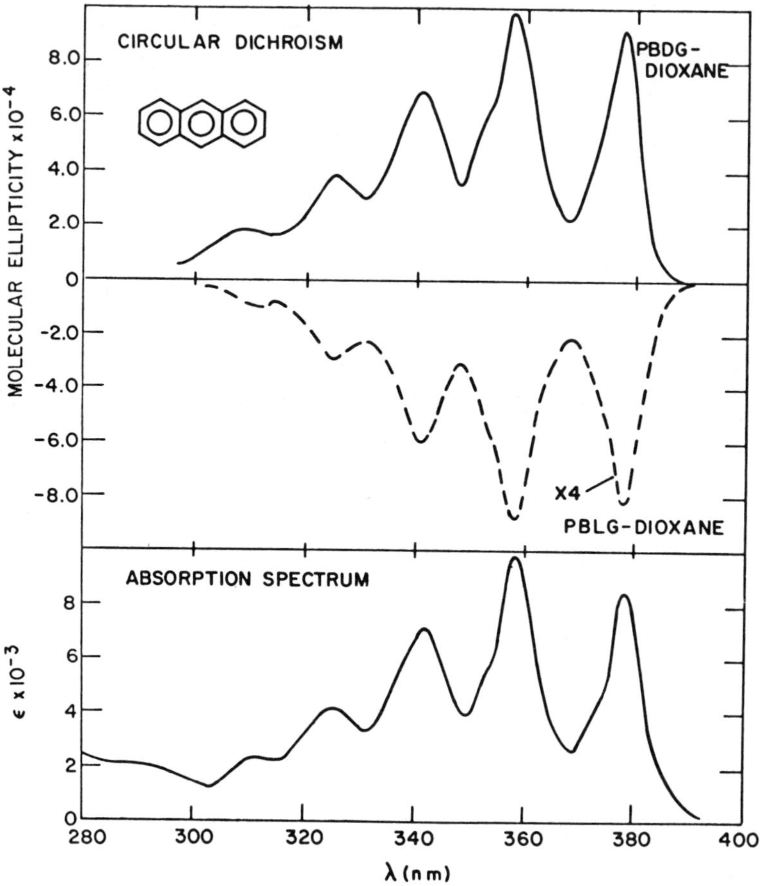

FIG. 10. Circular dichroism (upper) and absorption (lower) spectra of anthracene in (---) a lyotropic cholesteric mesophase composed of PBLG-dioxane (18:82 wt%) (MW PBLG = 125,000, pitch ≈ 18μm); (——) a lyotropic cholesteric mesophase composed of PBDG-dioxane (20:80 wt%) (MW PBDG = 120,000, pitch ≈ 21 μm). (Reprinted from Ref. 41, p. 7883, by courtesy of the American Chemical Society.)

E. Recent Developments

Pollmann et al. [44] have reported recently that achiral molecules dissolved in cholesteric mesophases exhibit circularly polarized fluorescence. The degree and sign of the circularly polarized emission was found to depend on the chirality of the cholesteric structure as well as on the direction of the electric transition moment within the emitting molecule analogous to LCICD.

In the early 1970s it had been observed that chiral molecules dissolved in nematics produced cholesteric mesophases which exhibit LCICD within the absorption bands of host molecules [12,21]. The sign of the induced CD was found to be dependent on the chirality of the solute [12]. More recently, Gottarelli et al. [45] have extended the above observations to other chiral solutes and have proposed a mechanism to account for the induction to a cholesteric mesophase via sandwich-type complex formation between chiral solute and host [46].

IRD spectropolarimetry has been also employed to probe the effect of dissolving chiral solutes in nematic N-(p-methoxybenzylidene)-p-n-butyl aniline. It was concluded from this study that a chiral additive which is not mesomorphic itself always induces the same handedness in a "host nematic" [47]. These same workers also characterized the absolute configuration of a series of chiral sulfoxides using a sequence of ligand parameters analogous to the Cahn-Ingold-Prelog nomenclature [48]. It was found that the ordering by atomic number used by the Cahn, Ingold, and Prelog rule was not suitable. The "effective volume" of the ligands was used to establish a priority scheme with the ligand of largest "effect volume" given the highest priority. Thus far over 100 chiral compounds have been found to fit into this scheme [48]. The simplicity of the LCICD technique to determine absolute configurations is predicted to be an important application of liquid crystal phenomena.

III. SUMMARY

The LCICD sign for ordered chromophores in cholesteric mesophases is dependent on the following:

1. Chirality of the cholesteric helix
2. Linear birefringence of the cholesteric matrix
3. Position of λ_0 relative to λ_{ab}
4. Linear dichroism of the solute

The LCICD intensity, on the other hand, is dependent on the pitch of the mesophase, its mean refractive index, and the magnitude of the relative birefringence of the matrix.

One may use the LCICD spectrum to determine the existence of a cholesteric mesophase, which we define as a chiral array of nematiclike layers. The sign of LCICD for an optical transition of known polarization and alignment, where λ_0 is known, provides the chirality of the mesophase [49]. The sign of the order parameter S can be readily found provided the polarization of the optical transition is known, or the polarization of the optical transition can be evaluated if the sign of the order parameter for the solute is known from some other experimental method. An achiral molecule whose alignment in the cholesteric mesophase is known may serve as a probe to the internal structure of the mesophase and be employed to measure relaxation processes. Figure 11 presents an example of such a study where the CD spectrum of anthracene dissolved in a lyotropic PBLG-dioxane cholesteric mesophase was monitored as a function of time after the application of a pressure pulse to the front face of the sample cell [50] which presumably caused a pitch variation in the material. The time dependence of the LCICD sign and intensity provides valuable information concerning the energetics of relaxation processes within cholesteric mesophases.

Liquid crystal-induced circular dichroism (LCICD) studies on a series of mono- and disubstituted benzene derivatives show both spectroscopic and conformational effects as a function of aromatic ring substitution. The sign of the LCICD within the $1L_b \leftarrow 1A$ electronic transition in monosubstituted benzenes was found to depend on the ortho-para directing ability of the substituent [51] similar to that found in magnetic circular dichroism studies. This variation in CD sign is attributed to the change in the magnetic moment contribution to the CD. LCICD spectra of some ortho-, meta-, and para-disubstituted benzenes generally show the para isomer to exhibit oppositely signed CD within the $1L_b \leftarrow 1A$ transition from less symmetric ortho and meta derivatives. This para effect has been attributed to a variation in the alignment of the optical transition moment between the isomers in a cholesteric mesophase of a single chirality [50,51].

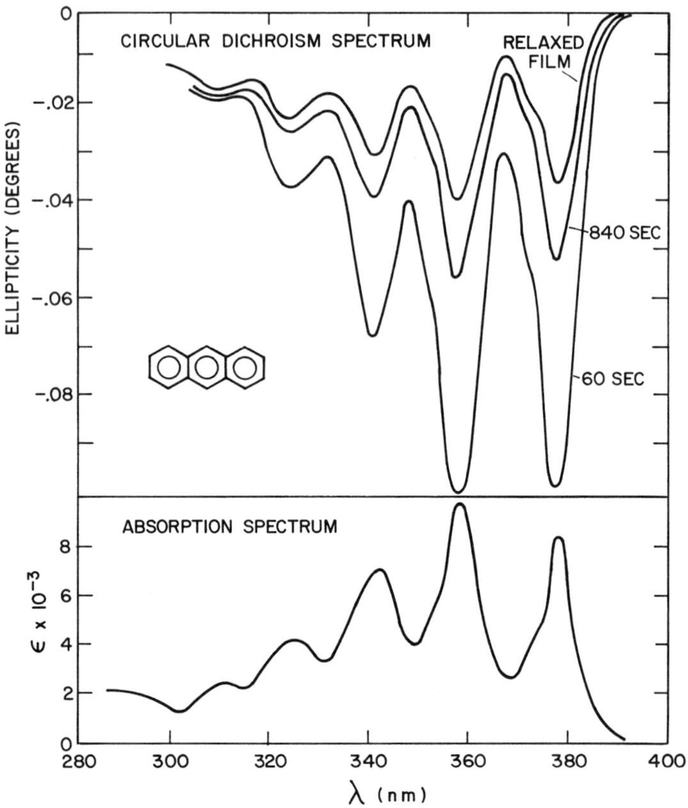

FIG. 11. Circular dichroism (upper) and absorption (lower) spectra of anthracene in a lyotropic cholesteric mesophase composed by PBLG-dioxane (18:82 wt5) (MW PBLG = 125,000, pitch $\approx 18 \mu m$) as a function of time after a pressure pulse was applied to the front face of the sample cell.

The investigation of the origin of LCICD has brought to light the interrelation between linear and circular dichroism effects in noninteracting ordered chromophores. The extrinsic circular dichroism observed for chromophores ordered in a helical fashion within cholesteric mesophases can be employed to probe the detailed nature of the host structure in both thermotropic or lyotropic mesophases.

LCICD studies have provided a unique opportunity to understand more completely the optical properties of chirally arranged anisotropic phases. Several questions still remain unanswered. The origin of extrinsic CD in unordered chromophores exposed to cholesteric mesophases has yet to be explained.

It is the author's opinion that the study of the effects of molecular ordering on optical as well as physical and chemical properties via the investigation of liquid crystalline phenomena is an area of emerging scientific importance with a bright future.

REFERENCES

1. K. Mislow, Introduction to Stereochemistry, W.A. Benjamin, Inc., New York, 1965.
2. E.L. Eliel, Stereochemistry of Carbon Compounds, McGraw-Hill, New York, 1962, pp. 59, 79.
3. N.K. Chaudhuri and M.A. El-Sayed, J. Chem. Phys., 47:1133 (1967).
4. E.R. Blout and L. Stryer, Proc. Natl. Acad. Sci. USA, 45:1591 (1959).
5. L. Stryer and E.R. Blout, J. Amer. Chem. Soc., 83:1411 (1961).
6. E.R. Blout, Biopolym. Symp., No. 1, 397 (1964).
7. S.F. Mason and A.J. McCaffery, Nature (London), 204:468 (1964).
8. B. Bosnick, J. Amer. Chem. Soc., 89:6143 (1967).
9. L.D. Hayward and R.N. Totty, J. Chem. Soc., Chem. Commun., 676 (1969).
10. R.A. Berg and B.A. Haxby, Mol. Cryst. Liq. Cryst., 12:93 (1970).
11. P.N. Schatz and A.J. McCaffery, Quart. Rev. (London), 23:552 (1969).
12. F.D. Saeva, Mol. Cryst. Liq. Cryst., 23:171 (1973).
13. J.J. Wysocki, J. Adams, and W. Haas, Phys. Rev. Cryst., 20:1024 (1968).
14. E. Sackmann, S. Meiboom, L.C. Snyder, A.E. Meixner, and R.E. Dietz, J. Amer. Chem. Soc., 90:3567 (1968).
15. C. Mauguin, Bull. Soc. Fr. Mineral. Cristallogr., 34:91 (1911).
16. H. de Vries, Acta Crystallogr., 4:219 (1951).

17. J.L. Fergason, Mol. Cryst., 1:293 (1966).
18. S. Chandrasekhar and K.N. Srinivasa Rao, Acta Crystallogr., A24:445 (1968).
19. F.D. Saeva and J.J. Wysocki, J. Amer. Chem. Soc., 93:5928 (1971).
20. G. Friedel, Ann. Phys. (Paris), 18:273 (1922).
21. K.J. Mainusch and H. Stegemeyer, Z. Phys. Chem. Neue Folge, 77:210 (1972).
22. H. Stegemeyer, Angew. Chem., 84:720 (1972).
23. B. Schrader and E. Korte, Angew. Chem., 11:226 (1972).
24. H. Stegemeyer and K.J. Mainusch, Chem. Phys. Lett., 16:38 (1972).
25. I. Chabay, Chem. Phys. Lett., 17:283 (1972).
26. I. Chabay, E.C. Hsu, and G. Holzwarth, J. Chem. Soc. Chem. Commun., 15:211 (1972).
27. R.J. Dudley, S.F. Mason, and R.D. Peacock, J. Chem. Soc. Chem. Commun., 1084 (1972).
28. F.D. Saeva, J. Amer. Chem. Soc., 94:5135 (1972).
29. E. Sackmann and J. Voss, Chem. Phys. Lett., 14:528 (1972).
30. G. Holzwarth and N.A.W. Holzwarth, J. Opt. Soc. Amer., 63:324 (1973).
31. G.S. Ranganath, S. Chandrasekhar, U.D. Kini, K.A. Suresh, and S. Ramaseshan, Chem. Phys. Lett., 19:556 (1973).
32. F.D. Saeva, P.E. Sharpe, and G.R. Olin, J. Amer. Chem. Soc., 95:7656 (1973).
33. E. Sackmann and H. Mohwald, J. Chem. Phys., 58:5407 (1973).
34. G. Holzwarth, I. Chabay, and N.A.W. Holzwarth, J. Chem. Phys., 58:4816 (1973).
35. S.F. Mason and R.D. Peacock, Chem. Phys. Lett., 21:406 (1973).
36. H. Falk, O. Hofer, and H. Lehner, Monatsh. Chem., 105:169 (1974).
37. S.F. Mason and R.D. Peacock, J. Chem. Soc. Chem. Commun., 712 (1973).
38. P. Doty, A.M. Holtzer, V.H. Bradbury, and E.R. Blout, J. Chem. Soc. Chem. Commun., 76:4493, (1954).
39. A. Elliott and E.J. Ambrose, Disc. Faraday Soc., 9:246 (1950).
40. C. Robinson, Mol. Cryst., 1:467 (1966).

41. F.D. Saeva and G.R. Olin, J. Amer. Chem. Soc., 95:7882 (1973).

42. T. Ooi, R.A. Scott, G. Vanderkooi, and H.A. Scheraga, J. Chem. Phys., 46:4410 (1967).

43. M.F. Perutz, Nature (London), 167:1053 (1951).

44. P. Pollmann, K. J. Mainusch, and H. Stegemeyer, Z. Physik. Chem. Neue Folge, 103:295 (1976).

45. G. Gottarelli, B. Samori, S. Marzocchi, Tetrahedron Lett. 1981 (1975).

46. G. Gottarelli and B. Samori, Chem. Phys. Lett., 40:308 (1976).

47. E. Korte, B. Schrader, S. Bualek, and H. Krabbe, Angew. Chem. Int. Ed., 16:790 (1977).

48. H. Krabbe, H. Heggemeier, B. Schrader, and E. Korte, Angew. Chem. Int. Ed., 16:791 (1977).

49. F.D. Saeva, Mol. Cryst. Liq. Cryst. 18:375 (1972).

50. F.D. Saeva and G.R. Olin, unpublished observations.

51. F.D. Saeva, P.E. Sharpe, and G.R. Olin, J. Amer. Chem. Soc., 95:7660 (1973).

Chapter 7

A STRUCTURAL INTERPRETATION OF THE RHEO-OPTIC PROPERTIES OF THE CHOLESTERIC MESOPHASE

John M. Pochan

Xerox Corporation
Webster, New York

I. INTRODUCTION

Liquid crystals are states of matter which exhibit rheological and optical properties of the liquid and crystalline states, respectively [1]. Optical birefringence is a characteristic of most textures of the various mesophases and many times is used to identify them. Rheologically, the mesophases can be quite complicated, exhibiting viscosities that vary from 10^3 poise or greater (the smectics) to approximately 1 poise (the nematics). Early work on the rheological properties of liquid crystals was limited and has been reviewed [2-5]. In these works, it was shown that the mesophases were non-Newtonian and usually exhibited abrupt changes in viscosity at mesophase-mesophase and mesophase-isotropic transitions. Flow activation energies were presented, and it was shown (particularly in the case of nematics) that external perturbations such as electric or magnetic fields could affect measured viscosity. The nematic phase in particular has been quite well studied with experimental and theoretical work presented in the recent literature [6]. It is the purpose of this paper to review recent rheo-optic and rheological work in the area of cholesteric liquid crystals and to present, hopefully, a unified interpretation of the structure of this mesophase during shearing.

Unified models of molecular orientation have been presented to interpret the cholesteric optic response with no shearing perturbation [7-13]. An early paper by Fergason [14] indicated that the reflective properties of cholesteric liquid crystals can be influenced by shear. Since reflective color changes were noted with shear, it might naturally be expected that the correlated helical structure of the mesophase is perturbed in some fashion during the shearing process. None of these earlier works addressed this question.

Early rheological work was conducted on pure materials (usually a derivative of the steroid cholesterol) where effects of temperature and shear on the mesophase viscosity were presented [15-27]. As noted above, the cholesteric mesophase exhibited non-Newtonian flow, and the viscosity was Arrhenius activated. Sharp viscosity maxima were also exhibited at the cholesteric-isotropic transition [28], when the sample was cooled from the isotropic state. This maximum is inversely dependent on shear and has been associated with relative amounts of "broken" and "unbroken" cholesteric structure in dynamic equilibrium near this transition point [27].

Of particular interest in the early work was (1) the observance of a high shear rate viscosity limit which correlated with an extension of the isotropic fluid viscosity into the liquid crystal temperature range, and (2) the observance of color changes in sheared cholesterics. The former indicates that the high shear limit of the ordered cholesteric structure is the isotropic fluid. This is unlike nematic materials where the nematic near the isotropic transition exhibits a lower viscosity than the isotropic liquid [29,30]. The latter effect indicates that at least in the low shear rate region, the helical correlation of the mesophase maintains its integrity. To reach the high shear limit, the helical structure must break up and produce the high shear "isotropic" fluid. Practical considerations plus a desire to obtain the structural response of the mesophase during shearing led us to the design of a series of rheo-optic experiments to ascertain the desired features. It was felt that since supramolecular structure controlled the circular dichroic reflective properties of the cholesteric mesophase, the effects of shear upon these properties would indicate the structural response and might provide a useful probe of the intermolecular forces controlling the helical structure [8,12,13,31-33]. Experiments were designed so that this mesophase could be visually observed during all experiments which included electroviscosity as well as normal viscosity measurements. The following is a review of this work to date. Whenever possible, anomalous effects observed by others will be explained in terms of our model.

Details of the experimental techniques, instrumentation, compounds, etc., used in the research presented by coworkers and myself are in the literature and will not be presented here [34-39]. To provide perspective, a set-up of the optical system that provided data is shown in Fig. 1.

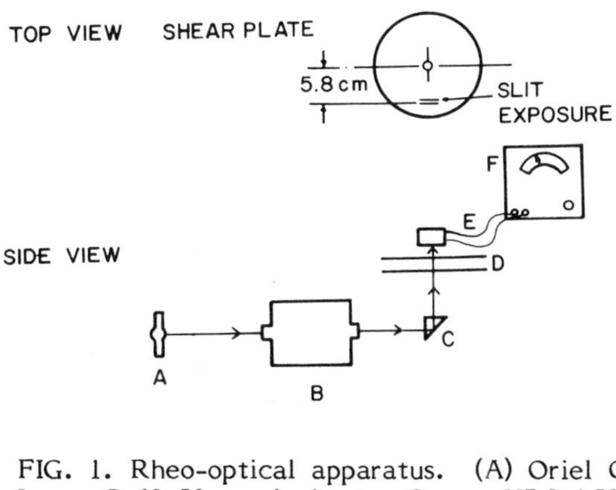

FIG. 1. Rheo-optical apparatus. (A) Oriel Optics Corporation Universal Lamp C-60-53 employing an Osram HBO 450 W/2 Xenon arc operated at 450 W; (B) Jarrell-Ash 0.25-m monochromator; (C) total reflection prism; (D) Pyrex shear plates containing liquid crystal sample; (E) Eppley Labs 16-junction circular thermopile; (F) Keithley Instruments, Inc., Model 150 A Microvoltammeter. (Reprinted from Ref. 22, pg. 1193, by courtesy of American Institute of Physics.)

II. DISCUSSION

A. Cholesteric Mixtures

Figure 2 is our generalized rheological curve for a cholesteric material [34,35]. Visual characteristics are also noted. Three regions (I, III, and V) of response are observed, two non-Newtonian regions with an intermediate Newtonian region. Transition regions II and IV couple these flow regions. In region I, if the pitch band is in the visible, a reversible color [7] change is noted with changing shear. The shift is to shorter wavelengths than P_0 (the zero shear pitch) and dependent upon the applied shear rate. At the cessation of shear, the relaxation to the unsheared state appears exponential in nature [35]. Because of the return to the zero shear pitch conditions and therefore the normal color, this region is called perturbed Grandjean. Region III can be described optically as a dynamic scattering region with a perturbed Grandjean reflective spectra superimposed. Rheologically, the viscosity of this Newtonian region is maintained into the low shear rate perturbed Grandjean region once transition region II has been attained. These data indicate a permanent structural change has taken place. Once shear rate II is attained, the perturbed Grandjean region can be obtained via anisotropic shear or regeneration via thermal melting. At the cessation of shear, region III exhibits the microscopic texture of the normal focal conic structure; therefore, the region is termed dynamic focal conic.

Region V is also unique in that it is a texture that is created via an irreversible shear-induced transition [34]. Optically, the system becomes almost transparent and is reminiscent of the thermally created homeotropic texture of the cholesteric mesophase. This region has been named the shear-induced homeotropic texture to distinguish it from the thermally induced texture. This region is also characterized by the observation of a normal force during shear [37]. Once this homeotropic texture has been generated the lower shear states can, as in the case of the dynamic focal conic state, only be induced via anisotropic shear or melting.

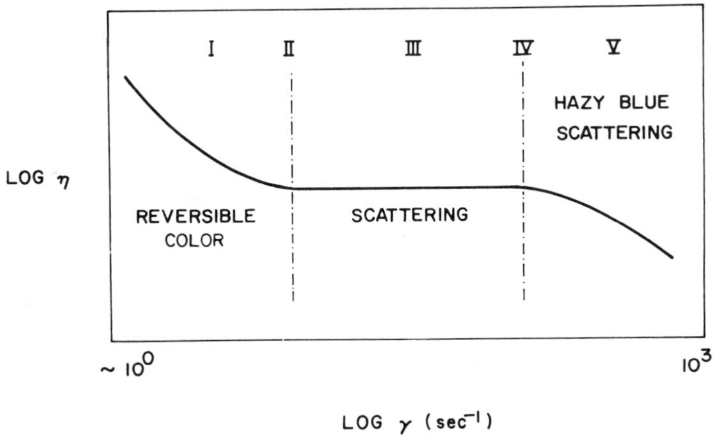

FIG. 2. Generalized rheological response curve for cholesteric materials.

Each of these regions, as well as the transition regions connecting them, have provided a picture of micro- and macrostructure of the sheared cholesteric state. In the following sections, each shear region will be individually discussed. Following that, cholesteric systems comprised of cholesteric-nematic and cholesteric polymer mixtures will be discussed in terms of the cholesteric mixture findings.

1. Region I

The unique rheo-optic properties of this region suggested the experiments necessary to ascertain structural response to shear. Electromagnetic interactions at a oblique angle to the cholesteric helix have been theoretically discussed [31], and in the limit of orthogonality Adams et al. [9] have experimentally shown the cholesteric state is in the normal focal conic structure. Our approach was to monitor the reflective spectrum as a function of shear (see Fig. 1). This is shown for a cholesteric mixture in Fig. 3. Note: Transmitted light is monitored in these experiments. The reflected spectrum increases in intensity, broadens in distribution, and shifts to shorter wavelengths with increasing shear rate. At intermediate shear rates (17.2 and 54.7 sec^{-1}), the spectrum appears to be composed of a

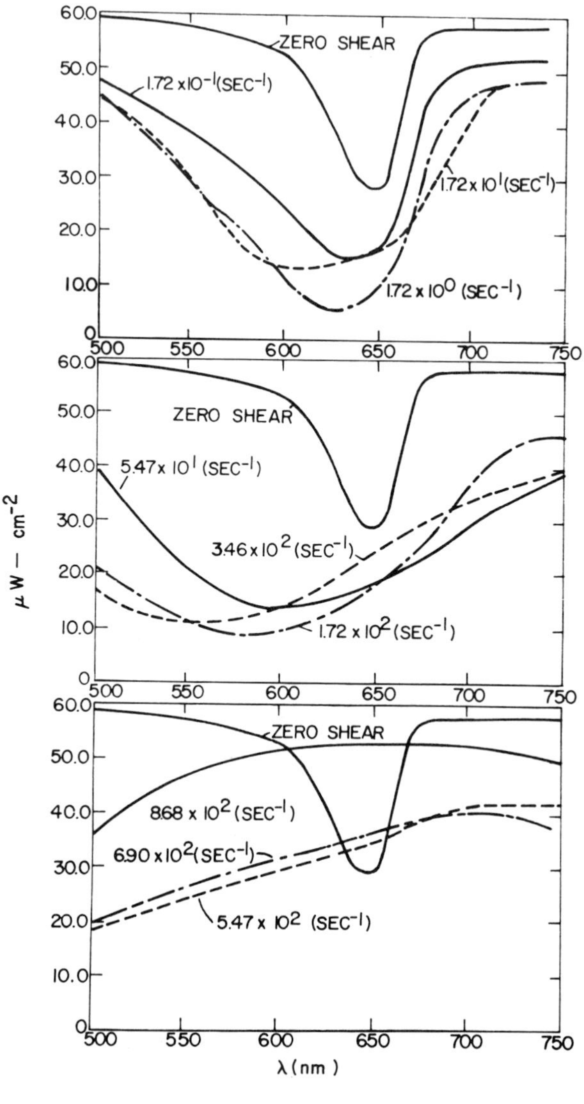

FIG. 3. Transmitted light intensity as a function of wavelength from 0 sec^{-1} to 868 sec^{-1} shear rate. Sample is a 27.2:72.8 wt% mixture of cholesteryl chloride in cholesteryl oleyl carbonate. (Reprinted from Ref. 35, p. 1195, by courtesy of the American Institute of Physics.)

doublet; one peak at the zero shear pitch, the other at shorter wavelengths of that reflection. This intermediate shear data qualitatively indicated that the sheared cholesteric was a layered structure composed of unperturbed and perturbed Grandjean structure. This was verified in a series of thickness experiments [38], the basic results of which are shown in Fig. 4. In those experiments the reflective spectrum of a sheared cholesteric material was monitored as a function of sample thickness with the shear rate held constant at 132 sec^{-1}; a shear rate in the dynamic focal conic region of the spectrum. In a 100- μm sample, this shear was sufficient to shift the reflective spectrum below our lowest wavelength of observation (450 nm). These experiments showed that between 50- and 16-μm thicknesses the zero pitch reflective band returned, indicating that: (1) a certain correlation length is needed to obtain the zero shear pitch band, and (2) surface interactions must stabilize the normal Grandjean texture during shearing. The latter effect is a function of shear, i.e., the homeotropic texture is generated at high shear rate demonstrating a complete structure breakdown of the shear state. More will be said of this later. A limit on the correlation length of the former effect was set at ≈ 10 μm.

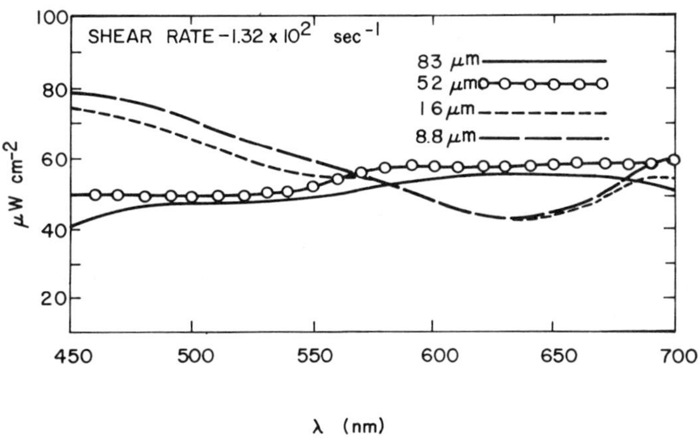

FIG. 4. Transmitted light intensity as a function of wavelength and sample thickness at 1.32 x 10^2 sec^{-1} shear rate. (Reprinted from Ref. 38, p. 5155, by courtesy of the American Institute of Physics.)

The preceding study was conducted when the system was in region III of Fig. 2. In order to analyze the reflective shear spectrum of region I, a model similar to that of Adams et al. [9] was used. The reflective model is shown in Fig. 5. Our original assumption was that

$$\lambda_0 = 2nP_0 \text{ for } \theta = 0^0 \tag{1}$$

i.e., the normal Grandjean reflection. For the sheared state, it was assumed the helical model tilted, and the equation relating tilt angle to reflective pitch is:

$$\lambda(\dot{\gamma}) = 2nP_0 \cos\theta = \lambda_0 \cos\theta \tag{2}$$

where λ_0 = normal Grandjean reflection wavelength

$\lambda(\dot{\gamma})$ = reflective wavelength at a given shear rate

$$\theta = \cos^{-1}\left(\frac{\lambda(\dot{\gamma})}{\lambda_0}\right) \tag{3}$$

and the tilt angle can be calculated from the minimum in reflected intensity as a function of shear. An assumption of the above equation is that the helical structure of the cholesteric mesophase does not distort during shear. This assumption was validated via comparative spectra where it was shown that little reflective shift to wavelengths larger than λ_0 is observed until very high shear rates. This type of shift would be expected with any distorted helical model [35].

The intensity of the reflective shift was also explained on the basis of the tilted model. Baessler and Labes [40] reported an infrared (IR) reflective coefficient per helical cycle of $\alpha_H = 0.03 \pm 0.01$ and defined the total optical density (OD) as

$$OD = \alpha_H N \tag{4}$$

where N = d/z, d is the sample thickness, and z is the pitch of the cholesteric mesophase. Since Baessler and Labes did not consider circular dichroic interactions in their model (an interaction which permits only one component of circularly dichroic light to be reflected from the helical

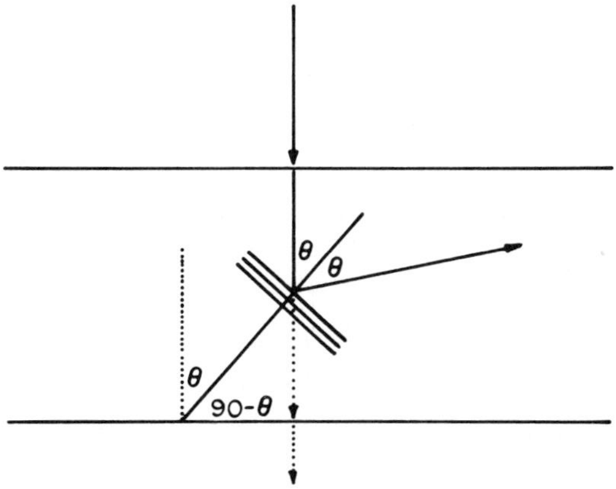

FIG. 5. Reflective model for a tilted cholesteric helix. Incident, reflected, and transmitted light shown by arrows. (Reprinted from Ref. 35, p. 1194, by courtesy of the American Institute of Physics.)

structure), a value of $\alpha_H = 0.06$ was used in our calculations. On a helical cycle basis OD = α_H, and for a given polarized component

$$\alpha_H = \log \frac{0.5I_0}{I_t} \tag{5}$$

where I_t is the amount of transmitted light and I_0 the incident radiation. With Eq. (5), I_t and thus, the amount of reflected light per helical cycle can be calculated. For each unit pitch dimension in the mesophase the amount of reflected and transmitted light can be calculated; therefore, the total interaction of light with a given thickness of normal Grandjean texture can be determined. If the cholesteric helices are considered rigid rods whose length is much greater than their cross-section, any tilting of the helix will provide less interactive length for electromagnetic radiation. This factor was considered in our calculations. At any given tilt angle θ the total interactive length of radiation with the helix will be $\ell \cos\theta$, and any helical length calculated as necessary to produce a certain percentage of reflective light will be increased by $1/\cos\theta$. Taking the above factors into account, computer programs generated tables of total sample length required at a given tilt angle to produce a given reflective efficiency.

The spectrum of Fig. 3 was then reanalyzed by assuming that a zero shear reflection always occurs (as shown above, this is a valid assumption) and subtracting this from the spectrum. The data are then replotted as intensity versus tilt angle, and from these plots and the computer-generated efficiency tables, plots of tilt population versus tilt angle are generated as in Fig. 6. The data show an initial increase in distribution, a leveling at intermediate shears, and then a continued increase in population. As expected, the distribution shifts to higher tilt angle with increasing shear. These plots actually serve as a check of the proposed parameters for Eqs. (4) and (5). The integrated thickness under the curves cannot be greater than the experimental thickness of our samples. The chosen value of $\alpha_H = 0.06$ fits this criterion, but must be considered an

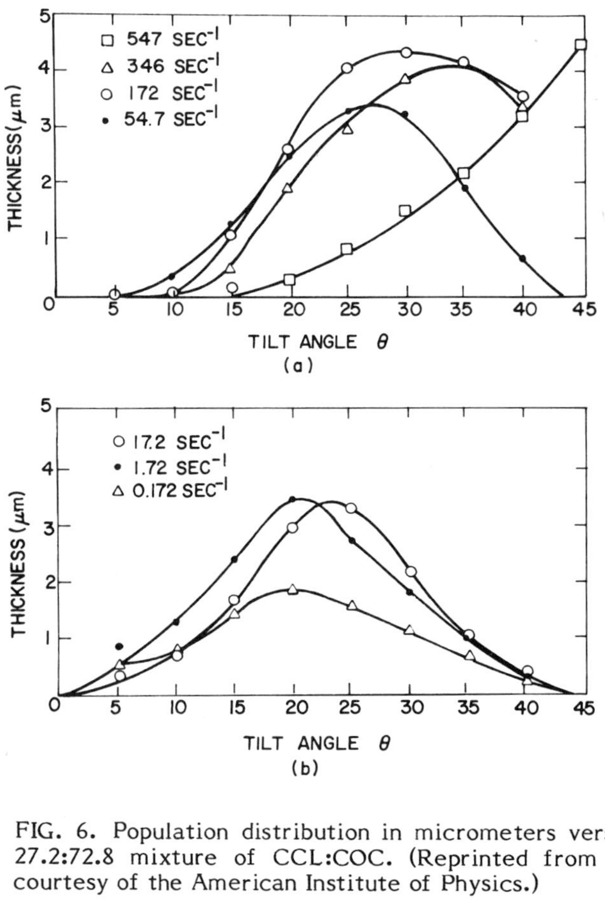

FIG. 6. Population distribution in micrometers versus θ in degrees for a 27.2:72.8 mixture of CCL:COC. (Reprinted from Ref. 35, p. 1198, by courtesy of the American Institute of Physics.)

upper bound. There is no way of experimentally accertaining the true value, unless very accurate thickness experiments can be performed.

Figure 7 is a plot of the maximum in the tilt distribution versus shear rate superimposed with the viscosity data. It is apparent that reflective data is obtained in the perturbed Grandjean (I) as well as the dynamic focal conic regions (III). The angular dependence is a function of shear and exhibits a large and nonlogarithmic effect near the dynamic focal conic to shear-induced homeotropic regions. The optical analysis of the data indicates that the initial Grandjean texture of the cholesteric mesophase is perturbed by shearing with the helical structure tilting with little or no distortion. This tilted structure is maintained into the dynamic focal conic viscosity region. The implication of these findings is that in region I of Fig. 2 the molecular interactions within the helical correlation are stronger than any applied shear energy.

2. Regions II and III

Figure 7 suggests to us that different structures are producing the results in regions I and III. This postulate was verified by optical observations during shearing as well as rheological information obtained in the transition region II. Rheologically, we have shown that a permanent shear-induced change in viscosity occurs when the perturbed Grandjean-to-dynamic focal conic transition occurs. If the torque response of the rheogoniometer (which is directly related to the viscosity of the medium) is

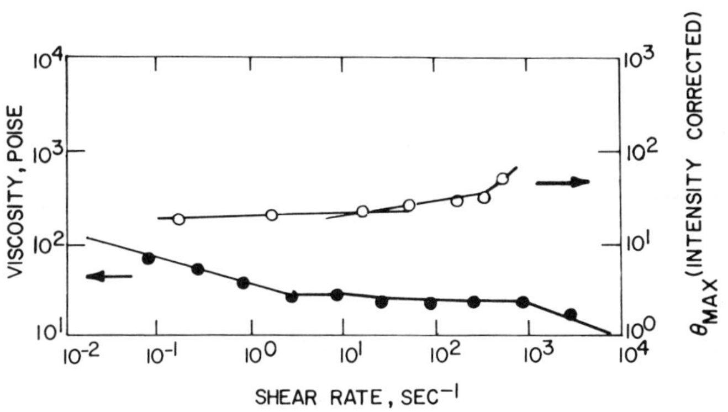

FIG. 7. Log-log plot of the viscosity (poise) versus shear rate (per second) and θ_{max} (degrees) versus shear rate (per second) for a 27.2:72.8 mixture of CCL:COC. (Reprinted from Ref. 35, p. 119, by courtesy of the American Institute of Physics.)

monitored as a function of time in the transition region II, curves such as those shown in Fig. 8 are observed [36]. The liquid crystal exhibits yield behavior in the transition region (curves 2 and 3) indicating changes in the fluid are taking place. Curve 5 is interesting, repeating conditions of curve 2 in which a large yield behavior was observed. After the transition has taken place, the torque value of curve 5 corresponds to the equilibrium torque obtained in curve 2 at long times, indicating that the transient behavior associated with any structural change has already occurred and is no longer observable, even at low shear rates.

These data, coupled with the rheo-optic data, indicate the following. It is known that the normal focal conic structure has the helical axis of the cholesteryl mesophase perpendicular to the normal alignment of the Grandjean texture. This state is obtained at the cessation of shear in region III. Helical response to shear in this region must be as shown in Fig. 9. In region I, the helical directrix describes a continuous function through the thickness of the samples. Minimum tilt occurs near the surface constraints of the system [38] with maximum tilt occurring somewhere

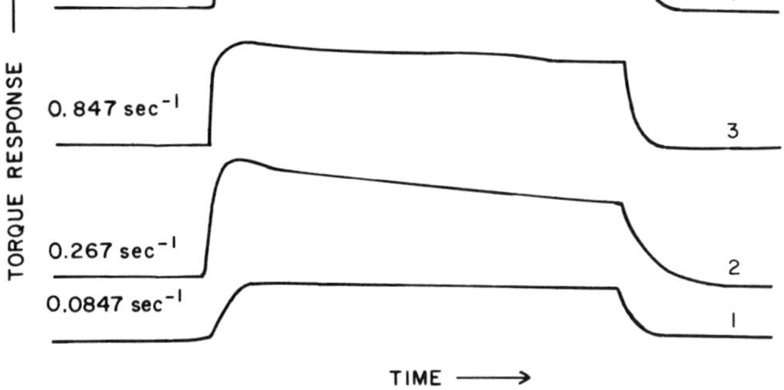

FIG. 8. Rheogoniometer torque traces as a function of time. Curve 5 is a repeat of condition for curve 2 after establishment of dynamic focal conic texture; 23:77 wt% CCL-COC mixtures. Note: Regions of Fig. 8 are labeled. (Reprinted from Ref. 37, p. 452, by courtesy of Plenum Press.)

within the body of the sample, possibly at the sample center. Near transition region II, the maximum tilt angle of the helix becomes unstable (this angle is approximately 26°); the continuity of the structure cannot be maintained, and realignment of a portion of the sample occurs with the director parallel to the shear vector in the sample. The realignment is stable in region III. It is impossible to tell from our experiments whether the helices are aligned during shear or if there is a region of complete disruption which relaxes to the normal focal conic structure at the cessation of shear. The total amounts of perturbed Grandjean and dynamic focal conic structures are a function of shear in region III and time in region II. This shear dependence is evident if the background intensity of the reflective spectrum is observed (see Fig. 3). With increasing shear, the background intensity is relatively constant in the perturbed Grandjean texture, decreases with increasing shear in the dynamic focal conic region, and increases as transition region IV is surpassed. These data indicate that the thickness of the scattering dynamic focal conic region increases with increasing shear until the homeotropic transition occurs, at which time visible scattering ceases.

Up to region IV of Fig. 2, our picture of the sheared cholesteric mesophase is one of a layered structure containing three or five layers. In the perturbed Grandjean region, the cholesteric is made up of two relatively unperturbed Grandjean layers near the surface of the constraints and a perturbed (or tilted) Grandjean layer coupling them. In this configuration, the helix director is continuous through the sample thickness. The actual thickness of the layers is a function of shear rate and surface interactions that stabilize the normal Grandjean configuration, i.e., if the stabilization energy is quite high, the unperturbed Grandjean layer could be quite thick.

In the dynamic focal conic region, the liquid crystal is composed of five layers: the two unperturbed layers adjaccent to the shear platens, two perturbed Grandjean layers adjacent to these layers, and a central dynamic focal conic layer. The perturbed Grandjean layer is characterized by non-Newtonian flow, while the dynamic focal conic region is Newtonian. The dynamic focal conic layer must control the viscosity characteristics of this shear region.

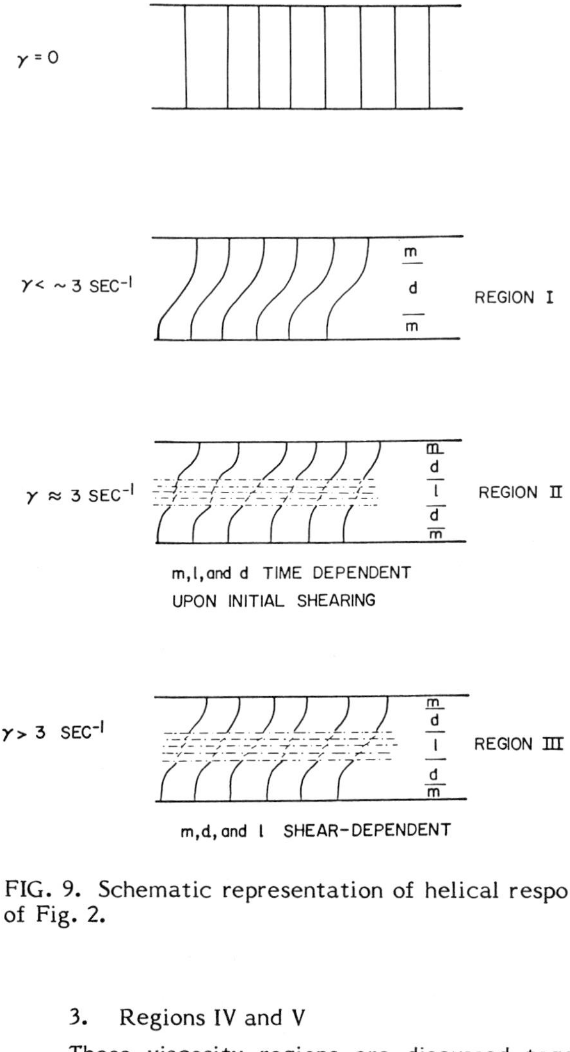

FIG. 9. Schematic representation of helical response in regions I, II, and III of Fig. 2.

3. Regions IV and V

These viscosity regions are discussed together, as region IV is a critical region marked by abrupt changes in properties. Rheologically, a transition to non-Newtonian flow is observed, and optically, the sample becomes transparent (possibly possessing a hazy blue scattering). When the shear rate has increased beyond the critical shear needed for the transition, the phase transforms in a few seconds. In the case of parallel glass platens, an abrupt boundary is formed. This boundary has led to a calculation of the critical shear necessary for the transition as $\approx 1.1 \times 10^3\ \text{sec}^{-1}$ for all cholesteric materials studied [36]. Once formed, this phase is stable (even though the shear is lowered below the critical value) until anisotropic shear or melting of the liquid crystal occurs. This is a factor which distinguishes this shear-induced homeotropic texture from the thermally induced texture [1]. The thermally induced texture is produced by cooling from the isotropic melt and usually exists only as a transient state prior to formation of the normal Grandjean or focal conic textures. The shear-induced state is stable under the conditions listed above. The implication of some structural difference is obvious.

Of more significance is the generation of normal stresses in the liquid crystal in the homeotropic region. This is the first instance of such forces being measured in liquid crystals. These forces usually result from hydrodynamic interactions of anisotropically structured materials rotating or tumbling in the fluid. Because of our proposition that the dynamic focal conic to homeotropic transition occurred because of breakdown of the helical structural correlations of the focal conic structure to smaller rodlike helical segments [36], a theory based on hydrodynamic interaction of rotating rigid rods by Kotaka [41] was used to analyze our data. In this theory, the normal stress differences are written in terms of the rotary diffusion constant D and the shear rate $\dot{\gamma}$.

$$P_{11} - P_{22} = CKT \left(\frac{42}{35}\right) \left(\frac{\dot{\gamma}}{6D}\right)^2 - \frac{2775}{1925} \left(\frac{\dot{\gamma}}{6D}\right)^4 + \cdots\cdots \tag{6}$$

where C = number of rods per unit volume

K = Boltzmann's constant

T = absolute temperature

$P_{11} - P_{22}$ = normal force measured experimentally

For rods with sufficiently large aspect ratio D can be written as:

$$D = \frac{3KT \ln (L/b)}{\pi \eta_0 L^3} \tag{7}$$

where η_0 is the Newtonian viscosity of the fluid containing the rods, L is the rod length, and b is a characteristic distance along the rod. Assumptions were made concerning certain values of Eqs. (6) and (7), namely that $C = \rho(L/b)$ where ρ is the measured density of the material

[42] and that b is 5 Å, a distance approximately equal to the interplanar distance in the cholesteric helix. Terminating Eq. (6) at the first term, substituting appropriately, and integrating led to the data in Fig. 10 for effective helical length in the phase. Note: Pure cholesteryl oleyl carbonate (COC) data is also shown. It is seen that the helical segments decrease in length with increasing shear and appear in the limit of high shear to be approaching a molecular length. Variations in b caused minor variations in the calculated rod length. The hazy blue scattering nature of this phase indicates rod lengths of several hundred to several thousand angstroms. This data is consistent with the observation and we feel corresponds to pieces of helical rods tumbling in a Newtonian fluid. The rod sizes are similar but slightly smaller than those observed for the thermally induced homeotropic texture [44]. The shear energy needed to produce this transition is sufficient to disrupt all previous cholesteric structure, as no reflective spectra are observed.

Because of the parallel plate configuration used in these experiments, the dynamic focal conic and homeotropic textures could be observed simultaneously. By relating the torque response of the rheogoniometer to

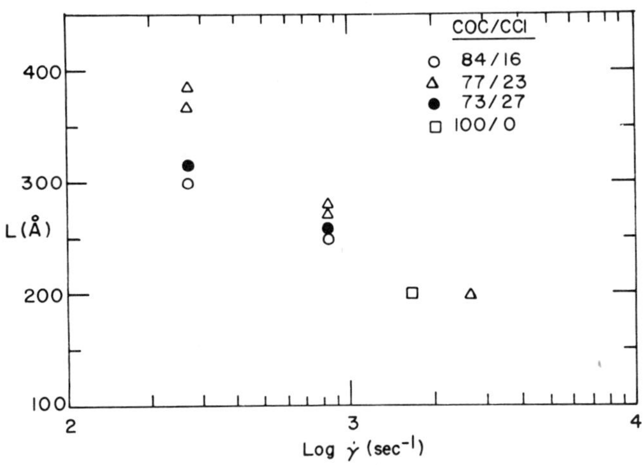

FIG. 10. Shear rate dependence of rod segment lengths calculated on the basis of Kotaka hydrodynamic interaction model. COC-CCL compositions listed in figure. (Reprinted from Ref. 37, p. 3600, by courtesy of the American Institute of Physics.)

the measurable transition boundary and assuming both phases are Newtonian, a value of 6.5 poise was attained for the viscosity of the homeotropic texture [37]. This value is used in calculations involving Eq. (6).

The normal force analysis thus leads to a picture of the dynamic homeotropic state as consisting of correlated lengths of helical fragments acting as 200- to 400- Å rods tumbling under steady shear. The study showed that a critical energy is needed to overcome the molecular cohesive forces in the cholesteric helix. At low shear rates, the helices act as rigid rods whose length is greater than their pitch. At higher shear rates, the correlation forces are overcome, resulting in fragmentation of the helix. It appears that in the limit of very high shear under isothermal conditions, the isotropic liquid would be attained. To complete the picture of the sheared state, Fig. 9 would also have to include a model ($\gamma \geq 1.1 \times 10^3$ sec^{-1}) showing fragmented helical structure.

Electroviscous experiments have also been conducted to ascertain the effect of field on this shear-induced transition. It has been shown that application of an electric field to a cholesteric material causes textural changes [44] (Grandjean → focal conic → uniaxial nematic). The latter transition is the well-known cholesteric-to-nematic phase transition. We have postulated that the dynamic focal conic region contains one layer in which the correlated cholesteric helical structure aligns parallel to the shear field. As in the static case, if an electric field is applied, this state would be expected to be destabilized with the resultant shear-induced transition to the homeotropic phase shifting to lower shear rates. The transition boundary should thus shift to lower shear rate. The change in shear energy upon application of the electric field could thus be equated to the change in shear required to produce the transition. In finalized form, the first-order equation relating these quantities is:

$$\gamma_{ce} = \frac{\omega}{\ell} R_2 = \frac{\bar{n}|E|}{\eta} \langle \mu \rangle + C \tag{8}$$

where γ_{ce} is the critical shear rate of the transition, ℓ is the sample thickness, ω is the angular velocity of a transparent glass platen (see Fig. 1), R_2 is the radius of the transition boundary, $\bar{n}$ is the population density of dipoles per unit volume, η is the viscosity of dynamic focal conic state, $\langle \mu \rangle$ is the Boltzmann average dipole distribution in the dynamic focal conic state, and |E| is the absolute value of the applied electric field. η had been previously measured [37]. R_2 was measured by observing the transition through tin oxide-coated transparent glass platens. With all other quantities measureable, values of γ_{ce} versus |E| can be plotted. If Eq. (8) is correct, values of the interaction moment $<\mu>$ for a given helix can be calculated. Because of symmetry arguments, the value for $<\mu>$ would be zero in an unperturbed helical system. In fact, this quantity can only be nonzero [39] if a helical distribution occurs which allows an average alignment of molecular dipoles with the applied field.

Figure 11 shows a plot of γ_{ce} versus |E| in which a linear relationship occurs at high field. From this linear region (which occurred in a variety of samples) values of $<\mu>$ can be calculated. The value obtained for five different samples of 28:72 wt% cholesteryl chloride (CCL)/cholesteryl oleyl carbonate (COC) was 0.16 Debye while that for a 14:86

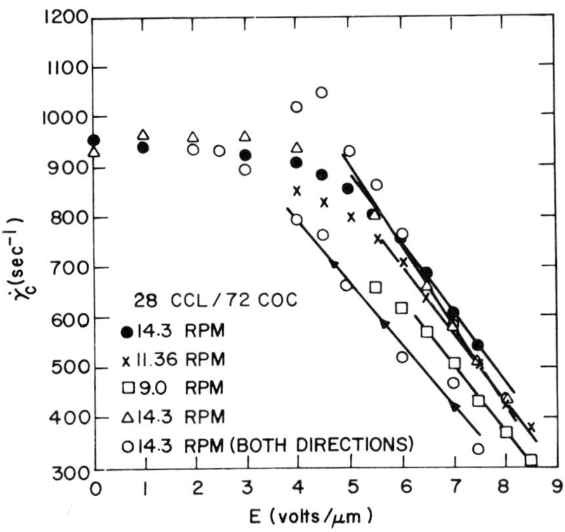

FIG. 11. Critical shear rate (γ_{ce}) versus applied field |E| for the focal conic to homeotropic transition in a 28% CCL mixture in COC. (Reprinted from Ref. 39, p. 98, by courtesy of Gordon and Breach Science Publishers.)

CCL:COC mixture was 0.09 Debye. This data indicates that a finite distortion of the helix occurs before the structure fragments to the shear-induced homeotropic texture. The electric field required was less than the minimum of approximately 8 V/μm needed to induce the cholesteric nematic phase transition in nearly compensated cholesterics [44] and indicates a combined shear-electric field destabilization of the structure.

B. Cholesteric-Nematic Mixtures

In order to complete the study on cholesteric rheology, systems containing noncholesteric materials were also studied to ascertain whether their structural response was similar to the pure cholesteric materials. These systems included cholesteric-nematic mixtures [45] and cholesteric-polymer mixtures [46]. In the case of cholesteric-nematic mixtures, it is well known that cholesteric molecules impart optical activity to the nematic moieties in the mixtures [47-52]. It is of interest to know whether the induced helix of these mixtures respond rheologically as the pure cholesteric systems. The spectral response of these mixtures is quite different from that in the pure systems (see Fig. 12). It is seen that at equivalent shear rates the reflective spectra for the mixed systems exhibit almost no change until shear rates greater than $\approx 17\ \text{sec}^{-1}$. The spectrum then exhibits behavior similar to pure cholesteric mixtures, i.e., shifting to lower wavelengths with increased shear. The marked intensity increases observed in pure cholesterics are not observed. The population distribution of helices, calculated as previously described, is shown in Fig. 13. The thinness of the perturbed layers is noteworthy and shows that the lower viscosity nematic material lessens the cohesive molecular interactions in the cholesteric helix permitting possible molecular slippage along the planes comprising the helix. This effect is also noted in the reflective spectra where evidence of helical distortion (not observed in the pure cholesterics) is seen [45]. It appears from this data that the dynamic focal conic region is easily formed and that small layers of Grandjean or perturbed Grandjean exist only near the surfaces of the sample container. The dynamic focal conic-to-shear-induced homeotropic phase transition is

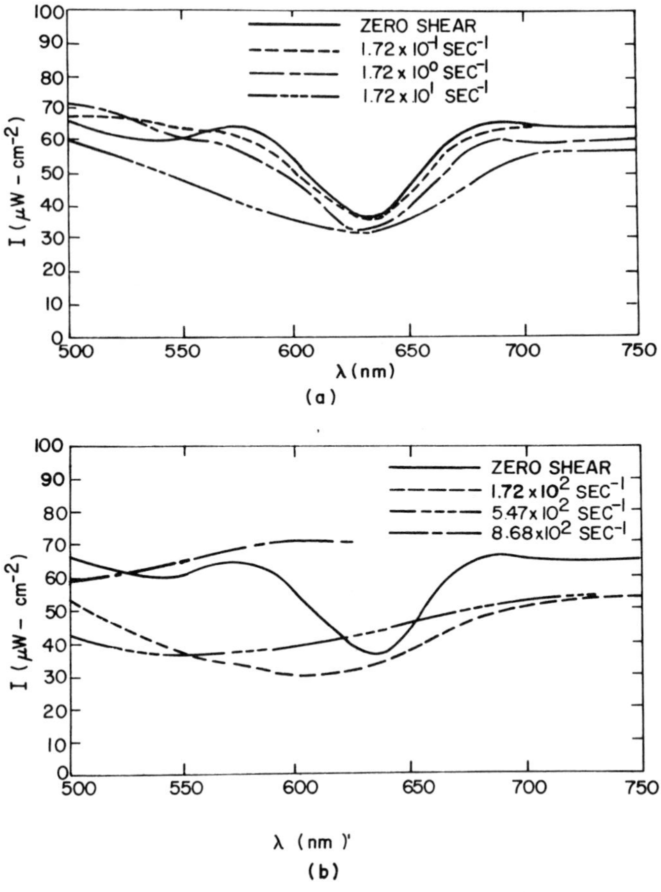

FIG. 12. Transmitted light intensity (μW | cm^2) as a function of wavelength for 55:45 wt% mixture of MBBA-COC. (Reprinted from Ref. 45, p. 2836, by courtesy of the American Institute of Physics.)

still observed, indicating that a critical shear is also needed in these systems before complete disruption of the cholesteric helical structure occurs. The structure proposed for the sheared mixtures is identical to that previously stated; the sheared mesophase consists of a five-layered structure: Grandjean near the surfaces, perturbed Grandjean adjacent, and a central dynamic focal conic state. The Grandjean and perturbed

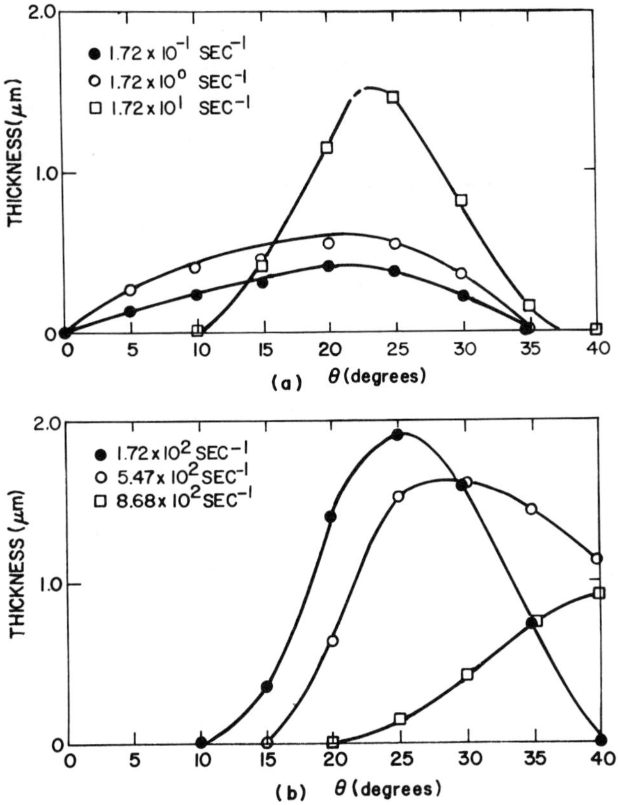

FIG. 13. Thickness of a given tilt versus tilt angles as a function of shear rate for a 55:45 wt% mixture of MBBA-COC. (Reprinted from Ref. 45, p. 2838, by courtesy of the American Institute of Physics.)

Grandjean are markedly thinner than in the pure mixed system while the dynamic focal conic is much thicker. Although electric field experiments were not performed, the critical shear for the dynamic focal conic to homeotropic texture was $\approx 400\ \text{sec}^{-1}$, indicating a compositional effect on this transition. It again appears that the cohesive molecular forces in the helix are less than in the pure systems and thus require less shear energy to distort or break down.

C. Cholesteric-Polymer Mixtures

The cholesteric-polymer mixtures [46] exhibited different characteristics from the pure or nematic mixed components. The polyisobutylene (PIB) used in the experiments was a low molecular weight (≈ 1,300) oily material that was compatible with pure room temperature cholesteric. After long periods of time (weeks), phase separation of these systems could be observed. At low percentages of PIB (3% by weight), the systems exhibit cholesteric optical properties. Rheologically, this system does not exhibit the irreversible transition between regions I and III of Fig. 2 that pure cholesterics show. Rather, this non-Newtonian-to-Newtonian transition becomes completely reversible. At PIB loadings greater than 3%, a peak in the viscosity versus shear data is observed near the perturbed Grandjean to dynamic focal conic structure (see Fig. 14). This anomalous increase can be explained in terms of the tilted helix distribution of the system (see Fig. 15). Analysis of the distribution was conducted as described above. In comparing Figs. 14 and 15, the anomalous increase in viscosity occurs at approximately the same shear that the tilt population of the helices narrows indicating a change in the flow mechanism. The population is not markedly affected below $\approx 5.5\ \mathrm{sec}^{-1}$. This region of shear rate is one in which the presence of PIB promotes the Grandjean texture and the reflected maximum intensity and population of tilt angles are not substantially altered. At a critical shear stress which decreases with increasing PIB concentration, a portion of the perturbed Grandjean structure breaks down, and the helical structure is forced into a focal conic layer, giving rise to the more or less abrupt shift in helix tilt angle population and the sharpening of tilt angle distribution. This abrupt flow change may also result in plug flow for the system with the viscosity determined by the breakdown or changeover region of the structure. The anomalous peak was also shown to be a function of sample history in that consecutive shear scans through the region resulted in the disappearance of the transition. It is interesting to note that the addition of small amounts of PIB to COC did not greatly alter the viscosity of the system. The study of PIB:COC gave the first indication of induced cholesteric helical structure in a polymeric material [46,48].

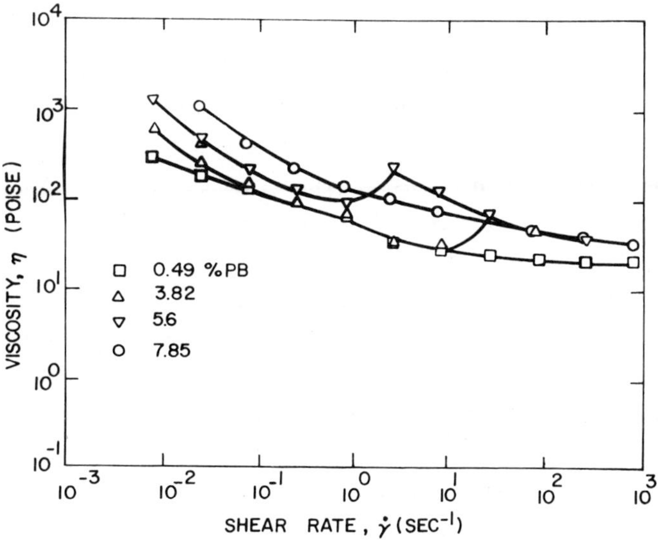

FIG. 14. Log viscosity (poise) versus log shear rate (per second) for various PIB:COC mixtures. (Reprinted from Ref. 46, p. 5799, by courtesy of the American Institute of Physics.)

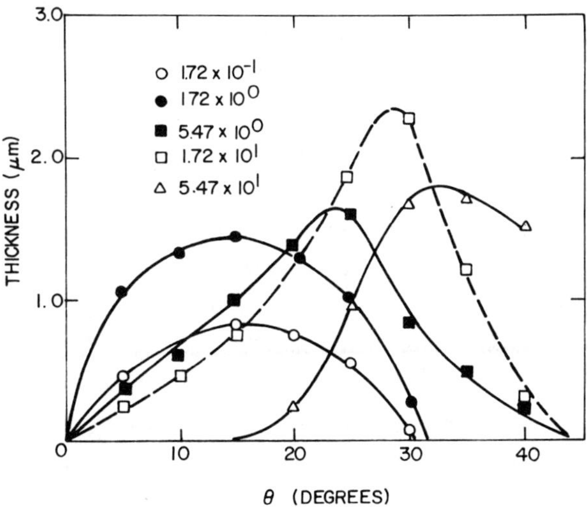

FIG. 15. Population distribution of tilted helices as thickness of helices at a given tilt angle versus tilt angle. Distribution shown as a function of shear; 4.2:95.8 PIB:COC. (Reprinted from Ref. 46, p. 5798, by courtesy of the American Institute of Physics.)

III. SUMMARY

This report has presented a structural interpretation of the sheared cholesteric mesophase. This has been done via investigations of pure cholesteric systems, mixed cholesteric systems, cholesteric nematic mixtures, and cholesteric polymer mixtures. In general, rheological behavior such as shear stress, effects of composition pitch, etc., have not been reported and will be the subject of future publications. Rheo-optic data and viscosity data have been used to provide a picture of the sheared cholesteric mesophase as being a complex structure whose configuration changes with shear conditions. The four general helical configurations of this mesophase that have been detected, are, in the order of increasing shear: (1) the normal Grandjean, (2) a perturbed Grandjean which is described by a tilted helical distribution, (3) a dynamic focal conic state (which is thought of as having the helical conformation orthogonal to the normal Grandjean texture), and (4) a dynamic homeotropic state (which consists of correlated lengths of helical molecular segments whose length is less than the characteristic pitch length of the helix). The former three configurations have been shown to exist simultaneously under proper shear conditions. The latter is formed via critical phenomena and exists separately. In the case of pure or mixed cholesteric materials, the cholesteric helix was shown to react to applied shear as a rigid rod, tilting with little or no distortion. In the case of cholesteric nematic mixtures helical distortion was apparent. These effects provide indications of the molecular interactions occurring in the helices and show that these interactions are not as strong in mixed systems where one component is noncholesteric. Analysis of rheo-optical data indicates that initially (in the nonshear case) the liquid crystal exists in the normal Grandjean texture. As shear is imposed, the helices tilt. The material next to the container walls remains relatively unaffected. This mixed structure exhibits non-Newtonian flow characteristics, and the helical directrix is continuous through the sample thickness. As some shear rate, the tilted configuration becomes unstable and a central dynamic layer is formed in which it is thought the helical correlations are parallel to the lines of flow in the system. In this case, the helical directrix is not continuous between walls.

The system is now Newtonian (an indication that the dynamic focal conic structure is Newtonian) and consists of five structural layers: normal Grandjean, perturbed Grandjean, and dynamic focal conic. The thickness of these layers is shear dependent. At some critical shear, the entire structure becomes unstable and the helices fragment to product a dynamic homeotropic texture. This transition is facilitated by an electric field. The size of the helical fragments is a function of shear and approaches molecular limits at high shear, indicating an essentially isotropic structure. This latter conclusion was attained because of normal forces generated within the system. Once generated, the dynamic focal conic structure and the homeotropic structures are stable unless anisotropic shear is applied or sample melting occurs. This work has led to the first liquid crystal normal force measurements and the first attainment of the viscosity of the homeotropic texture.

Addition of the nematic materials as well as nonmesomorphic materials such as polymers show, basically, the same characteristics with minor changes. Of importance is the fact that helical stabilization is not as prevalent (as witnessed by the small tilt distribution areas), and certain rheological transitions in pure cholesterics become reversible or are not observed at all. Others [53] have also observed sheared cholesteric systems and arrived at similar molecular configurations to the one presented here.

A comparison of our data with other forms of perturbational effects is in order. Our results are strikingly similar to those obtained during the electric field-induced cholesteric-nematic phase transition, the exception being the final structured state. Electrically, the transitions proceeds as Grandjean → focal conic → uniaxial nematic, while rheologically only the first step occurs. Through the transitions, the systems are analogous as the energy of the system is being minimized; in the electric field case by partial alignment of molecular dipoles with the applied field, while rheologically, less resistance of the helical structure to flow causes helical alignment with the shear fields. Energy minimization occurs beyond the first texture transition; however, the results are strikingly different. The electric field-induced uniaxial nematic is stable, but increasing shear on the cholesteric structure causes continued structural breakdown until an apparent isotropic state is attained. The critical phenomena in both cases are unrelated. Electrically, it is the energy required to totally overcome the helical correlation energy while rheologically it is an energy required to fracture the helical structure but not to "unwind" the helix. While our results are similar to the observed electric field effects, the results cannot be assumed to be analogous.

The shear results do not correspond to the effects of pressure upon pitch [54-55] where increasing pressure caused pitch to decrease. In that case, it was felt that pressure caused anharmonic vibrations of molecules in the helix leading to increased pitch. The pressures and shear stresses in those experiments were much higher than those of our experiments.

The experiments discussed leave many questions unanswered and provide the impetus for future experimental work. For instance, the effect of pitch and composition on the observed properties has not been determined. If there is a pitch dependence in any of the effects observed, it was within the limits of error of our measurement. Even more importantly, pitch is usually intimately related to composition in mixtures [56] and the two effects may be inseparable. Much of our experimental work was conducted on mixed liquid crystal systems, and the application of our analysis techniques (with the exception of the combined electric field-shear experiments and normal force measurements) to pure systems has not been done. Shear effects on chiral nematic material would be interesting, although initial data [45] indicate these systems would be nematiclike in nature. The simplified electromagnetic interaction theory used to extract perturbed structure (tilted helices) [35] should be tested with repeat experiments using incident circular dichroic light. This experiment is of the utmost importance for total understanding of the observed effects. Surface effects have not been analyzed. It is known that surface treatment can affect texture structure in liquid crystals [57]. The platen structures used in these experiments appear to stabilize the Grandjean structure. The question arises as to what rheological and rheo-optical effects would be observed if a surface treatment which stabilized focal conic structure had been used (would the first texture change not be observed, or would other effects be observed?). Thermal effects on the transition have not been studied to any great extent and should provide information on stability, activation, etc., of these unique systems.

Our model has proposed a centrosymmetric structure for the sheared phase. In the cholesteric-polymer mixture section, a plug flow was alluded too. This flow may also occur in pure and mixed cholesterics resulting in deformation behavior shown in Fig. 16 as opposed to those in Fig. 9. The same results could be obtained from both experiments, and therefore, the proposed model should be verified. Indeed, a rich experimental field has been exposed.

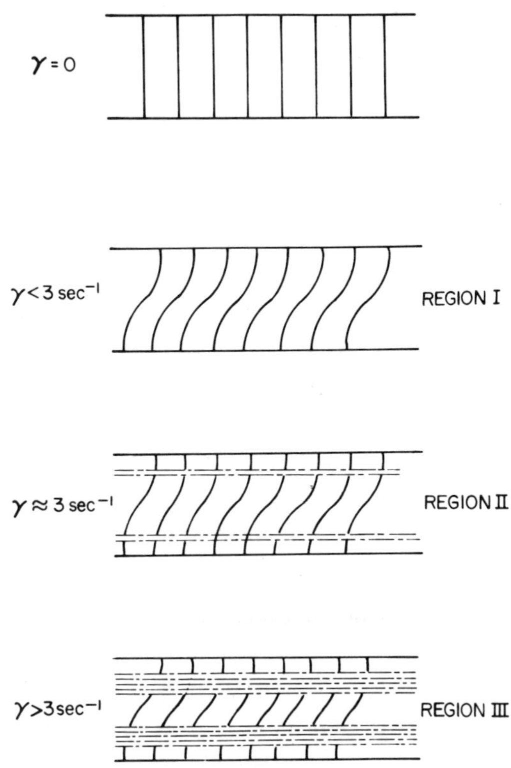

FIG. 16. Schematic representation of alternate helical response in region I, II, and III of Fig. 2.

This structural interpretation of the sheared cholesteric mesophase presents a unified picture as to effects of shear and combined shear-electric field. In many instances, first-order theories have been developed which, although rudamentary, adequately explain our observations. It is hoped that a coherent fluid dynamic model may someday be evolved against which these results may be tested.

ACKNOWLEDGMENTS

The author wishes to acknowledge coworkers D. G. Marsh, P. F. Erhardt, and W. C. Richards. A special note of gratitude is expressed to D. D. Pochan for help in the preparation of this manuscript.

IV. REFERENCES

1. G.W. Gray, Molecular Structure and the Properties of Liquid Crystals, Academic, New York, 1962.
2. R.S. Porter and J.F. Johnson, J. Appl. Phys., 34:51 (1963).
3. A.Saupe, "Recent Results in the Field of Liquid Crystals," Ange W. Chem. Intg. Ed., 1: (1968).
4. G.H. Brown, J.W. Doane, and V.O. Neff, "Structure and Physical Properties of Liquid Crystals," Crit. Rev. Solid State Sci., 1: (1970).
5. R.S. Porter and J.F. Johnson, Rheology, Vol. 4 (F. Eirich, ed.), John Wiley and Sons, New York, 1968.
6. Proceedings of the International Liquid Crystal Conference, Kent State University, Kent, Ohio, 1972. Published in Mol. Cryst. Liq. Cryst. (1973, 1974).
7. H.L. de Vries, Acta Crystallogr., 4:219 (1951).
8. J.E. Adams, W. Haas, and J. Wysocki, J. Chem. Phys., 50:6, 2458 (1969).
9. J.L. Fergason, Mol. Cryst. Liq. Cryst., 1:293 (1966).
10. J. Adams and W. Haas, Mol. Cryst. Liq. Cryst., 11:229 (1970).
11. R. Dreher, C. Meier, and A. Saupe, Mol. Cryst. Liq. Cryst., 13:17 (1971).
12. R. Nityananda, Mol. Cryst. Liq. Cryst., 21:315 (1973).
13. T.C. Lubensky, Phys. Rev., 6:452 (1972).

14. J.L. Fergason, Sci. Am., 76: (1964).
15. C.G. Kartha and R.K.L. Agarwall, J. Phys. Soc. Jap., 28:470 (1970).
16. C.G. Kartha and A.R.K.L. Padmini, J. Phys. Soc. Jap., 31:3 (1971).
17. R.S. Porter, K. Sakamoto, and J.F. Johnson, Mol. Cryst. Liq. Cryst., 8:443 (1969).
18. E.M. Barrall II, R.S. Porter, and J.F. Johnson, J. Phys. Chem., 71: (1967).
19. W. Ostwald, Trans. Faraday Soc., 29:1002 (1933).
20. W. Ostwald and H. Malss, Kolloid. Zh., 63:6 (1933).
21. A.S.C. Lawrence, Trans. Faraday Soc., 29:1080 (1933).
22. R.S. Porter, E.M. Barrall, and J.F. Johnson, J. Chem. Phys., 45: (1966).
23. K. Sakamoto, R.S. Porter, and J.F. Johnson, Mol. Cryst. Liq. Cryst., 8:443 (1969).
24. R. Schench, Z. Phys. Chem., 27:167 (1898).
25. M. March and A.V. Tobolsky, Physical Chemistry of High Polymeric Systems, John Wiley and Sons, New York, 1950, p. 229.
26. E. Bose, Phys. Z., 8:347 (1907).
27. T. Yamada and E. Fukada, J. Appl. Phys. Jap., 12:68 (1973).
28. R.S. Porter, C. Griffen, and J.F. Johnson, Mol. Cryst. Liq. Cryst., 25:131 (1974).
29. P. Martonoty and S. Candau, Mol. Cryst. Liq. Cryst., 14:243 (1971).
30. R.S. Porter and J.F. Johnson, J. Chem. Phys., 66:51 (1963).
31. D. Taupin, J. Phys. (Paris), 30:C4-32 (1969).
32. D.W. Berreman and T.J. Scheffer, Phys. Rev. Lett., 25:577 (1970).
33. J.L. Fergusin, in Liquid Crystals, (G.H. Brown, G.J. Dienes, and M.M. Labes, eds.), Gordon and Breach, New York, 1966.
34. J.M. Pochan and P.F. Erhardt, Phys. Rev. Lett., 27:2 (1971).
35. J.M. Pochan and D.G. Marsh, J. Phys. (Paris), 57:1193 (1972).
36. J.M. Pochan, P.F. Erhardt, and W.C. Richards, in Liquid Crystals and Ordered Fluids (J.F. Johnson and R.S. Porter, eds.), Plenum, New York, 1974, p. 449.
37. P.F. Erhardt, J.M. Pochan, and W.C. Richards, J. Chem. Phys., 57:9 (1972).
38. J.M. Pochan and D.G. Marsh, J. Chem. Phys., 57:5154 (1972).

39. J.M. Pochan, P.F. Erhardt, and W.C. Richards, Mol. Cryst. Liq. Cryst., 24:89 (1973).
40. H. Baessler and M.M. Labes, Mol. Cryst. Liq. Cryst., 6:519 (1970).
41. T. Kotaka, J. Chem. Phys., 30:1556 (1959).
42. J.M. Pochan and H.W. Gibson, J. Amer. Chem. Soc., 94:5573 (1972).
43. R. Stein, University of Massachusetts, private communication, 19xx.
44. J.J. Wysocki, J.E. Adams, and D.J. Oleckna, in Liquid Crystals and Ordered Fluids (J.F. Johnson and R.S. Porter, eds.), Plenum, New York, 1970.
45. D.G. Marsh and J.M. Pochan, J. Chem. Phys., 58:7, 2835 (1973).
46. D.G. Marsh, J.M. Pochan, and P.F. Erhardt, J. Chem. Phys., 58:12, 5795 (1973).
47. F. Saeva and J. Wysocki, J. Amer. Chem. Soc., 93:5928 (1971).
48. F. Saeva, see Chap. 6.
49. J.M. Pochan and D.D. Hinman, J. Phys. Chem., 78:12, 1206 (1974).
50. F.D. Saeva, in Liquid Crystals and Ordered Fluids (J.F. Johnson and R.S. Porter, eds.), Plenum, New York, 1974, p. 581.
51. J. Adams, G. Dir, and W. Haas, in Liquid Crystals and Ordered Fluids (J.F. Johnson and R.S. Porter, eds.), Plenum, New York, 1974, p. 421.
52. F.D. Saeva, J. Amer. Chem. Soc., 94:5135 (1972).
53. D.F. Ciliberti, G.D. Dixon, and L.C. Scala, Mol. Cryst. Liq. Cryst., 20:27 (1973).
54. P. Pollmann and H. Stegmeyer, Chem. Phys. Lett., 20:87 (1973).
55. B. Boethcher, Chem. Z., 96:214 (1972).
56. H.W. Gibson, J.M. Pochan, and D.D. Hinman, in Liquid Crystals and Ordered Fluids (J.F. Johnson and R.S. Porter, eds.), Plenum, New York, 1974, p. 593.
57. L.T. Creagh and A.R. Kmetz, Mol. Cryst. Liq. Cryst., 24:59 (1973).

Chapter 8

LIQUID CRYSTALS AND BIOLOGICAL MEMBRANES

Dennis Chapman

Department of Biochemistry and Chemistry
Royal Free Hospital School of Medicine
University of London
London, England

I. INTRODUCTION

An important topic of modern molecular biology is the structure and function of cell membranes. This topic is now the subject of much discussion [1]. The components of cell membranes include lipids (e.g., phospholipids), proteins, metal ions, sometimes cholesterol, and of course, water. One question which has recieved a great deal of attention is how these components are organized and mutually interact with each other. Conversely, while membranes are of interest for their biological and medical significance there is also great interest in the possibility of using the principles involved in their electrical properties for the design of industrial transducer and telecommunication devices.

A considerable amount of analysis of the lipid components of cell membranes has now been carried out, and it has been shown that various classes of lipids are present, e.g., lecithins (phosphatidylcholines), phosphatidylethanolamines, and phosphatidylserines. Typical structures of some phospholipids are shown in Fig. 1. Associated with each lipid class are a variety of fatty acids of differing chain length and unsaturation. At present much work is being carried out to analyze the protein components of membranes. Studies are also being made of the mutual effect of the lipid and protein components.

Important to the subject of this book is the fact that the lipids which occur in biological membranes have liquid crystalline properties. They appear to exhibit long-range order with a certain amount of short-range disorder.

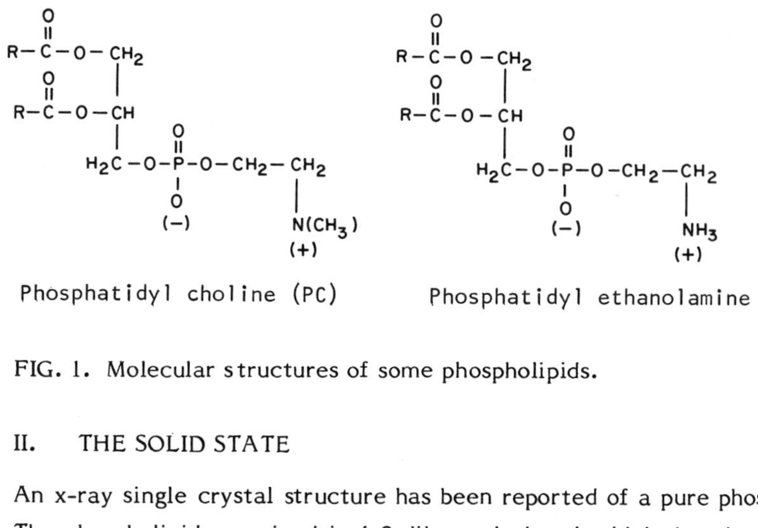

FIG. 1. Molecular structures of some phospholipids.

II. THE SOLID STATE

An x-ray single crystal structure has been reported of a pure phospholipid. The phospholipid examined is 1,2-dilauroyl phosphatidylethanolamine [2]. The molecular conformation of this lipid crystallized from acetic acid is shown in Fig. 2. The mean planes containing the lipid chains are essentially parallel to one another; the phosphodiester moiety has a double gauche conformation. The intermolecular packing produces the classic lipid bilayer structure, adjacent lipid bilayers being separated by acetic acid molecules of crystallization. The hydrocarbon chain packing can be considered either as a quasihexagonal type or as a complex orthorhombic subcell arrangement.

The capillary melting points of a number of pure phospholipids have been determined and are shown to be quite high, e.g., the values of the capillary melting points for the diacylphosphatidylethanolamines are about 200 °C, while for the phosphatidylcholines these capillary melting points are about 230 °C. It is important to note that these values are independent of both the chain length and the degree of unsaturation of the fatty acid residues associated with the phospholipid. These capillary melting points can be compared with those of other compounds having long chains. The capillary melting points of fatty acids containing the same length of hydrocarbon chain are much lower, e.g., stearic acid melts at 69.7°C. On the other hand, the capillary melting point of sodium stearate is ≈ 300°C. The high values of the capillary melting points of phospholipids are, therefore, consistent with the occurrence of ionic linkages associated with the polar groups of the phospholipid. The higher capillary melting points of the sodium soaps, compared with the phospholipids, suggest that there is possibly greater ionic character associated with the polar groups of these molecules than with the diacylphosphatidylcholines and phosphatidylethanolamines.

III. THE MESOMORPHIC STATE

A. Liquid Crystals Formed by the Action of Heat (Thermotropic Mesomorphism of Lipids)

In addition to the capillary melting point, other phase changes have been shown to occur with phospholipids at lower temperatures; for example, when a pure phospholipid, dimyristoylphosphatidylethanolamine, containing two fully saturated chains, is heated from room temperature up to the capillary melting point, a number of thermotropic phase changes occur (i.e., phase changes caused by the effect of heat). This was first shown by infrared spectroscopy [3], then by thermal analysis [4], and has now been studied by a variety of other physical techniques [5]. Optical studies of thin films show that dimyristoylphosphatidylethanolamine at room temperature is birefringent when viewed between crossed polars. On heating,

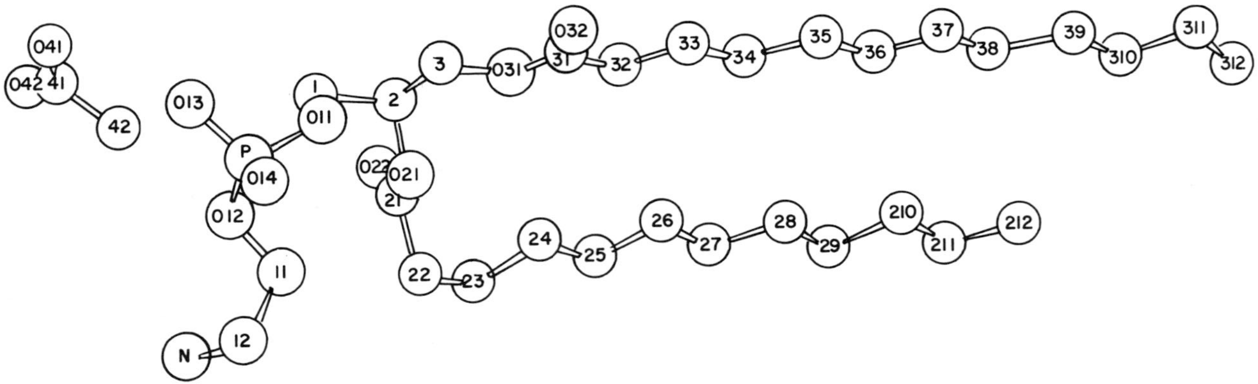

FIG. 2. The molecular conformation of 1,2-dilauroyl-DL-phosphatidylethanolamine: acetic acid (Hitchcock et al. [2]).

three processes occur: first, some loss of birefringence at the first transition temperature at ≈120 °C, then a small increase at 135°C, and finally a pronounced overall loss of birefringence near the melting point of 200 °C. Above 120°C, pressure on a cover glass with a needle causes the material to flow. When the temperature of the phospholipid reaches ≈120 °C, the infrared (IR) absorption spectrum undergoes a remarkable change and loses all the fine structure and detail which was present at lower temperatures. The spectrum is then similar to that obtained with a phospholipid dissolved in a solvent such as chloroform. The IR spectra of a phospholipid at different temperatures are shown in Fig. 3.

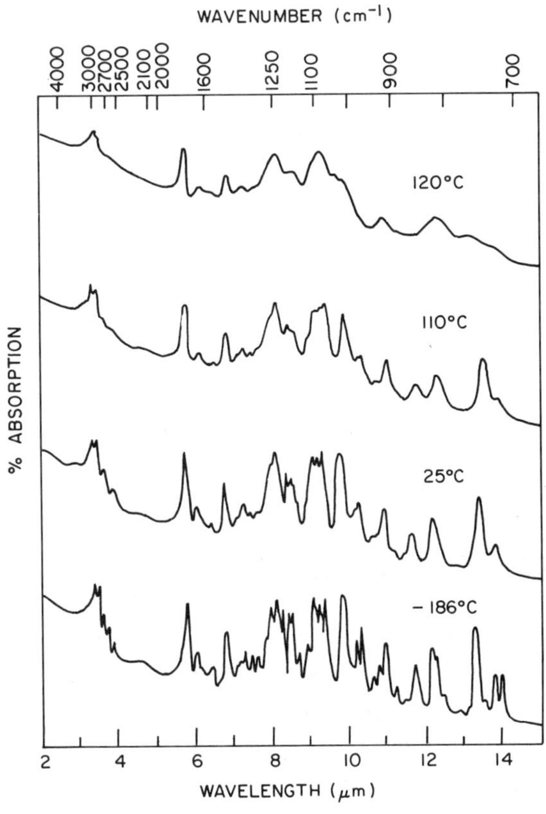

FIG. 3. The infrared spectra of 1,2-dilauroyl-DL-phosphatidylethanolamine at different temperatures. (From Chapman et al., Ref. 5.)

Differential thermal analysis (DTA) shows that a marked endothermic transition (absorption of heat) occurs at about 120°C [3]. An additional heat change occurs at ≈135°C, and only a small heat change is involved near the capillary melting point of the lipid. This behavior is similar to that which occurs with compounds which form liquid crystals, e.g., ρ-azoxyanisole or cholesteryl propionate, which give nematic and cholesteric liquid crystalline phases, respectively.

The x-ray long spacings show a dramatic reduction to some two-thirds of their original value at the first transition temperature with a further small reduction at the second transition temperature. The x-ray short spacings change at the first transition point from sharp diffraction lines to a diffuse spacing at ≈4.6 Å. Nuclear (proton) magnetic resonance (PMR) studies show a gradual reduction in line width from about 15 gauss at liquid nitrogen temperature until, at the first transition temperature, there is a sudden reduction in the line width (to ≈ 0.09 gauss). PMR spectra of dimyristoylphosphatidylethanolamine are shown in Fig. 4. These show that molecular motion increases gradually as the temperature increases until, at the transition temperature, a considerable increase in the molecular motion takes place [6], as revealed by an appreciable reduction in line width.

The main conclusions from these various studies are that: (1) Even with the fully saturated phospholipid at room temperature, some molecular motion occurs in the solid. This is evident from the PMR spectra and from the IR spectra taken at liquid nitrogen and at room temperatures. (Note the difference between the IR spectra at -186°C and room temperature shown in Fig. 3). (2) When the phospholipid is heated to a higher temperature, it reaches a transition point at ≈120°C, a marked endothermic change occurs, and the hydrocarbon chains in the lipid "melt" and exhibit a very high degree of molecular motion. This is evident both in the appearance of the IR spectrum and also in the narrow PMR line width at temperatures of 120°C and higher. On the one hand the broad diffuse appearance of the IR spectrum is consistent with the chains flexing and twisting and with a "break-up" of the all-planar trans configuration of the methylene groups of the chains. (The IR spectrum can be regarded as giving a "rapid snapshot" showing the condition and organization of the

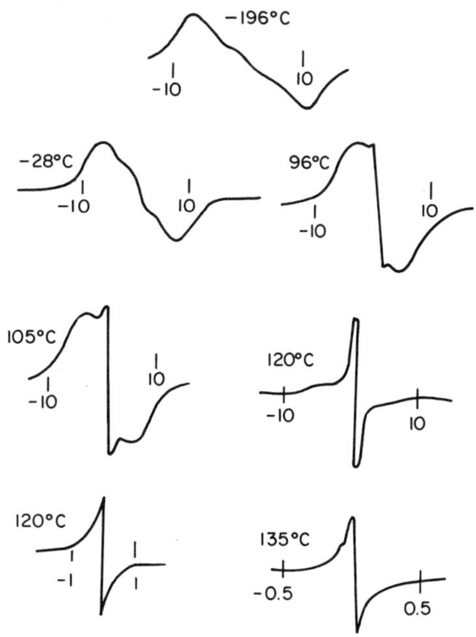

FIG. 4. PMR derivative absorption spectra of 1,2-dimyristoyl-DL-phosphatidylethanolamine at various temperatures. The abscissae are marked in gauss. (From Chapman and Salsbury, Ref. 6.)

phospholipid molecule within a period of 10^{-15} sec. The PMR spectrum, on the other hand, is affected by slower motions. It provides an impression of any molecular motion, i.e., rotation, translation, or diffusion, of the phospholipid molecule occurring within a much longer time period, some 10^{-7} sec. The fact that the phase transition is concerned primarily with the hydrocarbon chains of the phospholipid is confirmed by the x-ray data.

When phospholipids contain shorter chains, or unsaturated bonds, those marked endothermic phase transitions occur at lower temperatures. The changes in the temperatures at which these transitions occur parallel the trends in the melting points of the related fatty acids. The transition temperatures are high for the fully saturated long-chain phospholipids. They are lower when there is a trans double bond present in one of the chains and lower still when there is a cis double bond present. This trend in the transition temperatures also confirms that this phase transition is primarily associated with a melting of the hydrocarbon chains of the phospholipid, while this in turn is related to the dispersion forces between the chains.

Only one main melting of the chains occurs even when there are two different types of chain present in the phospholipid. The transition temperatures for different phospholipid classes differ even though they contain exactly the same fatty acid residues. While natural phospholipids from erythrocyte or mitochondrial membranes contain large numbers of unsaturated cis double bonds and, therefore, in the dry condition have endothermic transition temperatures either near or below room temperatures, the highly saturated derivatives exhibit transition temperatures much higher than room temperature. With mixtures of phospholipids we might expect the transition temperatures to occur over a wider range of temperature than with a single phospholipid.

When a crystalline phospholipid approaches the first thermotropic transition temperature we can imagine that the molecular motion of the chains increases until, at the first transition point, lateral expansion of the crystal lattice is forced to take place. This expansion allows even greater chain mobility, possibly involving cooperative motion of the chains with rotation about C-C bonds occurring. A similar effect probably takes place with polar long-chain molecules in a monolayer at an air-water interface dependent upon the density of the chain packing.

A very simple theoretical system was devised [7] to provide some insight for such situations. The model system is restricted to a set of simple chains in two dimensions and with the chains confined to lie on a two-dimensional hexagonal lattice. The end-to-end distance of each chain (corresponding in a real system to the distance from the polar group to the methyl group) and other properties were determined as a function of the density of packing of the chains using two simple potential functions and the Monte Carlo computational method. Typical configurations of sets of simple chains on a two-dimensional hexagonal lattice are shown in Fig. 5. At the highest density only the fully extended configuration occurs. At lower densities other configurations are allowed. As the density of the chains decreases, the end-to-end distance of the chains suddenly falls, consistent with the occurrence of a cooperative phase transition.

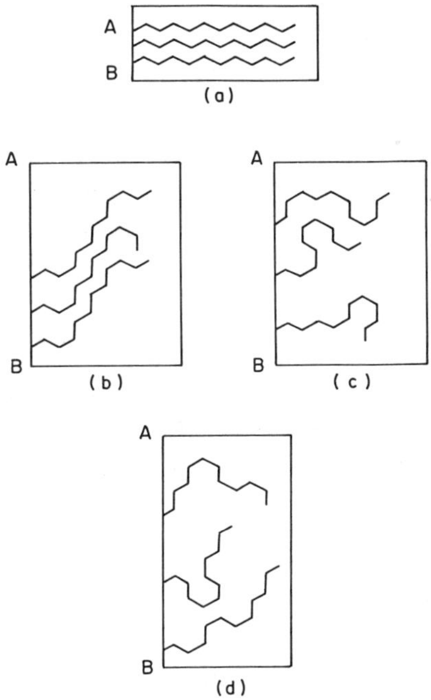

FIG. 5. Typical configurations of sets of chains on a two-dimensional hexagonal lattice as the interchain separation is increased. Separation increases from (a) to (d). (From Whittington and Chapman, Ref. 7.)

An important conclusion which we obtain from these phase transition studies is that, near the endothermic transition temperature, a given phospholipid can be in a highly mobile condition with its hydrocarbon chains flexing and twisting. The more unsaturated the chain, the lower the temperature will be at which this occurs. This is a fundamental property of the phospholipid, and we can expect this chain mobility in whatever situation the phospholipid occurs unless, for special reasons, this motion is somewhat inhibited. We can envisage that inhibition of chain motion by interaction with other molecules would provide one of these special reasons, e.g., cholesterol or protein (see later discussion). In other circumstances, due to less perfect packing arrangements, we might expect, at a particular temperature, even greater mobility of the chains of the lipid and indeed of the whole lipid molecules themselves.

B. Liquid Crystals Formed in the Presence of Water (Lyotropic Mesomorphism of Lipids)

Small amounts of water can have important effects upon the mesomorphic behavior. Thus, the diacylphosphatidylcholines (lecithins) exhibit an additional liquid crystalline form between the first transition temperature and the capillary melting point [8]. The intermediate liquid crystalline form is found to exhibit x-ray spacings consistent with a cubic phase organization.

When phospholipids are examined in the presence of increasing amounts of water, the various physical techniques, such as microscopy, PMR spectroscopy, or DTA, show that as the amount of water increases, the marked endothermic transition temperature for a given phospholipid falls. The transition temperature does not fall indefinitely; it reaches a limiting value independent of the water concentration (Fig. 6). We can understand this if we regard the effect of water as leading first to a "loosening" of the ionic structure of the phospholipid crystals. This in turn affects the whole crystal structure and causes a reduction, up to a certain limit, of the dispersion forces between the hydrocarbon chains. Large amounts of energy are still required to counteract the dispersion forces between the chains, and quite high temperatures are still required to cause the chains to melt. These limiting transition temperatures parallel the melting point behavior of the analogous fatty acids, becoming lower with increasing unsaturation. This further reduction of the endothermic transition temperature by water means that the natural phospholipids extracted from biological membranes usually exhibit this crystalline-to-liquid crystalline transition many degrees below the biological environmental temperature.

There are a number of important features associated with the transition temperature for the lipid when it is in the presence of water. The first of these is that the ability to disperse the lipid in water increases markedly above this transition temperature. Only those phospholipids which give transitions when placed in water below or near to room temperature spontaneously form myelin figures. Fully saturated phospholipids which have high transition temperatures do not form myelin figures

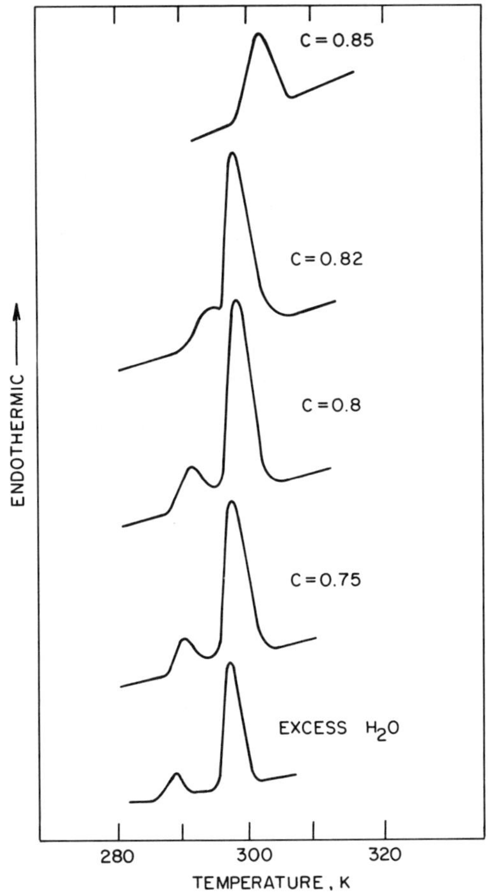

FIG. 6. DSC curves for the DML-H_2O system as a function of lipid concentration c = wt lipid/(wt lipid + wt water).

at room temperature. However, if the temperature is raised to the transition point, these phospholipids form myelin figures. In the presence of an excess of water, phospholipids, such as 1,2-dipalmitoylphosphatidylcholine, spontaneously form myelin figures at 42°C. The behavior of lipids in water can be conveniently summarized by means of the phase diagram shown in Fig. 7.

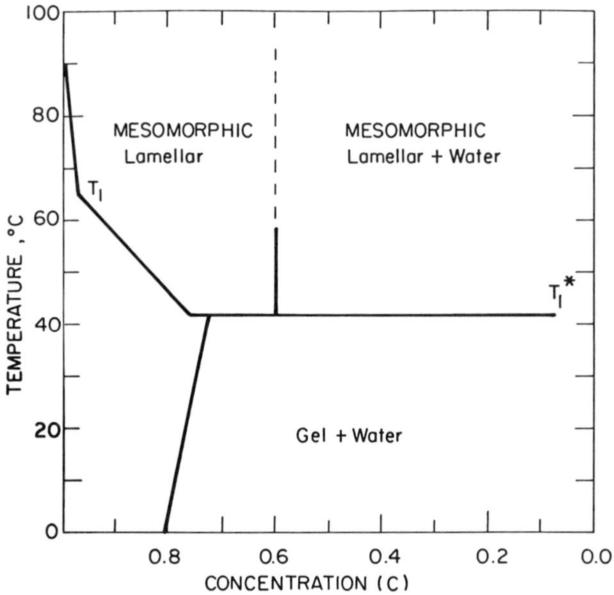

FIG. 7. Phase diagram of the 1,2-dipalmitoyl-L-phosphatidylcholine-water system. (From Chapman et al., Ref. 8.) In the gel phase water separates the double layers of lipid, and the hydrocarbon chains are packed in an organized crystalline manner, probably in an hexagonal arrangement. The cubic phase occurs at high temperatures and in the presence of only small amounts of water. It therefore occurs close to the left-hand side of the phase diagram.

Recently, further studies have been carried out on lecithin-water systems [9]. The various structures which can occur below the thermotropic phase transition temperatures were examined. The β form is obtained with heterogeneous lipids and has lipid chains vertical to the plane of the lamellae. The β′ form is observed with pure lipids of identical chains but has the chains tilted with respect to the normal in relation to the lamellae. Both forms are packed with rotational disorder in a two-dimensional hexagonal lattice.

Upon heating, the pure lecithins, e.g., dipalmitoyl lecithin, transform from the β′ to the β form just prior to the fluid condition. (The fluid form is termed the α-form according to the Tardieu convention.) Tardieu et al. [9] also show that with the pure lecithins, as water is added to the anhydrous lipid (dipalmitoyl lecithin), the thickness of the lipid leaflet decreases as the amount of water increases. This is associated with an increased tilt of the chains with respect to the lamellae. The dimensions of these phases are shown in Table 1.

The lecithin-water system has received much attention because of the common occurrence of these lipids in animal cell membranes. A range of physical techniques have now been applied to the study of these systems. For example, NMR spectroscopy has been applied to the study of the lipid phase transition. The first study using wide-line NMR spectroscopy was by Veksli et al. [10]. This showed that a marked line-narrowing occurs at the transition temperature associated with the anisotropic but rapid molecular motion of the long-chain protons. Also, just prior to the

TABLE 1. Dimensions of the phases Lβ and Lβ'[a]

	T' (°C)	$\bar{\nu}$ (cm^3/g)	c	d (Å)	d_1 (Å)	S ($Å^2$)	θ (°)
Egg-PC	0	0.932	(0.97)	60.6	(58.8)	(41.4)	0
'βC18-C-10-PC	0	0.935	0.99	52.0	51.5	40.8	0
Mitochondria	-20	0.902	0.94	60.6	56.6	40.7	0
diC16-PC	20	0.948	0.94	57.7	54.0	42.7	17
DiC16-PC	20	0.948	0.89	57.0	50.7	45.6	26
diC16-PC	20	0.948	0.82	60.5	49.2	46.8	30
diC16-PC	20	0.948	0.80	61.5	48.6	47.4	31
βdiC16-PC	20	0.948	(0.75)	64.0	(47.4)	(48.6)	(33)
diC18-PC	20	0.958	0.90	63.0	56.5	44.6	24
diC18-PC	20	0.958	0.83	63.0	52.0	48.4	32
diC18-PC	20	0.958	0.78	63.0	48.5	52.0	38
diC18-PC	20	0.958	(0.71)	67.0	(47.0)	(53.4)	(40)

[a]θ is the tilt of the chains with respect to the normal to the lamellae. The dimensions in parentheses are doubtful; egg-PC is hygroscopic and consequently a low water content is difficult to determine with accuracy; the phase Lβ' at high water content might be accompanied by some water in excess. (From Ref. 9.)

transition temperature a line narrows which is associated with the rapid molecular motion of the $N(CH_3)_3$ groups. This study also showed that even below the transition temperature there is restricted rotational and oscillatory modes of reorientation with a line width considerably reduced from the rigid lattice value. Deuterated lipids have also been examined [11] using deuteron magnetic resonance. This work has been recently extended by Seelig and Seelig [12] to a study of the order parameter along the lipid chains. Related studies on oriented deuterated soap systems have also been carried out by Charvolin et al. [13]. Electron spin resonance (ESR) spectroscopy using spin labels has also been used to study the lecithin-water phase transition [14,15].

Studies by Salsbury et al. [16] have also been made of water organization associated with the lipid molecules using deuteron magnetic resonance. As dipalmitoyl lecithin is heated through the phase transition temperature the splitting of the deuteron doublet decreases sharply up to the transition temperature. The splitting then increases with further increase in temperature. These effects are probably due to the increasing mobility of the polar group up to the transition temperature. The further increase in splitting may be associated with a redistribution of the bound water.

Fluorescence studies [17], triplet probes [18], dilatometry [19], freeze-fracture electron microscopy [20], laser Raman spectroscopy [21], and light scattering [22] are among the many techniques applied to the study of phase transitions of lipid-water systems.

The concept that lipid mobility was so great above the transition temperature that lipids could diffuse readily along the length of the bilayer [5] was confirmed and diffusion coefficients have been measured in lipids and membranes by means of spin label methods ($D = 1.8 \times 10^{-8}$ cm^2/sec) [23,24,25].

1. Types of Lyotropic Phase

Above the first transition temperature, when sufficient water is present, phospholipids can form several different types of phase organization, e.g., some phospholipids give lamellar- and hexagonal-type structures. Recent studies have shown that the lecithins appear to exhibit only a lamellar-type structure over a large range of water concentration. Other phospholipids, such as phosphatidylethanolamines and samples of brain lipid, appear to be able to exist in both the hexagonal and lamellar types of organization [26], depending upon the concentration in water (see Fig. 8). Studies by Gulik-Kryzwicki et al. [27] have shown that four phases can be distinguished with a mitochondrial lipid extract. The composition of the extract from beef heart mitochondria was 34% lecithin, 29% phosphatidylethanolamine, 10% phosphatidylinositol, 20% cardiolipin, 2% cholesterol, and 5% neutral lipids. Hexagonal phases were observed, as well as various lamellar phases. As yet no relationship between these phases and their biological function has been made. Nevertheless, these phase properties have to be considered and held in mind in discussions of lipid behavior.

Studies have been made of large monodomain phospholipid bilayer liquid crystals [28]. The technique involves taking the sample of lipid-water between two treated glass slides clamped together with a spacer. The spacer (either glass or Teflon) determines the sample thickness. The glass slides are treated with a polymerizable surfactant to orient the bilayers near the glass surfaces. Assembly of the sample is made in dry nitrogen. The sample is heated to the high end of the (smectic A) lamellar phase. The lipid is then in its most fluid state, i.e., viscosity is minimum,

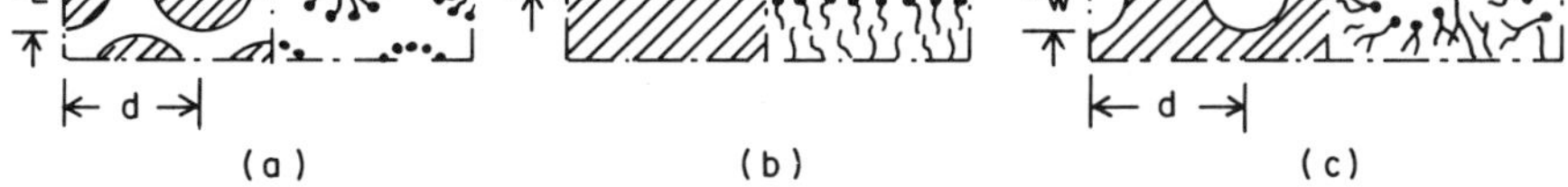

FIG. 8. Structures of some high-temperature phases of lipid-water systems; schematic representation of a cross-section of the rods and of the lamellae. The hydrophobic group is represented by a dot, the paraffin chains by a wriggle: (a) hexagonal I: (b) lamellar; (c) hexagonal II. (From Luzzati et al., Ref. 26.)

diffusion constants and dislocations are at a maximum. Annealing of the distorted lamellar structure to its lowest energy configuration is obtained by slow heating (about 0.5°C/min) to a temperature of about 125°C (for dipalmitoyl lecithin monohydrate). Orientation is obtained most readily at low water concentration with arrays of 1 cm^2 in area and up to 500-μm thickness obtainable with dipalmitoyl lecithin. Orientation of the lipid and the lipid-containing cholesterol and also monovalent effects due to divalent ions have been obtained.

Optical birefringence, light scattering, and x-ray studies have been made with the monodomain phospholipid systems. The light scattering data were used to provide some information about the elastic properties of the various lipid bilayers present.

2. Hydrocarbon Chain Organization

The detailed organization of the hydrocarbon chains of lipid mesophases both in anhydrous and aqueous systems have been variously described as liquidlike and chaotic [29], extended and ordered [30], significantly ordered [31], and partly ordered with some rotational isomerism [32]. The interpretation of the x-ray data associated with the short spacings is not unequivocal [33].

The diffuse diffraction band at 4.6 Å from randomly oriented bilayers is similar to that obtained from liquid paraffins. This diffraction is too complex to be analyzed in terms of a detailed model for the organization of the hydrocarbon chains, and it is yet to be shown that it corresponds to a unique mode of packing of the chains. Warren [34] has discussed the origin of the diffraction band from liquid paraffins in terms of a cluster of seven hexagonally packed chains in the all-trans conformation. This model was confirmed by Pierce [35] and by Zachariasen [36] who calculated the radial distribution of chains in the liquid. These workers, however, assumed that liquid paraffin is an isotropic structure, and this is clearly not valid in the case of lipid bilayers.

Calorimetric data has been used to examine the hydrocarbon chain motion above the phase transition temperature of anhydrous lecithins [32]. The heat involved in the transition is found to be about 95% of the total heat of fusion.

The total entropy per methylene group is the same for all long-chain compounds in the crystalline form at their chain melting point. The entropy gain during the transition from β crystal to isotropic liquid for n-alkanes, triglycerides, and fatty acids is 2.6 eu per CH_2 group, whereas for the crystal-to-liquid crystal transition for lecithins the equivalent figure is 1.1 eu (entropy unit). Thus, in the liquid crystalline state, the chain fluidity is about one-half that found in liquid n-alkanes at the transition temperatures. It was suggested that this may arise by inhibition of rotation about the carbon-carbon bonds due to the presence of the neighboring chains.

Calorimetric studies [32] with the fully hydrated lecithins showed that they have greater chain mobility above the transition temperatures than do the anhydrous or monohydrate liquid crystals but still less than that which occurs in the n-alkane melt so that some order is retained and the chains are never in a chaotic arrangement [37].

Below the transition temperature the area per lipid molecule is about 48 Å while above this temperature (at 40% H_2O) the area per molecule is some 70 $Å^2$ (see Chapman et al. [8] and Phillips and Chapman [38]). These dimensions are also consistent with the concept that in the lipid systems examined there is still some order normal to the plane of the membrane.

The recent deuterium magnetic resonance studies of Seelig and Seelig [39] comparing NMR and spin-label studies, and that of Charvolin et al. [40] provide good evidence on the order parameter which shows that within the lipids which they examined the order parameter is essentially constant along the chain except for the region near the center of the bilayer (Fig. 9).

It should also be noted that as the lipid is heated to higher temperatures above the transition temperature the disorder in the chain will increase.

3. Monolayer Properties

Chapman [41] and Phillips and Chapman [38] pointed out the correlation which exists between the monolayer properties at the air-water interface of lipids and the properties of the aqueous dispersions of lipids. The condensed monolayer corresponds to the "gel-type" structure and the

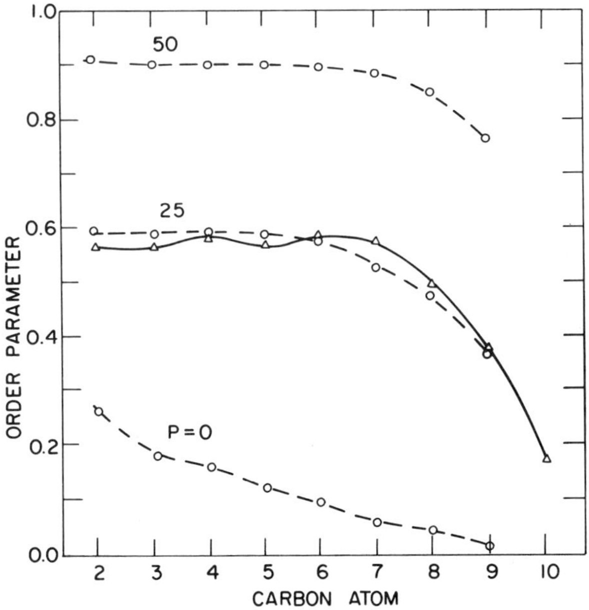

FIG. 9. Order parameter, at 28°C, shown as a function of a position along the chain. Dashed lines, calculation for different values of lateral pressure (in dynes per centimeter); solid lines, data by Seelig and Neiderberger. (After Marcelja, Ref. 42.).

expanded monolayer to the "fluid condition" above the lipid transition temperature, so that similar thermotropic phase changes occur with the monolayers.

Marcelja [42] has used this correlation between the monolayers and lipid-water dispersions to calculate the lateral pressure in the hydrocarbon chain region. He estimates this to be of the order of 20 dynes/cm for each half of the bilayer.

4. Phase Separation

The first studies on phase separation of lipid water systems were discussed by Ladbrooke and Chapman [43] who reported studies of binary mixtures of lecithins using calorimetry [44]. These authors examined mixtures of distearoyl and dipalmitoyl lecithin (DSL-DPL) and also distearoyl lecithin and dimyristoyl lecithin (DSL-DML). With the DSL-DPL mixtures the phase diagram shows that a continuous series of solid solutions are formed below the T_c line. It was concluded that compound formation does not occur and that with this pair of molecules having only a small difference in chain length, cocrystallization occurs.

With the system DSL-DML monotectic behavior was observed with limited solid solution formation. Here, the difference in chain length is already too great for cocrystallization to occur so that as the system is cooled migration of lecithin molecules occurs within the bilayer to give crystalline regions corresponding to the two compounds.

The use of spin labels such as TEMPO to prove phase separation of mixed lipid systems has been reported by Shimshick and McConnell [45] on similar lipid mixtures.

5. Effects Due to Foreign Molecules

a. Cholesterol Effects. In 1968 a paper was published [46] describing studies on lecithin-cholesterol-water interactions by differential scanning calorimetry (DSC) and x-ray diffraction. The 1,2-dipalmitoyl-L-phosphatidylcholine-cholesterol-water system was studied as a function of both temperature and concentration of components. This particular lecithin was used because it exhibits the thermotropic phase change in the presence of water at a convenient temperature (41°C). The addition of cholesterol to the lecithin in water lowers the transition temperature between the gel and the lamellar fluid crystalline phase and decreases the heat absorbed at the transition. No transition at all is observed with an equimolar ratio of lecithin with cholesterol in water. This ratio corresponds to the maximum amount of cholesterol which can be introduced before cholesterol precipitation occurs. These effects are not specific for dipalmitoyl lecithin; unsaturated lecithins and the lipid extract of human erythrocyte ghosts exhibit similar behavior. The addition of cholesterol to the lecithin-water system in the gel phase causes a reduction in the cohesive forces between adjacent hydrocarbon chains of the lecithin; this leads to a fluidization of these chains. The ordered array of hydrocarbon chains in the gel phase is disrupted by cholesterol. The presence of water was also shown to be of prime importance in this interaction.

The hydrocarbon chains of the lecithin in water have less thermal motion in the presence of cholesterol than they have in the absence of cholesterol above the T_c line, and they have more thermal motion than in the absence of cholesterol below the T_c line. The requirement for water has been stressed. X-ray evidence [46] indicates a lamellar arrangement, and at 50% cholesterol an additional long-spacing pattern occurs due to the separation of crystalline cholesterol. These results may be interpreted in terms of penetration of the lipid bilayer by cholesterol. In the lamellae of aqueous lecithin, the chains are hexagonally packed and tilted at 58°. It can be envisaged that penetration will be facilitated when the chains are vertical. This causes an increase in the x-ray long spacing. At concentrations of cholesterol greater than 7.5% the long spacing then decreases. Above this critical concentration a reduction occurs of the cohesive forces between the chains producing chain fluidization. The PMR of egg yolk lecithin shows that the addition of cholesterol produces effects which depend on the state of the lipid [47]. At temperatures in excess of the transition temperature, the addition of cholesterol reduces the signal due to the methylene protons of the hydrocarbon chains, i.e., the motion of these is inhibited. It has also been demonstrated that addition of cholesterol to sonicated lecithins broadens the choline residue signal to a lesser extent than it does that of the alkyl chains (Darke et al. [48]). Thus, the chain protons are severely immobilized. Linewidth measurements clearly demonstrate that the fluidity of the chains in the presence of cholesterol is intermediate between that of aqueous lecithin below and above its transition temperature. Even in aqueous lecithin below its transition temperature the choline group still retains appreciable mobility.

Phillips and Finer [49] have discussed the variation with composition of the cooperativity of the lipid thermotropic transition in mixed dipalmitoyl lecithin cholesterol bilayers. In bilayers containing less than equimolar amounts of cholesterol, discrete regions of 1:1 complex separate out leaving lecithin molecules at the boundaries of these regions for which cooperative motions are not possible and clusters of free lecithin molecules which freeze at the normal transition temperature.* Phillips and Finer comment on suggestions of Rothman and Engelman [50] concerning 2:1

*It has recently been suggested that cholesterol molecules may be randomly arranged in the plane of the lipid bilayer [71].

lipid-cholesterol complexes and also some calorimetric studies of Hinz and Sturtevant [51].

b. Cations and Anions. Metal ion interactions have been known for some years to affect the thermotropic phase transition of soap systems. The thermotropic phase transition of stearic acid occurs at 114°C with the sodium salt and at 170°C with the potassium salt. These phase transitions can be linked to the monolayer characteristics. The precise nature of the interaction between ions and phospholipids is still open to doubt. There is some evidence that charge neutralization is the prime interaction of charged phospholipids with divalent cations [52,53]. Divalent cations were found to increase the transition temperature and the monovalent cations to fluidize the bilayer. Some authors believe that the primary effect of the cation on lecithins may be on the aqueous portion of the lipid bilayer [54-56]. A recent study of an extensive range of salts with lecithin bilayers indicates that the anion present (Fig. 10) has a very large effect in determining the state of fluidity of the bilayer, the results obtained being best explained by a thermodynamic treatment based on relative association constants. Trauble [57] has shown that pH can affect lipid transition temperatures, particularly lipids such as the phosphatidylethanolamines.

c. Protein and Polypeptides. Various studies have now been made of the effects of polypeptides and proteins on phase transition temperatures. Chapman et al. [58] have studied phosphatidylserine interactions with cytochrome C lysozyme and polylysine and have shown shifts of transition.

The "electrostatic" effect of polypeptides and proteins on lipid fluidity may also be important in transient and structural situations. It is clearly relevant where comparisons are made between transitions which occur with membranes where electrostatic interaction of lipid and protein is thought to occur, compared to those of the extracted lipids.

This effect on lipid fluidity by interaction of protein with the polar groups of lipids may also occur in other important systems, such as the concanavalin A-glycolipid system and antigen-antibody interactions and could be of considerable biological importance.

With gramicidin A the heat involved in the main lipid endothermic transition is markedly reduced in a somewhat similar manner to that observed with cholesterol. It seems reasonable to conclude, therefore, that

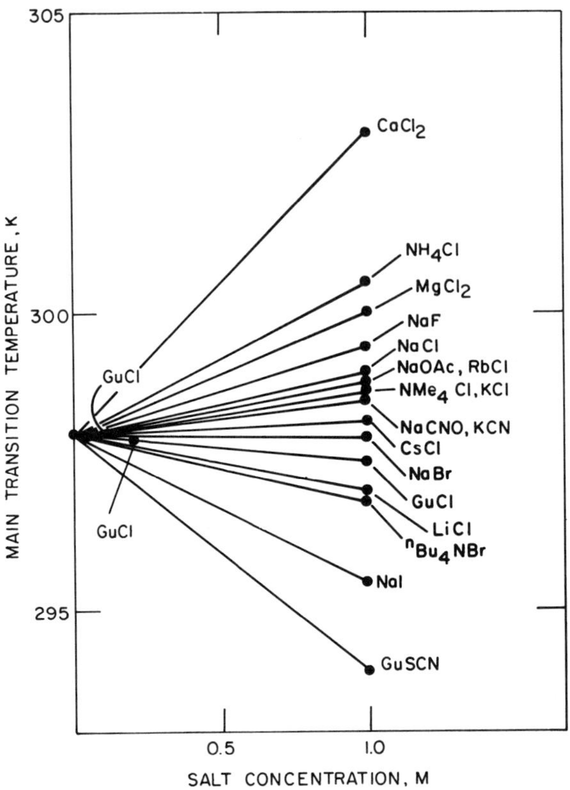

FIG. 10. The dependence of the main endotherm transition temperatures of DML on stoichiometric salt molarity for alkaline, alkaline earth, and guanidinium (Gu) salts; c = 0.5.

as with cholesterol the molecule is interdigitated among the lipid chains preventing chain crystallization from occurring.*

Chapman and Urbina [59] suggested that lipid chain crystallization can cause protein molecules to be squeezed out of that portion of the bilayer. Consistent with this idea is the evidence that lipid phase separation can cause protein aggregation [60]. When alveolar membranes are grown at 28°C the component protein molecules are randomly distributed, whereas chilling the membrane to 5°C causes aggregation.

*Recent studies of intrinsic protein interactions in lipid bilayers show that a small disordering effect can be produced upon the lipid chains (see review [71]).

6. Permeability Characteristics

Studies have been made of permeability characteristics for water and various other molecules above and below the lipid phase transition temperature, e.g., the kinetics of water permeability through lipid systems has been studied and related to lipid fluidity. Thus, there is a marked increase in water permeability as the lipid chains become more unsaturated. This is also the case with nonelectrolytes such as glycerol and erythritol [61]. The self-diffusion rate of $^{22}Na^+$ through lecithin bilayers also shows a marked increase at the transition temperature to the liquid crystalline form [62].

The effect of cholesterol on the lipid chains when the lipid is in the fluid condition is to decrease water permeability [63]. Black lipid membranes also show the same effect. These results are consistent with a reduction in the fluidity characteristics of the lipid chains. On the other hand, the presence of cholesterol is to enhance the rate of water permeability of liposomes derived from saturated liposomes [63].

7. Theoretical Treatments

A number of theoretical studies have now been made of the thermotropic phase transition of lipids and lipid-water systems. Whittington and Chapman [64] examined the motion of chains in the hexagonal form of long-chain molecules. In later studies they also examined [7] the way in which the end-to-end distance of chains (which are fixed at one end) varies as the distance between the chains is increased. They used two different potential functions and restricted the chains to a two-dimensional hexagonal lattice. This study emphasized the cooperative nature of the twisting and movement of the CH_2 groups of adjacent chains leading to the phase transition and also above it in the more fluid state. Rothman [65] and Nagle [66] have also emphasized this cooperative nature of chain movements.

Marcelja [42] has used a molecular field approximation and examined different statistical average over all conformations of a single chain in a field due to neighboring molecules. This method has been extended to describe the properties of lipid monolayers and bilayers. The theoretical pressure-area diagrams are in good agreement with the experimental data of Phillips and Chapman [38]. The order parameter for hydrocarbon chains as a function of temperature, lateral pressure, and position along the chain is also calculated and compared with the experimental. A comparison between calculated and experimental order parameters (using NMR methods) is shown in Fig. 10. This shows the order parameter at 28°C as a function of a position along the chain for different values of lateral pressure. Temperatures and latent heats of the thermal phase transition were also calculated and are of the correct magnitude.

C. Relevance to Biology

When we consider the relevance of the liquid crystalline behavior of phospholipids to membrane structure and function we have to remember that the interaction with protein has still to be explored. However, even with this limitation we can make a few general comments.

1. Model Membranes

The spontaneous formation by phospholipids in water, above their transition temperature, of globules containing bilayers of lipid separated by aqueous layers has led to their use as model membranes. The argument is that as real membranes contain bilayers of lipid then here is a natural system whose constitution can be chosen to mimic the lipid characteristic of a real membrane. Ions can then be trapped in the aqueous compartments and the permeability of these ions across the lipid bilayer studied. Admittedly, there is no protein associated with the system but this can be included in later studies. Using this model membrane system various effects which occur with real membranes can be mimicked, e.g., anesthetic and drug effects.

A second model membrane system also uses the liquid crystalline behavior of phospholipids. In this system a phospholipid above its transition temperature is painted with a brush across an orifice separating two aqueous compartments. It spontaneously thins down and forms a single bilayer. This model membrane can be used to study ion permeability, water flow, electrical resistance, and excitability characteristics. The model membrane is fairly stable, and some of this stability is probably associated with the flow characteristics of the phospholipid in its liquid crystalline phase.

2. Natural Membranes

Despite their apparent complexity, natural membranes have been successfully examined using the same techniques which have been applied to simple lipid-water systems.

3. Calorimetric Studies

We have seen that lipids in water may show a marked thermal transition associated with a phase change from gel to liquid crystal, but that the presence of cholesterol can modify or remove this transition.

Studies by Steim et al. [67] were made with a membrane which does not contain cholesterol. This is the membrane of Acholeplasma laidlawii. With this membrane, a strong endothermic transition is indeed observed. The transition temperature corresponds with the transition temperature of the extracted lipids dispersed in water. Comparison of the heat changes associated with the transitions observed with the membrane and with the lipid extract indicate that appreciable amounts of lipid contribute to this membrane transition.

Studies have also been made with membranes which contain cholesterol, e.g., myelin and red blood cell membranes. The studies with the myelin membrane isolated from white matter of ox brain show [68] that:

1. With wet myelin, thermal transitions are not detectable. In this case the cholesterol and other lipids appear to be organized in a single phase. The organization of cholesterol in the membrane appears to prevent the lipids from crystallizing.
2. To maintain the organization of the lipid in myelin a critical amount of water appears to be required. This water is unfreezable at 0°C and may correspond to "bound" water.
3. On drying the myelin, the cholesterol and other lipids crystallize and precipitate. Endothermic transitions associated with the cholesterol and the lipid can then be observed.

Studies of the red blood cell membrane also show no detectable thermal transitions associated with the lipid component. These results are understandable if there are regions of lipid bilayer present in the membrane with the cholesterol preventing lipid chain crystallization.

The growth medium of Escherichia coli is found to have marked effects on the thermal transition observed. When the fatty acid composition of the medium is altered the observed transition temperature shifts as in simple lipid-water systems; (Esfahani et al. [69]). For both systems, Arrhenius plots of growth rates are found to depend on the fatty acids in the growth medium (Fox and Tsukagoshi [70]).

The fluidity of membranes may be related to the transition temperatures of their components. Thus, membranes which contain phospholipids having little unsaturation will have less fluidity than those having phospholipids with much unsaturation. This control of fluidity may then be related to diffusional characteristics of molecules passing into and out of the cell.

In line with this idea it is interesting that poikilothermic organisms, i.e., those which alter their body temperature to correspond to their environment, apparently alter the unsaturation in their phospholipids [27]. The alteration may provide a more or less constant fluidity of the membrane structure. At high temperatures the chains are more saturated; at low temperatures, they become more unsaturated.

Some cell membranes, e.g., in some bacterial systems, contain lipids having saturated branched chains. In this case, the presence of the methyl groups reduces the dispersion forces and hence, the transition temperatures. It may be the case that in these biological structures it is more "convenient" to have branching rather than unsaturated double bonds in the membrane systems while at the same time retaining the appropriate fluidity of structure.

ACKNOWLEDGMENT

The author wishes to thank the Wellcome Trust for supporting this research.

REFERENCES

1. D. Chapman, in Biological Membranes, Academic, London and New York, 1968.
2. P.B. Hitchcock, R. Mason, K. Mark Thomas, and G.G. Shipley, Proc. Natl. Acad. Sci. USA, 71:3036, 1974.

3. P. Byrne and D. Chapman, Nature, (London), 202:987 (1964).
4. D. Chapman and D.T. Collin, Nature (London), 206:189 (1965).
5. D. Chapman, P. Byrne and G.G. Shipley, Proc. Roy. Soc., Ser. A, 290:115 (1966).
6. D. Chapman and N.J. Salsbury, Trans. Faraday Soc., 62:2607 (1966).
7. S.G. Whittington and D. Chapman, Trans. Faraday Soc., 62:3319 (1966).
8. D. Chapman, R.M. Williams and B.D. Ladbrooke, Chem. Phys. Lipids, 1:445 (1967).
9. A. Tardieu, V. Luzzati and F.C. Reman, J. Mol. Biol., 75:711-733 (1973).
10. Z. Veksli, N.J. Salsbury and D. Chapman, Biochim. Biophys. Acta, 183:434-446 (1969).
11. E. Oldfield, D. Chapman and W. Derbyshire, FEBS Lett., 16(2):102-104 (1971).
12. J. Seelig and A. Seelig, Biochem. Biophys. Res. Commun., 57:406-407 (1974).
13. J. Charvolin, P. Manneville and B. Deloche, Chem. Phys. Lett., 23:345-348 (1973).
14. M.D. Barratt, D.K. Green and D. Chapman, "ESR Studies of Nitroxide Probes in Lecithin-Water Systems," Chem. Phys. Lipids, 3:140-144 (1969).
15. W.L. Hubbell and H.M. McConnell, "Molecular Motion in Spin-Labelled Phospholipids and Membranes," J. Amer. Chem. Soc., 93:314-326 (1971).
16. N.J. Salsbury, A. Darke and D. Chapman, Chem. Phys. Lipids, 8:142-151 (1972).
17. J.M. Vanderkooi and B. Chance, FEBS Lett., 22:23-26 (1972).
18. K.R. Naqvi, J.P. Behr and D. Chapman, Chem. Phys. Lett., 26:440-444 (1974).
19. H. Trauble and D.H. Haynes, Chem. Phys. Lipids, 7:324-335 (1971).
20. A.J. Verkleij, P.H.J.T. Ververgaert, L.L.M. Van Deenen and P.F. Elbers, Biochim. Biophys. Acta, 288:326-332 (1972).
21. J.L. Lippert and W.L. Peticolas, Biochim. Biophys. Acta, 282:8-17 (1972).

22. P. Overath and H. Trauble, Biochemistry, 12:2625-2634 (1973).
23. R.D. Kornberg and H.M. McConnell, Proc. Natl. Acad. Sci., USA, 68:2564-2568 (1971).
24. E. Sackmann and H. Trauble, J. Amer. Chem. Soc., 94:4482-4497 (1972).
25. P. Devaux, C.J. Scandella and H.M. McConnell, J. Amer. Chem. Soc., 95:474-85 (1973).
26. V. Luzzati and F. Reiss-Husson, Nature (London), 210:1351 (1966).
27. T. Gulik-Krzywicki, E. Rivas and V. Luzzati, J. Mol. Biol., 27:8 (1967).
28. L. Powers and N.A. Clark, Proc. Natl. Acad. Sci. USA, 72:840 (1975).
29. V. Luzzati, In Biological Membranes (D. Chapman, ed.), Academic, New York, 1968, p. 71.
30. F.A. Vandenheuvel, J. Amer. Oil Chem. Soc., 40:455-471 (1963).
31. Y.K. Levine and M.H.F. Wilkins, Nature New Biol., 230:69-72 (1971).
32. M.C. Phillips, R.M. Williams and D. Chapman, Chem. Phys. Lipids, 3:234:244 (1969).
33. Y.K. Levine, Progr. Surface Sci., 3:279-352 (1973).
34. B.E. Warren, Phys. Rev., 44:969-973 (1933).
35. W.C. Pierce, J. Chem. Phys., 3:252-255 (1935).
36. W.H. Zachariasen, J. Chem. Phys., 3:158-161 (1935).
37. R.M. Williams and D. Chapman, in Progress in the Chemistry of Fats and Other Lipids (R.T. Holman, ed.), Pergamon, New York, 1970, pp. 1-79.
38. M.C. Phillips and D. Chapman, Biochim. Biophys. Acta, 163:301-313 (1968).
39. J. Seelig and A. Seelig, Biochem. Biophys. Res. Commun., 57:406-407 (1974).
40. J. Charvolin, P. Manneville and B. Deloche, Chem. Phys. Lett., 23:345-348 (1973).
41. D. Chapman, Ann. N.Y. Acad. Sci., 137:745-754 (1966).
42. S. Marcelja, Biochim. Biophys. Acta, 367:165-176 (1974).
43. B.D. Ladbrooke and D. Chapman, Chem. Phys. Lipids, 3:304-367 (1969).
44. M.C. Phillips, B.D. Ladbrooke and D. Chapman, Biochim. Biophys. Acta, 193:35-44 (1970).

45. E.J. Shimshick and H.M. McConnell, Biochemistry, 12:2351-2360 (1973).
46. B.D. Ladbrooke, R.M. Williams and D. Chapman, Biochim. Biophys. Acta, 150:333-340 (1968).
47. D. Chapman and S.A. Penkett, Nature (London), 211:1304-1305 (1966).
48. A. Darke, E.G. Finer, A.G. Flook and M.C. Phillips, J. Mol. Biol., 63:265-278 (1972).
49. M.C. Phillips and E.G. Finer, Biochim. Biophys. Acta, 356:199-206 (1974).
50. J.E. Rothman and D.M. Engelman, Nature New Biol., 237:42-44 (1972).
51. H.J. Hinz and J.M. Sturtevant, J. Biol. Chem., 247:3697-3701 (1972).
52. A.J. Verkleij, B. De Kruyff, P.H.J.Th. Ververgaert, J.F. Tocanne and L.L.M. Van Deenen, Biochim. Biophys. Acta, 339:432-437 (1974).
53. H. Trauble and H. Eibl, Proc. Natl. Acad. Sci. USA, 71:214-219 (1974).
54. M.H. Gottlieb and E.D. Eanes, Biophys. J., 12:1533-1543 (1972).
55. M. Ehrstrom, L.E.G. Eriksonn, J. Israelachvili and A. Ehrenberg, Biochem. Biophys. Res. Commun., 55:396-402 (1973).
56. D.V. Godin and T.W. Ng, Mol. Pharmacol., 9:802-819 (1974).
57. H. Trauble, in Biomembranes, Vol. 3 (F. Kreutzer and J.F.G. Slegers eds.), Plenum, New York, 1972, pp. 197-227.
58. D. Chapman, J. Urbina and K.M. Keough, J. Biol. Chem., 249:2512-2521 (1974).
59. D. Chapman and J. Urbina, FEBS Lett., 12(3), 169-172 (1971).
60. V. Speth and F. Wunderlich, Biochim. Biophys. Acta, 291:621-628 (1973).
61. J. De Gier, J.G. Mandersloot and L.L.M. Van Deenen, Biochim. Biophys. Acta, 150:666-675 (1968).
62. D. Papahadjopoulos, K. Jacobson, S. Nir, and T. Isac, Biochim. Biophys. Acta, 311:330-348 (1973).
63. R. Bittman and L. Blau, Biochemistry, 11:4831-4839 (1972).
64. S.G. Whittington and D. Chapman, Trans. Faraday Soc., 61:2656-2660 (1965).
65. J. Rothman, J. Theor. Biol., 38:1-16 (1973).
66. J.F. Nagle, J. Chem. Phys., 58:252-264 (1973).

67. J. Steim, M.E. Tourtellotte, J.C. Reinert, R.D. McElhaney, and R.L. Rader, Proc. Nat. Acad. Sci. USA, 63:104-109 (1969).

68. B.D. Ladbrooke, T.J. Jenkinson, V.B. Kamat, and D. Chapman, Biochim. Biophys. Acta, 164:101-109 (1968).

69. M. Esfahani, A.R. Limbrick, S. Knutton, T. Oka, and S.J. Wakil, Proc. Natl. Acad. Sci. USA, 68:3180-3184 (1971).

70. F.C. Fox and T. Tsukagoshi, in Membrane Research, Academic, New York, 1972, p. 145.

71. D. Chapman, J.C. Gomex-Fernandez, and F.M. Goni, FEBS Lett., 98:211-223 (1979).

Chapter 9

THERMODYNAMICS OF MESOPHASE TRANSITIONS

Edward M. Barrall II

IBM General Products Division
San Jose, California

I. INTRODUCTION

The melting of a crystalline substance is a familiar phase transition. The highly structured solid melts to the isotropic liquid phase at a well-defined temperature and with a characteristic heat of fusion.

In contrast, a considerable number of organic compounds, more than 3,000, do not melt directly from crystalline solid to isotropic liquid [1,2]. Instead, the substance passes through an intermediate phase, a mesophase.

In such a case, two phase transitions are involved: at a lower temperature, a transition from crystalline solid to mesophase, and at a higher temperature, a transition from mesophase to isotropic liquid.

The mesophase appears as a turbid liquid, and the transitions can be observed visually or by other common techniques. The term liquid crystal has been used interchangeably with mesophase. To a lesser extent, this state has been referred to as anisotropic or paracrystalline [1].

Mesophases are ordered on a molecular level, yet possess some of the mechanical properties of liquids. The ordered domains in mesophases contain typically about 10^5 molecules. However, the sense of arrangement of the molecules is not the same in all mesophases. Three basic types are recognized: nematic, smectic, and cholesteric mesophases. A single compound can exhibit more than one mesophase.

Figure 1 illustrates the general structural features of the three mesophases. The nematic type is the simplest. In pure compounds, e.g., p-azoxyanisole, the nematic mesophase is thought to consist of bundles of rodlike structures. The only structural restriction in nematic mesophases is that the molecules are parallel or nearly parallel within a bundle. In the smectic mesophase, molecules are arranged side by side in a series of stratified layers [2], e.g., diethyl p,p'-azoxydibenzoate. Molecules in the layers may be arranged in a regular or random side-by-side spacing. The long axes of the molecules are parallel to one another and essentially perpendicular to the base plane of the layer. The layer can be one or more molecules thick. In general, the smectic mesophase has greater order than the nematic. The third type is the cholesteric mesophase; it is formed principally by derivatives of cholesterol, e.g., cholesteryl benzoate, but not by cholesterol itself. The cholesteric mesophase resembles the smectic mesophase in that molecules are arranged in layers. Within each layer, however, the parallel arrangement of molecules is reminiscent of the nematic mesophase. The cholesteric layers are thin with the long axis of molecules parallel to the plane of the layers. A screw axis runs through the layers [2].

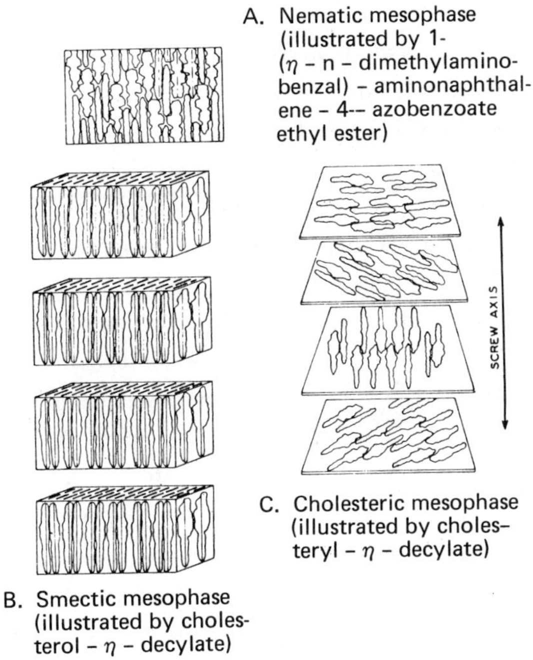

FIG. 1. Three types of thermotropic mesophases.

II. CALORIMETRY

A relatively new but widely applied focus in mesophase research is the calorimetric study of the phase transitions. From the molar heats of transition, q, the entropy change of transition is calculated from the familiar relationship $\Delta S = q/T$, where T is the absolute temperature. Only about a dozen mesophase-forming compounds had been subjected to such calorimetric study prior to 1960. Indeed, Brown and Shaw, in their exhaustive review of the literature prior to 1957, could find heat of fusion data on only four materials [1]. Thermodynamic data are now known on entire homologous series of mesophase-forming materials. Well over 600 materials have relatively complete temperature and entropic data available.

A knowledge of both the temperature and the heat of transition is necessary if the principals of physical analysis are to be applied to mesophase-forming systems. From this information the transition entropy may be calculated. This acts as the key for evaluating the type and degree of order present in many systems.

In the past it was customary to measure and record only the temperature of mesophase transitions. While this has been very profitable in a number of systems [3], temperature alone cannot furnish a sufficiently encompassing basis for prediction of mesophase behavior. Most attempts to predict mesophase transition temperatures, even in homologous series, using only a knowledge of the transition temperatures of other members, have not met with a high degree of success. A survey of the transition temperatures of any homologous series indicates important trends. However, "exceptions" appear. A list of transition entropies is a somewhat more regular indication of important thermal trends.

The recent proliferation of thermodynamic information is principally due to two causes: pressing need for the information and great improvements in thermal instrumentation. Liquid crystals are currently receiving attention in such varied areas as computer display and biomedical work [4]. Liquid crystalline materials when placed between conductors or in a strong magnetic field undergo dramatic changes in optical rotation and/or light scattering properties. It is this set of properties that have interested computer display developers and video display designers. The search for homologs with room-temperature liquid crystal ranges is a driving force in the large-scale study and synthesis of a broad range of materials. Every known mesophase type has been employed in some type of display. Progress has been so rapid that in only 3 years commercial calculator, wristwatch, and thermometric displays have appeared on the open market. The low voltage and minute current requirements make these displays compatible with the most sophisticated solid-state devices. It is projected that in 10 years liquid crystal displays will be competing closely with cathode ray and glow-type tubes for the indicator and television markets. Large industrial and governmental research departments are actively engaged in a wider range of liquid crystal research than would have been thought possible a few years ago. Not since the active days of Lehmann and Vorlander have so many mesophase materials been made. The question of thermodynamic properties naturally arises in these research efforts. Recent application of the methods of differential thermal analysis (DTA) and differential scanning calorimetry (DSC) has greatly facilitated the determination of temperatures, heats of transition, and heat capacity of various phases. The approach of classical calorimetry has not been neglected. At least one laboratory has made detailed studies of thermodynamic properties by conventional methods [5].

The use of recording differential techniques has greatly emphasized two items of importance. (1) The solid phase polymorphism of many mesophase forming materials is as complex as the mesophase behavior. Indeed, this statement may be made in general about organic compounds: solid phase polymorphism is very common. (2) Absolute purity and types of impurity play very important roles in the type, transition temperature, and range of mesophase formation. To some extent, this was realized by most early careful workers, but these methods did not drive the point home as effectively or dramatically as differential measurements.

The purpose of this paper is to outline the thermodynamic data which are presently available for homologous series of liquid crystal-forming materials. Materials will be listed only when calorimetric data are available. It is hoped that this outline will furnish a framework for the further study and systemization of the increasing body of information appearing monthly. It will become obvious to the reader that certain general statements relate the thermodynamic constants of all liquid crystal-forming materials. Without a doubt these general statements will be revised and grow more powerful as additional information from other sources is added in the future. A more complete listing of thermal properties is available as Vol. 2 of Ref. 3.

A. Techniques of Measurement

Arnold and coworkers have made a large number of measurements using classic adiabatic calorimetry [5,6]. However, the bulk of the thermodynamic data presently available has been obtained by dynamic calorimetry.

By the very nature of the instrumentation dynamic methods are much less accurate than adiabatic calorimetry. The expected accuracy is usually not much better than $\pm1\%$ and in some cases only $\pm10\%$. Nonetheless, the information is useful for general comparisons, so long as the limitations are recognized. Kreutzer and Kast [7] employed dual ice calorimetry which is intermediate in precision between adiabatic and scanning calorimetry.

The techniques of DTA have been described in detail elsewhere [8]. Briefly, a sample and reference material are heated at some linear rate. The absolute temperature and the differential temperature between sample and reference are recorded. The area beneath the differential curve is related to calories via calibration with a material of known heat of fusion. Since the area is due to a temperature difference, factors such as sample and instrument heat capacity are important. Temperatures, heats of transition, and heat capacity may be determined from the curves to about $\pm1\%$ given adequate calibration.

DSC also involves the comparison of the sample with an inert reference during a dynamic heating or cooling program. However, instead of a temperature difference being permitted, heat is added via the filaments, to keep the sample and reference in balance. To a first approximation, this method is a kind of scanning adiabatic calorimetry.

Obviously, both scanning methods are measuring the same phenomena. However, it is a serious logical error to confuse the two techniques. When adequately calibrated and carefully employed the two methods give comparable results. Indeed, under favorable circumstances DSC data compare closely to data from the highly precise calorimetry of Arnold (see Fig. 2) and calculated values. Serious errors may be expected for transitions which do not reach equilibrium quickly. Fortunately, most liquid crystal transitions are rapid.

Other methods for the evaluation of transition heat have been employed occasionally for liquid crystals. These methods will be mentioned in passing when particular compounds are discussed.

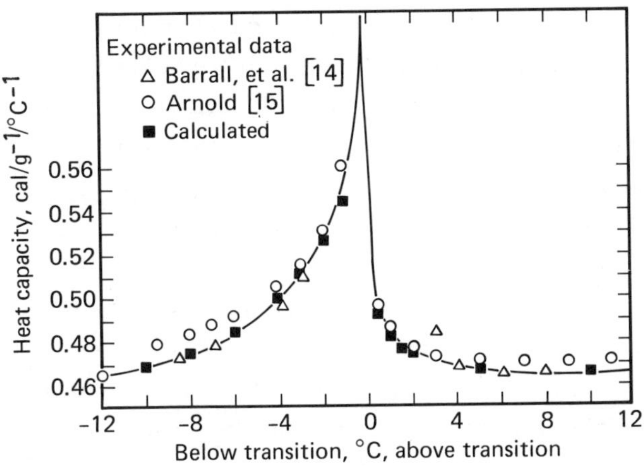

FIG. 2. A comparison of experimental (calorimetry and DSC) and calculated heat capacity data. Nematic-isotropic transition to p-azoxyanisole, ≈ 135°C, comparison of calculated and measured specific heats. (From Ref. 9.)

B. Thermodynamic Data

The data in this collection are gathered from various articles in the literature. The original workers reported the results in a variety of units, i.e., calories per gram, calories per mole, joules per mole, and degrees Kelvin and Celsius. For the sake of unity of presentation all the diverse units have been converted to calories per mole for heat of transition and calories per mole per degree Kelvin for entropy of transition. In some cases various authors have given several values for transition heat depending upon the point taken as the start of transition. The effect of purity is very large on the heat and temperature of transition [10,11].

When comparing transition heats measured by various methods or by the same method by different authors it is not uncommon to find variations as large as 10%. This does not imply less precision on the part of one worker over another or necessarily imply differences in sample purity. The usual source of the "error" is how much of the pretransition effect is included in the measured transition heat. Liquid crystal transitions are certainly not thermodynamically simple; McCullough [12] has identified at least six characteristic nonisothermal melting and solid-solid transitions on the basis of heat capacity-temperature curves of organic compounds. From the available heat capacity curves of liquid crystal-forming materials it is obvious that several of the above nonisothermal processes are represented in nematic and cholesteric systems [13,14].

However, it is one thing to recognize a problem and yet another to arrive at a satisfactory solution. For a nonisothermal process the "pretransition energy" must be included and properly belongs with the main transition energy, especially if entropy of transition is to be used to judge order changes in a system. Given two sets of measurements on samples of equivalent purity, the set involving the largest energy change is more likely to be correct. The statement about equivalent purity is the axis of the argument, for impurities acting as solvents can bring about nonisothermal melting in systems which should be isothermal. It is very useful to know not only the temperature of maximum transition rate (T_m) but also the temperature range considered in evaluating the heat of transition. For the present, this sort of information is unavailable in many cases. Hopefully, transition range will be included in more reports in the future. As for the present study, transition entropies will be reported directly as given in the literature with a minimum of speculation concerning possible pretransition effects.

A sampling of the various compound structures for which thermodynamic information is available is given in Table 1. In most cases at least 10 members of a homologous series have been made and studied. In series which exhibit properties of interest to liquid crystal displays or thermal indication as many as 25 members of the series have been studied. This list should be regarded as an <u>outline only</u> as the totals and types continue to change. The letters, A, B, C, etc., will be used where convenient in this text for reference.

A quick examination of Table 1 reveals that liquid crystal-forming materials are usually rod shaped and rigid (A to F) or have some unusual symmetry (G). The rigid rod-shaped materials may exhibit smectic and

TABLE 1. Some structures which yield liquid crystal families of known thermodynamic properties

A 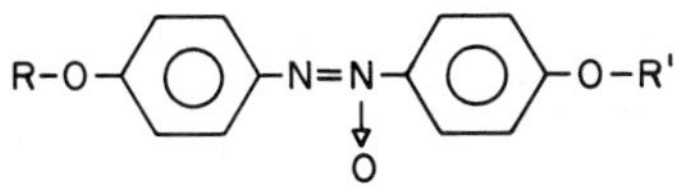

4,4'-Di-n-alkoxyazoxybenzenes

B 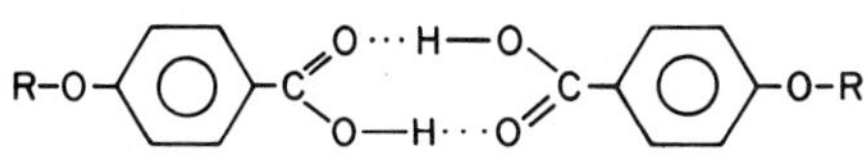

p-Alkoxybenzoic acids

C 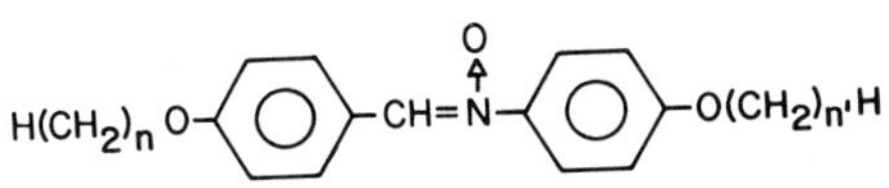

4,4'-Di-n-alkoxyphenylnitrones

D 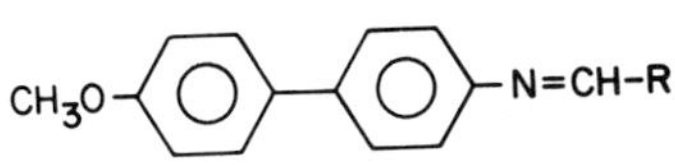

4-Benzylidene-amino-4'-methoxybiphenyls (R = aromatic or heterocyclic)

E 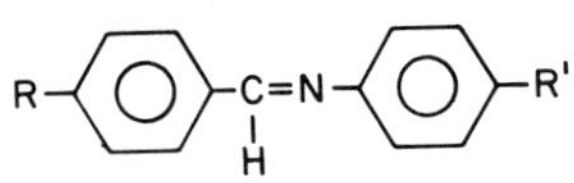

Schiff's bases

F 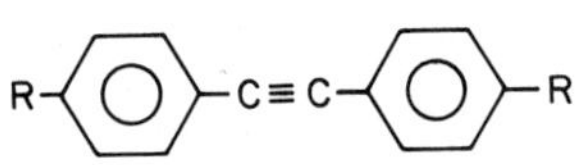

Diphenylacetylides

G 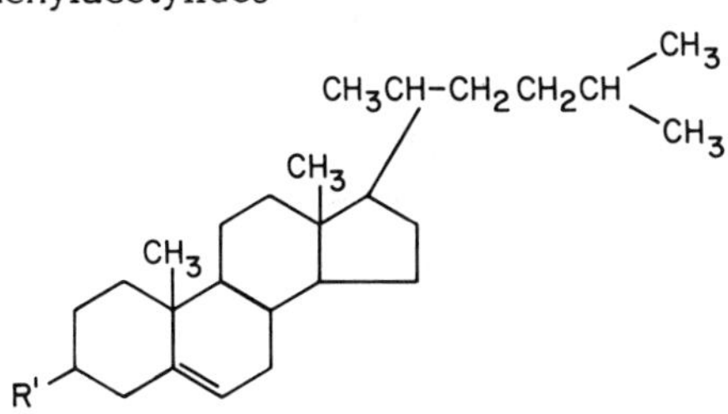

TABLE 1.continued.

	R'
Cholesteryl esters	$\mathrm{R{-}\overset{\overset{O}{\|}}{C}{-}O{-}}$
Carbonates	$\mathrm{R{-}O{-}\overset{\overset{O}{\|}}{C}{-}O{-}}$
S-Alkyl carbonates	$\mathrm{R{-}S{-}\overset{\overset{O}{\|}}{C}{-}O{-}}$
ω-Phenyl alkanoates	$\phi\mathrm{{-}(CH_2)_n{-}\overset{\overset{O}{\|}}{C}{-}O{-}}$
Thiocholesteryl-ω-phenylalkanoates	$\phi\mathrm{(CH_2)_n\ \overset{\overset{O}{\|}}{C}{-}S{-}}$
Halides	Cl-, Br-, I-

nematic mesophases in any combination, with the nematic form being the highest melting. A large number of smectic forms have been defined (see Fig. 3). Compounds belonging to the cholesteryl group (G) exhibit smectic and cholesteric mesophases. The nematic mesophase is excluded in pure materials since the cholesteric is a special case. To the present author's knowledge only a single smectic and single cholesteric transition is observable in G-type materials. That is, the transition sequence is never larger than solid, smectic, cholesteric, isotropic liquid. Two or three smectics are not observed as may be the case with any of the other materials.

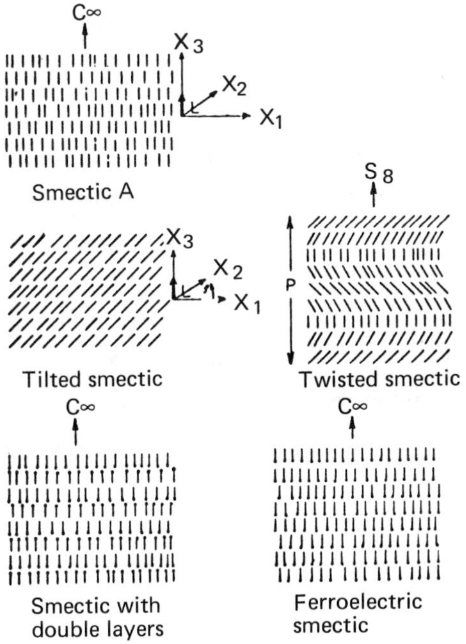

FIG. 3. Schematic representation of several smectic liquid crystal geometries. (From Ref. 15.)

C. Monotropism

Mesophases may also be monotropic, that is, form only on cooling the isotropic liquid. This is usually due to the extent of the crystalline order being great enough to preserve the true solid above the range of mesophase thermal stability [16,17]. When impurities are added to many monotropic systems, the melting of the solid is depressed sufficiently to make the mesophases nonmonotropic. Cholesteryl palmitate is a good case in point. In ultrapure material the transition path is as follows [18].

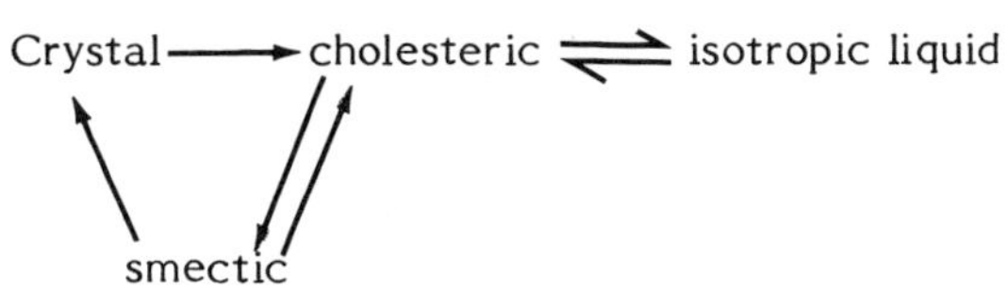

However, in the presence of a trace of antioxidant the melting of the crystal is depressed giving [19]:

Crystal $\rightleftharpoons$ smectic $\rightleftharpoons$ cholesteric $\rightleftharpoons$ isotropic liquid

Any or all the mesophases may be monotropic. Solid phase polymorphism may also determine monotropism, and the phase from which crystallization occurs determines the nature of the solid phase formed. Cholesteryl formate and propionate are instructive of this phenomenon.

Using a formate ester sample with a DSC purity of 99.86% (98.67% using the method of Davis et al [20]), the following transitions were noted using a polarizing microscope and hot stage [21]:

1. On heating, the solid conversion to the isotropic liquid was noted at 97.0 °C.
2. The stage temperature was lowered to 95.7°C, and a prismatic solid formed.
3. Rapid cooling of the melt on the slide produced a cholesteric mesophase.
4. By trial and error, a stage temperature, 57.2°C, was found at which the mesophase would neither melt to the isotropic liquid nor the above solid form rapidly.
5. After several changes of the "mosslike" cholesteric texture, the mesophase became isotropic at 59.6°C.
6. If the mesophase is cooled to below 50°C a new, spherulitic mesophase forms.

The above observations suggest the following path of phase transition:

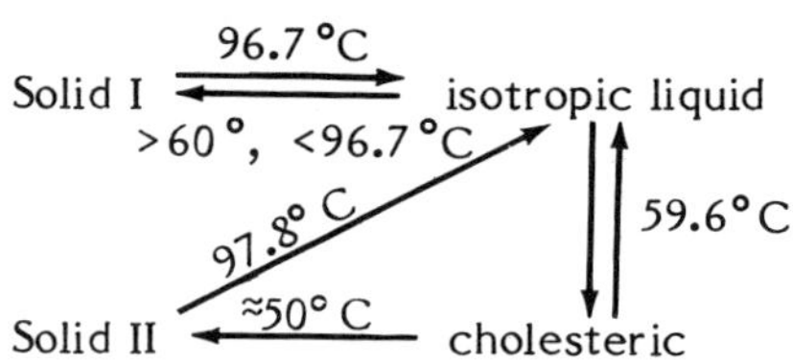

Note: Solid I melting followed by slow cooling indicates no mesophase. The enthalpy difference between solid I and solid II is at present not known. The present authors suggest that it is less than 1 kcal/mol.

Three sets of calorimetric data are available. Agreement between three independent studies of the solid → isotropic liquid transition is excellent, 2% in terms of ΔS. The difference between the two measurements of the very small mesophase transition is ∼9% in terms of ΔS.

The propionate ester is the first of the aliphatic ester series that all workers agree forms a mesophase, and that on heating as well as cooling. However, precise scanning calorimetry has demonstrated the existence of two solid phases with different melting points [22]. Microscopy has verified two solid forms: solid I, spherulitic and solid II, needlelike. The following recrystallization path was suggested [22]:

Solid I ⇌ cholesteric ⇌ isotropic liquid
≈ 88 °C ⇅ 98.0 °C
Solid II

No direct conversion from solid I to solid II without the mesophase was observed. Five sets of thermodynamic data and two sets of transition temperatures are known [23-27]. The variation in transition temperature is probably due to differing amounts of solids I and II in individual samples.

III. MATERIALS EXHIBITING THE CHOLESTERIC MESOPHASE

All compounds with structure G exhibit a cholesteric mesophase (hence, the name). This characteristic twisted or glide-plane structure has also been found in *d* or *l* nematic mesophase forms containing an asymmetric carbon atom in the side chain. The racemic *d-l* mixtures are normal nematics. In particular the compound which belongs to type E or Schiff's bases has been

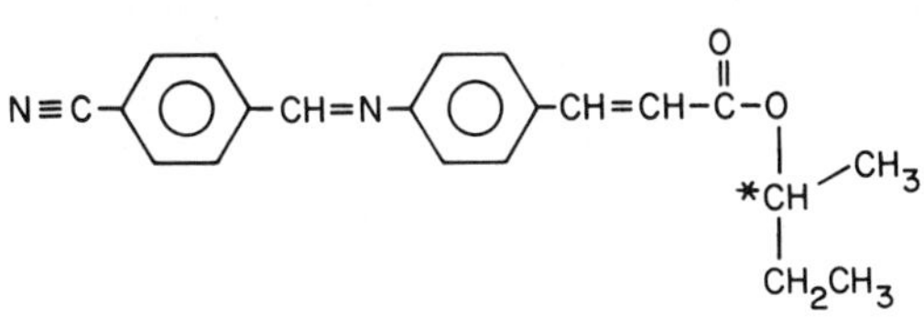

studied in both the *d* and *d-l* mixture form [28,29]. The *d* form shows a good cholesteric at 99.5 °C (monotropic) with an entropy of 0.27 cal mol^{-1}

K^{-1} [29]. The solid to isotropic liquid transition occurs at 99.6°C with a ΔS of 17.5 cal mol^{-1} K^{-1}. The d-l form is not monotropic with the solid-to-nematic transition at 100 °C and ΔS = 17.3 cal mol^{-1} K^{-1}. The nematic-to-isotropic liquid transition occurs at 109 ° C with a ΔS of 0.24 cal mol^{-1} K^{-1}. The differences in the mesophases are probably real.

A. Cholesterol Mesogens

More conventional, type G, cholesteric mesogens have been studied in great detail. For example, the n-alkyl acid esters have been evaluated up to 21 carbon numbers, and for most of the series by at least four workers [23-27]. The transition entropy for the solid-to-mesophase (or isotropic liquid in monotropic cases) process is shown as a function of acid chain carbon number in Fig. 4. The scatter below C_5 is probably due to polymorphism in the solid phase. The sharp discontinuity between C_8 and C_9 has been reported by most workers, and from x-ray diffraction studies appears to be due to a fundamental change in molecular packing [30]. Above C_{10} there is some evidence for even-odd alternation in entropy.

The same kind of plot for the mesophase transitions is given in Fig. 5. From C_1 to C_8 the cholesteric → isotropic liquid entropy increases in a relatively smooth manner. A sharp break in this trend occurs at C_9 when a lower but more linear trend is established up to C_{20}. Also, at C_9 the smectic mesophase appears. The entropy of the smectic to cholesteric transition increases even more rapidly with additional carbon numbers. This would be expected from the closer association of molecules in the smectic mesophase.

A detailed study using nonalkane acid esters has shown that the transition entropies of cholesteryl esters is sensitive only to the length of the ester group [31]. When the length of the ester in angstroms is plotted with entropy, such diverse materials as the benzoate, adipate, and tetrafluorobutyrate esters fall on the same smooth relationship [31]. Indeed, the ester functionality is not necessary to preserve. Marker's acid

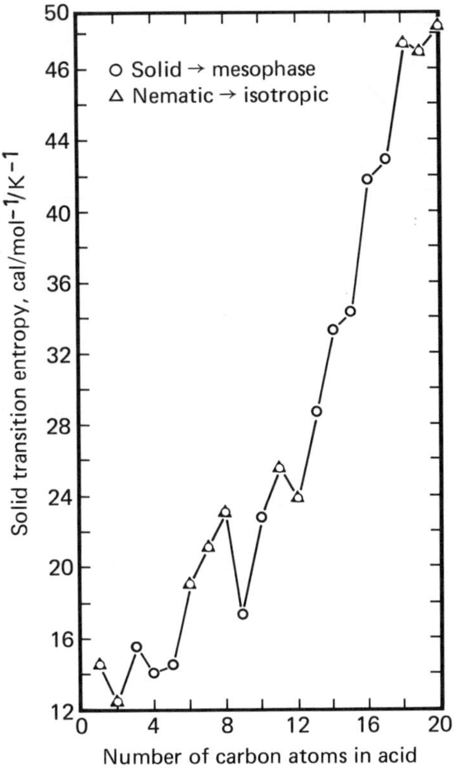

FIG. 4. Effect of acid carbon number on solid → mesophase or solid → isotropic liquid transition entropy of cholesteryl esters.

(the positional analog of cholesteryl formate) has almost identical solid and mesophase (cholesteric) transition entropies although hydrogen bonding in the acid translates the transition temperatures upward by more than 150°C [32].

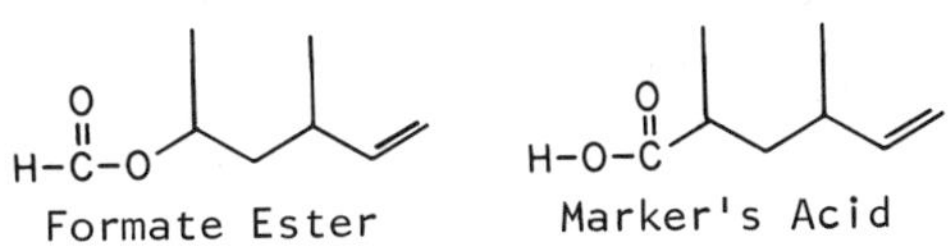

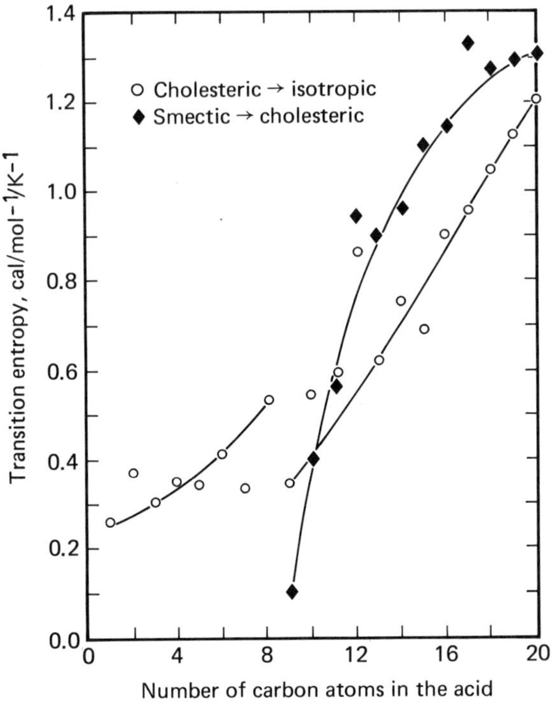

FIG. 5. Effect of acid carbon number on mesophase transition entropy of cholesteryl esters.

Indeed, for the cholesterol mesogens, geometry appears to be the only important term in defining the entropy change involved in any transition. Ennulat et al. have investigated the thio esters, carbonates, and thio carbonates of cholesterol [23,33-35]. Allowing for some small differences in chain length brought about by the thio-substitution, the analogs are almost identical to their oxygen counterparts. The thio-esters do exhibit the smectic mesophase at C_8.

B. Other Mesogens

These materials obviously form the bulk of the compounds investigated. They are characterized by a rodlike shape and more or less strong dipolar interactions. Although in Table 1 the central bridge between rings has been stressed, the end group substitution has profound influence on the temperature and heat of the various transitions.

Two extreme cases have been chosen as examples. The azoxy benzenes and azoxy cinnamates belong to group A of Table 1, but the cinnamate has much more polar end groups. The thermodynamic picture is correspondingly vastly different. Arnold has carried out precision calorimetry on the methyl through dodecyl values of the side chain [6] for the symmetrical dialkoxyazoxybenzenes as well as octadecyl. His values for transition heats are shown in Fig. 6. In addition, the first three members of the series, p-azoxyanisole (see Table 2), p-azoxypenetole, and p,p'-dihexyloxyazoxybenzene have been studied by a number of other workers.

Arnold's calorimetric data present a remarkably scattered picture for the transition from the solid to the mesophase (Fig. 6). No regular order is apparent. This is somewhat surprising in consideration of the regular order which is evident for the nematic → isotropic, smectic → nematic, and smectic → isotropic transitions in the same study. A possible explanation for such erratic entropy of transition values is offered in Table 2. Chow and Martire confirmed the existence of two separate and distinct solid phases of p-azoxyanisole ∽ 99.7 mol%, by scanning calorimetry [40]. Although solid II, the high-temperature form, is metastable at room temperature, it can exist in various mixtures with solid I. The formation of two or more solid phases is very common in mesophase-forming materials. In considering the entropy of the transition solid → mesophase it is necessary to use the <u>total</u> entropy, i.e., the summed entropy of transition for all crystal forms which are stable between 0 K and the transition temperature. Bondi has adequately demonstrated this for other organic transitions [42]. Therefore, the present authors suggest that most of the materials investigated by Arnold were complicated by solid phase polymorphism. The relationship between transition entropy and alkyl chain length should be regular if all crystal forms were accounted for. The collection of data in Table 2 for the solid-to-nematic mesophase transition of p-azoxyanisole is scattered for probably the same reason. It is interesting to note that the values of nematic → isotropic liquid transition as measured by

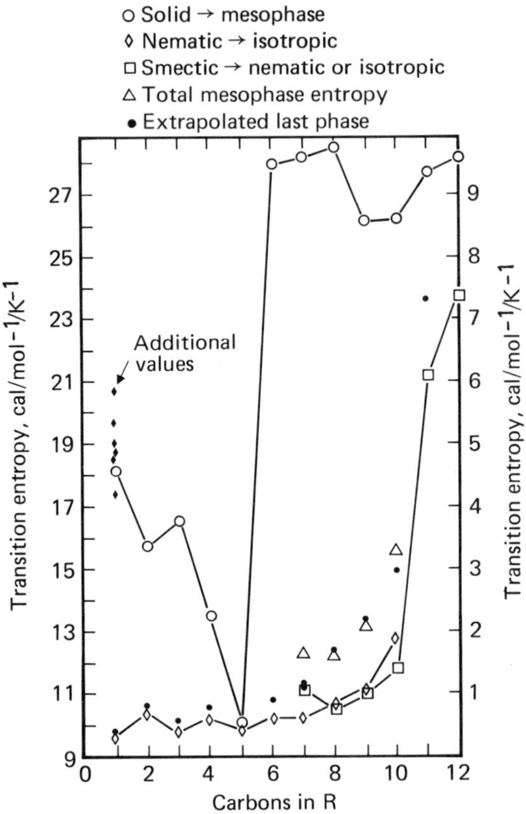

FIG. 6. Transition entropies for some p,p-n-dialkoxy-azoxybenzenes.

eight workers are in very close agreement. By elimination of the extremes the value for this transition is between 0.403 and 0.453 cal mol^{-1} K^{-1}. Thus, errors due to the small entropy of transition notwithstanding, the nematic isotropic transition is better defined than transition involving the solid phase, i.e., polymorphism does not complicate the nematic mesophase. In the calorimetric study, Fig. 6, the nematic isotropic transition shows even-odd alternation up to the C_8 side chain.

Chow and Martire measured the thermodynamic properties of the C_6 side-chain material, 99.6 mol% [40]. The entropy of the solid → nematic transition was found to be 29.4 cal mol^{-1} K^{-1} (heat of transition 10.4

TABLE 2. A Comparison of the Thermodynamic Data Available on the Transitions of p,p'-Azoxyanisole

Reference	Transition type	Transition Temperature (°C)	Transition Temperature (K)	Transition heat (cal/mol)	Transition entropy (cal mol^{-1} K^{-1})
36	Solid → nematic	117.6	390.6	7,440	19.0
	Nematic → isotropic	133.9	406.9	176	0.432
37	Solid → nematic	118	391	7,700	19.7
	Nematic → isotropic	132	405	176	0.434
38	Solid → nematic	117	390	7,280	18.7
	Nematic → isotropic	128 onset	401	178	0.444
	Nematic → isotropic	132 sharp break	405	178	0.440
37	Nematic → isotropic	132	405	176	0.434
37	Nematic → isotropic	132	405	183	0.453
7, 39	Nematic → isotropic	132	404	462	1.14
40	Solid → nematic	117.5	390.5	7,260	18.6
	Nematic → isotropic	134.2	407.2	181	0.444
	Solid II → nematic	104.4	377.4	5,630	14.9
	Nematic → solid II	76	349	7,230	20.7
	Solid II → solid I	61	334	1,030	3.09
6	Solid → nematic	118.2	391.2	7,067	18.07
	Nematic → isotropic	135.3	408.3	137.2	0.3360
	Nematic → isotropic	135.3 extrapolated	408.3	164.9	0.4039
41	Solid → nematic	117.5	390.5	6,800	17.4
	Nematic → isotropic	134.0	407.0	150	0.37

kcal/mol). This compares to Arnold's value of 27.92 cal mol^{-1} K^{-1}. The nematic → isotropic liquid transition was found by the same authors to have an entropy of 0.895 cal mol^{-1} K^{-1} (heat of transition 0.359 kcal/mol). This compares closely with Arnold's extrapolated value of 0.898 cal mol^{-1} K^{-1}. In addition, Chow and Martire noted the presence of a smectic mesophase very close to the solid transition [40]. They were unable to determine the heat of transition due to poor resolution on the DSC trace.

In general the results of DTA and DSC on the mesophase → isotropic liquid transition compare more closely with Arnold's "extrapolated" values of transition heat than with the quantities derived by more straightforward measurements. The extrapolated values included pretransitional heat capacity changes. The method of baseline construction in DTA and DSC accounts for the pretransition inclusion. Martin and Muller have discussed this phenomenon [38].

In the homologous series of p,p'-dialkoxyazoxybenzenes the alkyl side chains do not appear to contribute significantly to the order of the nematic system up to C_6. Below C_6 the alkyl side chains are probably not laterally associated and may determine which of two possible lateral arrangements is adopted by neighboring azoxybenzene nuclei in order to account for the odd-even effect. At C_6 the configurational geometry of the side chains becomes such that they are closely associated and contribute to the order of the system. Undoubtedly, the appearance at either C_6 or C_7 of a smectic mesophase is due to interalkyl chain order. Beyond C_{10} the interchain order is so elaborated that the stability range of the nematic mesophase is exceeded by the time the system has sufficient energy to separate the alkyl chains. Thus, the nematic mesophase vanishes from the phase diagram.

Between C_{12} and C_{18} a new type of order appears, and at C_{18} two smectic mesophases appear, the smectic B and smectic C of Saupe [43]. The C_{18} compound is one of the few materials giving two smectic mesophases for which calorimetric information is available. It may be characteristic of the smectic B → smectic C transition that it is smaller in entropy change than the smectic C → isotropic liquid transition. Logically, this is difficult to visualize. If the same chain segments are involved in both transitions, then the transition entropies should be approximately equal, not a factor of two different, 6.79 versus 13.9 cal mol^{-1} K^{-1}. It is possible to speculate that only one-half the alkyl chains are involved in the smectic B → smectic C transition. This sort of speculation is predicted on the fact that the smectic B modification is not a metastable form.

Arnold et al. presented a heat capacity versus temperature curve for the dioctadecyl material [44]. The smectic B form shows little pretransition heat capacity anomaly whereas the smectic C modification shows a relatively large pretransition effect (as does the solid). The heat capacity in the mesophase range is unusually high; well above that of both the isotropic liquid and the solid phases. This is analogous to the heat capacity phenomena noted for the smectic and cholesteric mesophase of cholesteryl myristate [13].

Arnold et al. [45] have prepared and thermally characterized a series of compounds with the general structure:

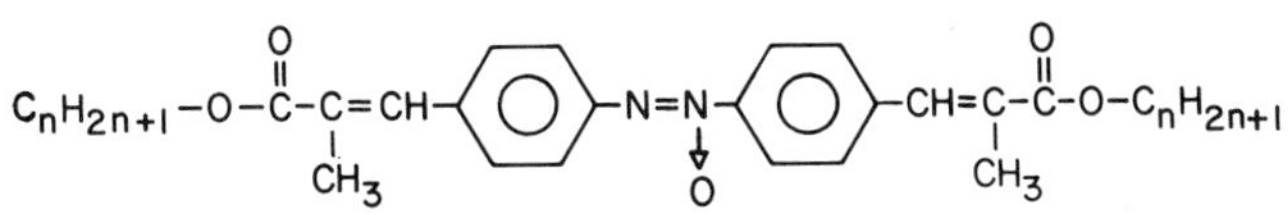

with n = 2, 3, 4, 5, 6, 7, 8, 9, and 11. These compounds represent a symmetrical form of the Schiff's base p-aminocinnamic acid esters described by Leclercq et al. [28] with a much increased dipole due to the nitroso bridge. This should be comparable to the previously discussed 4,4'-di-n-alkoxyazoxybenzenes [6]. The entropic data are shown in Figs. 7 and 8.

The ethyl and propyl diesters have two mesophases: smectic A and nematic. The nematic mesophase is not exhibited by the next four members of the series. Only the smectic A mesophase appears. From the octyl to dodecyl diester two mesophases appear: smectic A and smectic C. Complete heat capacity curves of the mesophase range of the didecyl and didodecyl esters have been presented [46].

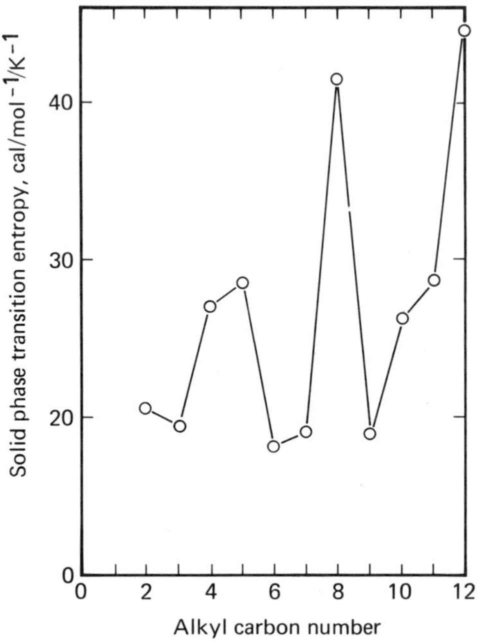

FIG. 7. Effect of alkyl chain length on the solid-to-smectic mesophase transition entropy for a series of 4-4'-azoxy-α-methyl cinnamic acid di-n-alkyl esters.

Figure 7 shows the solid-to-mesophase transition entropy as a function of ester carbon number. Although the scatter is bad, there is a general upward trend with increasing ester chain length. As with the p,p'-n-dialkoxy azoxybenzenes this scatter is probably due to solid phase polymorphy.

Figure 8 indicates that the mesophase entropies are somewhat more regular. Contrary to other smectic series which have been studied, there is little evidence of even-odd carbon number alternation. The series increases along a smooth curve from diethyl to didodecyl. The entropies of nematic, smectic A, and smectic B are shown individually and totaled in Fig. 8. The totaled form would be expected to be more significant when all mesophases are not monotropic. The entropy increase for this series is less rapid with increasing chain length than with the dialkoxy azoxybenzenes. However, it is more regular. The additional chain length in the cinnamic

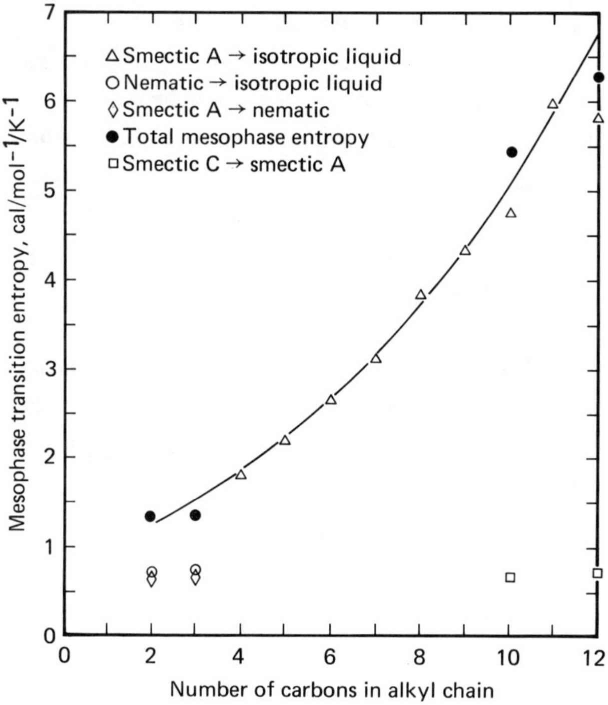

FIG. 8. Effect of the alkyl chain length on mesophase transition entropies for a series of 4,4'-azoxy-α-methyl cinnamic acid di-n-alkyl esters.

acid link probably does not contribute to the smectic order. If a contribution were made the entropy of the didodecyl compounds of azoxy-α-methylcinnamic acid and azoxybenzenes should be comparable, with the cinnamate the larger. This is not the case.

IV. CONCLUSIONS

From this brief summary of known mesophase thermodynamics it is apparent that much can be learned from this complex picture. In theory, transition entropy and heat capacity define molecular order and can be used to describe in both words and mathematics exactly how molecules behave in such diverse states as solid and liquid. X-ray studies have given us the outline of matter just as a city map give us the frame of a city. The frame is just that, not filled in with the patterns of flow and movement for which the city really exists. The same is true of a crystallographic approach to matter; movement of the units is neglected because of the static nature of the method. Thermodynamics gives a picture (albeit complex) of the movement of molecules and the effects of size, shape, and dipole interaction. Liquid crystal-forming materials have probably the most completely explored thermal picture (with the possible exception of the n-alkanes) of any group of chemicals. The body of data available on homologous series is very large and should be helpful in disentangling dipolar from other effects. The cholesteric series is almost a pure function of geometry as contrasted to nematic mesogens. Liquid crystals constitute an important interface, and it has been at interfaces at which the largest advances of our knowledge of the universe have been made.

RECENT DEVELOPMENTS

Since the final preparation of this chapter in 1975 the field of liquid crystal research has moved steadily onward. The earlier work remains true but important additions have been made and should be mentioned for the sake of completeness.

New Liquid Crystals

Three fundamentally new molecular types have been added to the roster of potential mesogens. These are:

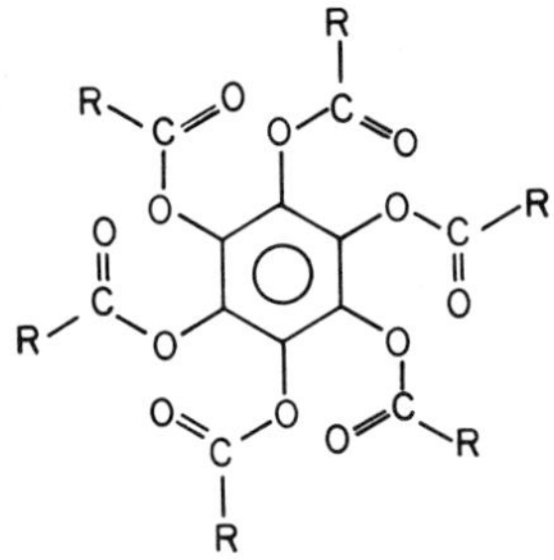

Disk Molecules

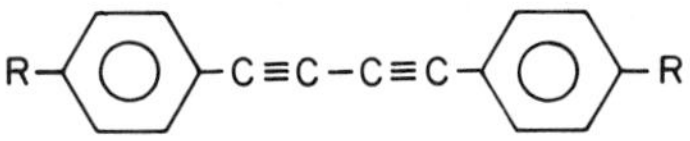

Diacetylenes

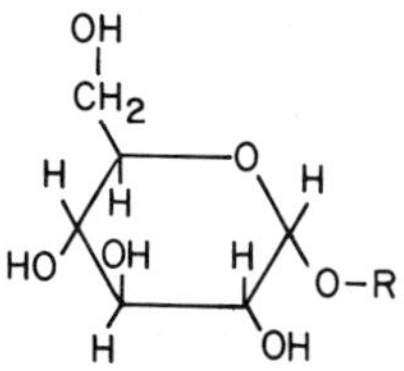

Glucopyranosides

Disklike molecules have been synthesized and the thermal properties investigated by Chandrasekhar et al. from n-butyl to n-nonyl [47]. The unique disk configuration with no internal "stiff" segment has been demonstrated by x-ray diffraction to form a very stable smectic phase. The molecules stack like coins with the stacks arranged to form a hexagonal close-pack array. The separation between parallel "coins" is random, about 6 Å, but the interstack separation is quite regular. It is possible that this smectic phase may be new and in Saupe's system would be classified SmI. Very similar disklike mesogens have been identified by Brooks and Taylor [48] and Zimmer and White [49] in petroleum and coal tars.

The diacetylene materials have been known for some time [50], but their mesogenic nature was not recognized until the di-n-alkyl and di-n-alkoxy homologs were synthesized [51]. These materials are potentially important in the mesophase polymerization of the diacetylene group. The series has been explored thermally up to n-pentadecyl [51].

The glucopyranosides were recognized some time ago as forming both lyotropic and thermotropic mesophases [52]. Recently, the thermal properties have been explored from n-heptyl to n-decyl. The mesophase was at first incorrectly identified as a cholesteric from microscopy. Later, x-ray analysis revealed a chiral smectic form [53]. The mesophase is unusually stable. Once transformed into the chiral smectic the solid phase does not form upon cooling until recrystallized from water or other solvent [53].

Developments in Smectic Mesogens and Mixtures

Display technology has produced very satisfactory nematic (twisted) displays suitable for calculators and digital watches. This development has been based on the work reported at the beginning of this chapter. Recently, laser-addressed smectic displays have received great attention as possible large-scale projection displays [54]. This has stimulated research into the smectic mesophase.

With the great improvement in our understanding of mesophase phase diagrams, it has been possible to construct multicomponent smectic-nematic mixtures with temperature ranges tailored to order [55-57]. For many mesophase systems, the cyanobiphenyls for example, the Schroeder-van Laar equation accurately predicts the solid-to-mesophase transition temperature to within a degree for multicomponent mixtures [56]. All that is required is simultaneous solution of as many of the following equations as there are components:

$$\ln X_A = \frac{\Delta H_A}{R}\left(\frac{1}{T_A} - \frac{1}{T}\right)$$

where X_A = mole fraction of component A

ΔH_A = molar heat of fusion of component A

T_A = crystal → mesophase temperature (K) of pure A

T = Eutectic point of the mixture (K).

Obviously, accurate values for ΔH_A and T_A are required.

In many cases the smectic → nematic and nematic → isotropic liquid transition temperature can be calculated from simple mixing equations. Cox has shown that the more complex behavior noted for some liquid crystal mixtures [58] may be predicted from equations originally derived for metal mixtures.

REFERENCES

1. G.H. Brown and W.G. Shaw, Chem. Rev., 57:1049 (1957).
2. J.L. Fergason, Sci. Am., 211:76 (1964).
3. G.W. Gray, Molecular Structure and Properties of Liquid Crystals, Academic, New York, 1962, pp. 131-133.

4. J.L. Fergason, T.R. Taylor, and T.B. Harsh, Electr. Technol., 85:41 (1970).

5. H. Arnold, Z. Phys. Chem. (Leipzig), 225:45 (1964).

6. H. Arnold, Z. Phys. Chem. (Leipzig), 226:146 (1964).

7. K. Kreutzer and W. Kast, Naturwissenschaften, 25:233 (1937).

8. E.M. Barrall and J.F. Johnson, in Techniques and Methods of Polymer Evaluation, Vol. I (P.E. Slade and L.T. Jenkins, eds.), Dekker, New York, 1966, pp. 1-40.

9. A. Torgalkar, R.S. Porter, E.M. Barrall, and J.F. Johnson, J. Chem. Phys., 48:3897 (1968).

10. R.D. Ennulat, in Analytical Calorimetry, Vol. I (R.S. Porter and J.F. Johnson, eds.), Plenum, New York, 1968, p. 219.

11. E.M. Barrall and M.H. Vogel, Thermochem. Acta, 1:127 (1970).

12. J.P. McCullough, Pure Appl. Chem., 2:221 (1961).

13. E.M. Barrall, R.S. Porter, and J.F. Johnson, J. Phys. Chem., 71:895 (1967).

14. H. Arnold and P. Roediger, Z. Phys. Chem. (Leipzig), 239:283 (1968).

15. G.H. Brown, J.W. Doane, and V.D. Neff, Crit. Rev. Solid State Sci., 303: (1970).

16. W.R. Young, I. Haller, and A. Aviram, Program Abstract, Third International Liquid Crystal Conference, August 24-28, 1970, Berlin, p. 136.

17. J.S. Dave and M.J.S. Dewar, J. Chem. Soc., 4305 (1955).

18. E.M. Barrall, R.S. Porter, and J.F. Johnson, J. Phys. Chem., 71:1224 (1967).

19. M. Vogel, E.M. Barrall, and C.P. Mignosa, Mol. Cryst. Liq. Cryst., 15:49 (1971).

20. G.J. Davis, R.S. Porter, and E.M. Barrall, Mol. Cryst. Liq. Cryst., 10:1 (1970).

21. E.M. Barrall and R.S. Porter, unpublished results.

22. M.J. Vogel, E.M. Barrall, and C.P. Mignosa, in Second Symposium on Ordered Fluids and Liquid Crystals (J.F. Johnson and R.S. Porter, eds.), Plenum, New York, 1970, p. 333.

23. R.D. Ennulat, Mol. Cryst. Liq. Cryst., 8:247 (1969).

24. E.M. Barrall, R.S. Porter, and J.F. Johnson, J. Phys. Chem., 71:1224 (1967).
25. G.J. Davis and R.S. Porter, Nat. Bur. Stand. (U.S.), Spec. Publ. 1973, p. 99.
26. P.J. Sell and A.W. Newman, Z. Phys. Chem. Neue Folge, 65:13 (1969).
27. G.W. Gray, J. Chem. Soc. (London), 3733 (1956).
28. M. Leclercq, J. Billard, and J. Jacques, Mol. Cryst. Liq. Cryst., 10:429 (1970).
29. E.M. Barrall, R.S. Porter, and J.F. Johnson, Mol. Cryst., 3:299 (1968).
30. J.H. Wendorff and F.P. Price, Mol. Cryst. Liq. Cryst., (1973), p. 129.
31. E.M. Barrall, J.F. Johnson, and R.S. Porter, Mol. Cryst. Liq. Cryst., 8:27 (1969).
32. W.R. Young, E.M. Barrall, and A. Aviram, in Analytical Calorimetry, Vol. 2 (J.F. Johnson and R.S. Porter, eds.), Plenum, New York, 1970, p. 113.
33. W. Elser, Mol. Cryst. Liq. Cryst., 8:219 (1969).
34. W. Elser, Mol. Cryst. Liq. Cryst., 2:1 (1967).
35. W. Elser and R.D. Ennulat, J. Phys. Chem., 74:1545 (1970).
36. E.M. Barrall, R.S. Porter, and J.F. Johnson, J. Phys. Chem., 68:2801 (1964).
37. R. Schenk, Kristalline Flüssigkeiten und Flüssigekristalle, W. Engelmann, Verlag, Leipzig, 1905, pp. 84-89.
38. H. Martin and F.H. Müller, Kolloid-Z. Z. Polym., 187:107 (1963).
39. K. Kreutzer, Ann. Phys., 33:192 (1938).
40. L.C. Chow and D.E. Martire, J. Phys. Chem. 73:1127 (1969).
41. M. Leclercq, J. Billard, and J. Jacques, C.R. Acad. Sci., Paris, Ser. C, 264:1789 (1967).
42. A. Bondi, Chem. Rev., 67:565 (1967).
43. A. Saupe, Mol. Cryst., 7:59 (1969).
44. H. Arnold, E.G. El-Jazairi, and H. Konig, Z. Phys. Chem., 234:401 (1967).
45. H. Arnold, D. Demus, H.J. Koch, A. Nelles, and H. Sackmann, Z. Phys. Chem. (Leipzig), 240:185 (1969).
46. H. Arnold, J. Jacobs, and O. Sonntag, Z. Phys. Chem. (Leipzig), 240:177 (1969).
47. S. Chandrasekhar, B.K. Sadashiva, and K.A. Suresh, Pramana, 9:471 (1977).

48. J.D. Brooks and G.H. Taylor, Çarbon, 3:185 (1965).

49. J.E. Zimmer and J.L. White, Mol. Cryst. Liq. Cryst., 38:177 (1977).

50. G. Wegner, Makromol. Chem., 154:35 (1972).

51. E.M. Barrall, B. Grant, and A.R. Gregges, in Liquid Crystals and Ordered Fluids, Vol. 3 (R.S. Porter and J.F. Johnson, eds.), Plenum, New York, 1978, pp. 19-39.

52. C.R. Noller and W.C. Rockwell, J. Amer. Chem. Soc., 60:2076 (1938).

53. E.M. Barrall, B. Grant, E. Samulsky, P.C. Moews, J.R. Knox, R. Gaskill, and J. Haberfeld, "The 1-N-Alkyl-α-d-Glucopyranosides, a New Series of Thermotropic Liquid Crystals," Mol. Cryst. Liq. Cryst., in press.

54. F. Kahn, J. Appl. Phys. Lett., 22:111 (1973).

55. G.W. Gray, K.J. Harrison, and J.A. Nash, Pramana Suppl., 1:381 (1975).

56. D.S. Hulme, E.P. Raynes, and K.J. Harrison, J. Chem. Soc. Chem. Commun., 98 (1974).

57. E.C. H. Hsu, J.L. Haberfeld, J.F. Johnson, and E.M. Barrall, Mol. Cryst. Liq. Cryst., 27:269 (1975).

58. R. Cox, "Liquid Crystalline Mixtures," Ph.D. Dissertation submitted 1977, U. Conn., Dept. of Materials Science.

Chapter 10

INFRARED AND RAMAN SPECTROSCOPY OF LIQUID CRYSTALS

Bernard J. Bulkin*

Hunter College of the
City University of New York
New York, New York

---

*Present Affiliation: School of Arts & Sciences, Polytechnic Institute of New York, Brooklyn, New York

I. INTRODUCTION

Infrared and Raman spectroscopy have been applied with increasing frequency to the study of liquid crystals. These techniques, which are fundamentally the study of molecular vibrations, are useful for probing a number of aspects of liquid crystal structure. Although to date the work has been neither as extensive nor as quantitative as that from magnetic resonance studies, it is likely that the studies discussed in this chapter are in the nature of "seed" investigations and that the technique will yield ever more quantitative data in the future.

It is useful to begin by considering the type of information about liquid crystals that one can hope to obtain from study of the vibrational spectrum. Following this we will discuss the various studies in the usual way, namely by dividing them into nematic, smectic, and cholesteric phases. Because of their close relationship to the thermotropic smectic phases, some studies on lyotropic liquid crystals will be included in that subsection. The instrumentation of infrared and Raman spectroscopy has been the subject of numerous texts and reviews and will not be treated in this article. In covering the literature, we have attempted to be representative and thorough, but not exhaustive. Where some work has not been discussed or referenced, it is usually because the aim of this chapter is to illustrate the types of studies which can be carried out and the types of information which can be obtained therefrom.

II. POTENTIAL OF VIBRATIONAL SPECTROSCOPY FOR THE STUDY OF LIQUID CRYSTALS

The vibrational spectrum extends over the energy range from 4,000-5 cm^{-1}. It is conveniently divided into two regions, that from 4,000-200 cm^{-1} where the vibrations of molecular (rather than ionic) substances are intramolecular (internal) in origin, and the region below 200 cm^{-1}, where intermolecular (external) vibrations are often observed. The dividing line is, of course, not a strict one. For certain light molecules with strong intermolecular forces the lattice vibrations may appear above 200 cm^{-1}. More important to our discussion here, however, is the fact that for higher molecular weight species with internal rotors, torsional frequencies are often observed below 200 cm^{-1}, indeed through the entire low-frequency region.

The spectrum can be further subdivided according to the technique used to probe the vibrations. There are several techniques available for probing inter- and intramolecular vibrations. Of these, infrared and Raman spectroscopy have been used most extensively. However, some work on neutron scattering, dielectric relaxation, and ultrasonic absorption has also been reported. The latter technique is primarily a lower frequency probe; hence, processes on the time scale of magnetic resonance are generally observed. The same is true of most of the dielectric relaxation measurements made to date, although such measurements could be made in the lower frequency end of the region described above.

It is often stated that infrared and Raman spectroscopy are complementary to one another. Infrared spectra are related to dipolar changes during a vibration, Raman spectra to polarizability changes. Thus, for certain high-symmetry molecules one can see a mutual exclusion between the two techniques, so that both are required for a complete study of the molecular vibrations.

For the complex low-symmetry molecules which form liquid crystals this is not generally expected to be the case. To date no molecules have been synthesized which have mesomorphic properties and possess any symmetry elements at all. Thus, in every case the formal intramolecular selection rules for all such molecules are those of C_1 symmetry, i.e., all vibrational modes allowed in both infrared and Raman spectra, with all Raman bands polarized.

This situation may be altered in several ways. First, complex molecules often have local symmetry. Thus, a methyl group attached to a complex molecular framework will frequently obey C_{3v} or C_s point group selection rules (Fig. 1). The important point is not the symmetry shown by molecular models in these complex cases, but rather that which is present in the potential energy distribution (PED). In some cases the derivation of these local symmetry selection rules can become quite complicated. The subject has recently been treated by Edgell [1].

A second important perturbation of the selection rules on internal vibrations is due to intermolecular forces. This is extremely important for liquid crystals as it is generally the intermolecular interactions which we wish to investigate. The systematic method by which such perturbations

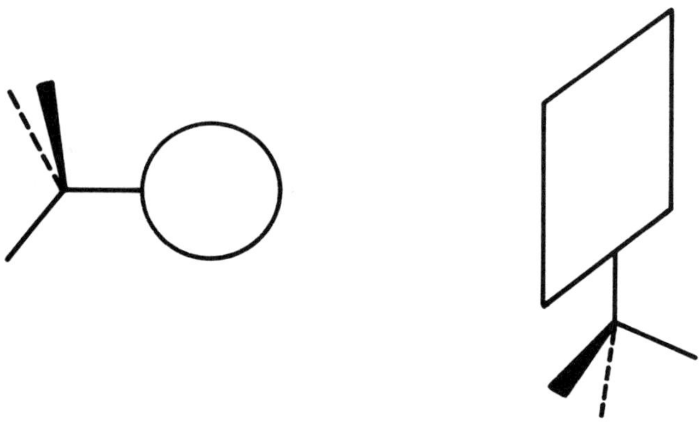

FIG. 1. Methyl group attached to "molecular" framework showing (left) C_s or (right) C_{3v} symmetry.

are expected to occur has been well studied for crystals [2,3]. The ideas can probably be extended to the study of liquid crystals as well. Briefly, this system proceeds in the following way: In an isolated molecule the molecular selection rules are observed. When the molecule is in a crystal (or crystal field) two other groups may be important in determining the selection rules. These are the site group (group formed by the symmetry elements which intersect the molecular site) and the factor group or unit cell group. The latter can be thought of as the group of all symmetry elements which interchange molecules in the crystalline unit cell. It is the space group of the crystal without the translation group. These factor groups are isomorphous with point groups. In this isomorphism, screw axes and glide planes must be equated to ordinary rotations and reflections. A correlation method exists for determining the effect of the site and factor groups on molecular selection rules [3].

This latter perturbation is important to our discussion of liquid crystals. The sensitivity of molecular spectra to restrictions on motional freedom in the crystalline phase, especially in the case of strong intermolecular forces, makes this an important starting point for the study of liquid crystals. To understand this, we need only recognize that in a crystal all external molecular rotations and translations are hindered, hence they are vibrations. Further, the factor group selection rules should be obeyed. In the liquid crystal, if our qualitative picture is correct, certain of these hindered rotations and translations become free, while others remain as vibrations. A change in intramolecular selection rules may be expected at the crystal-liquid crystal phase transition. In the low-frequency region a decrease in the number of external modes observed should be apparent.

It is interesting to point out here that the sample probably does not need to be aligned by external fields or other techniques to see these effects. This is because the vibrational selection rules are not very much affected by long-range order. This is especially true of the internal vibrations, where the intermolecular forces only provide a perturbation to the potential. This perturbation, in most cases, probably falls off as r^{-3} or steeper. A consequence of this is that one does not expect the vibrational spectrum selection rules to be very sensitive to phase transitions between liquid crystalline phases or even to the liquid crystal-isotropic liquid transition. We shall see that there are exceptions to this statement, for example, if one makes polarization measurements on highly aligned samples, measures frequency positions or integrated intensities quite accurately, or perhaps, if band shapes are studied. However the generalization that the effects will be small on internal vibrations nonetheless holds. By contrast, with intermolecular vibrations we will see that in the cases studied to date, the Raman spectrum has been somewhat sensitive to the transitions between liquid crystalline phases as well as to the transition to isotropic liquid.

In the previous paragraph we alluded to the types of phenomena one might study in the vibrational spectrum beyond the simple question of selection rules. These effects can be conveniently grouped under the headings of band intensities, positions of band maxima, shapes of absorption or scattering envelopes, and polarization of the bands. All these phenomena may contain information about the intermolecular ordering forces responsible for the liquid crystalline state. All are illustrated in specific studies below. In most cases, attempts have been made to study the temperature (and phase) dependence of one or more of these properties. In this there is a complication, however, which has not always been considered, namely, the temperature dependence which is not related to the presence of a liquid crystalline phase. Indeed, the temperature and phase dependences of vibrational spectra are, for the most part, not quantitatively understood even for relatively simple molecules. On the one hand, they seem to cotain information about the intermolecular forces in the various phases. On the other, they do not contain enough information for these forces to be extracted without the aid of additional data. We will see that this problem also plagues studies of liquid crystals. Before turning to a more detailed description of the work on liquid crystals, we wish to briefly comment on the measurement of the various quantities enumerated above.

To measure the location of frequency maxima is relatively straightforward, and since these positions give the most direct information about the potential that can be obtained from vibrational spectra it is probably the most valuable measurement that can be made. The greatest complication which occurs for liquid crystals is the problem of band overlap. The probability that this will occur is, of course, directly related to the number of atoms (N) in the molecule (3N - 6 vibration/molecule), so for mesomorphic materials, where 3N - 6 is usually very large, most of the vibrational modes are overlapped. This poses two problems: First, the presence of two bands in one envelope may shift the apparent positions of the maxima from those in the isolated bands. Second, if the vibrations are coupled in the potential energy distribution, as is likely if they are from vibrations which are due to physically proximate groups, then a change in the frequency of one will strongly alter the frequency of the other. Thus, resolution of an effect may be difficult.

Integrated band intensities suffer from this difficulty as well as others. To study the temperature dependence of intensities, measurements must be made well into the wings of the band. As the temperature increases, more intensity will often be shifted far from the band center. In general, this is difficult to measure because the signal to noise ratio in the wings will be poor. Thus, apparent decreases in band intensity should be viewed with caution. A further complication is due to "hot bands." These arise from transitions between populated fundamental levels and overtone or combination levels [4]. Because of anharmonicity, hot bands are often found on the low-frequency side of the main band. Their intensity will be temperature dependent according to the Boltzmann factor for the various levels. In complex molecules, where there are many low-lying vibrational levels ($kT \approx 200\ cm^{-1}$ at 300K), hot bands are an important contributor to the band envelope. It is interesting that neglecting them will have the opposite effect on measured intensity as neglect of a shift into the wings.

Band shape measurements are also affected by hot bands and overlapping bands. In addition, such aspects as spectrometer transfer function must be carefully considered. If, as is often the case, one attempts to extract information about rotational correlation from measurement of the band shape, then competing processes which also contribute to the shape must be considered. The problems have been discussed in the literature [5].

Finally, Raman and infrared polarization measurements should be the most straightforward to make, especially if they can be done digitally. In this manner polarization as a function of frequency can be recorded. With aligned samples we can expect vibrational polarization measurements to provide an accurate measure of order which is free of long-time averaging effects. The principal complication in such measurements comes from depolarization of the incident beam due to multiple scattering (refractive index changes) within the sample. Thus, if one observes the Raman spectrum of an unaligned liquid crystal, all bands will appear to have the same intensity independent of polarization of the incident light. This is because the polarization is effectively scrambled by the sample. While this effect is clear and recognizable, partial scrambling in a sample which is presumed to be aligned may be more difficult to compensate for. Some of the studies on nematics, discussed below, illustrate this problem quite well.

III. INFRARED AND RAMAN SPECTRA OF NEMATOGENIC MATERIALS

A. Vibrational Assignments

Although many nematic substances are known, vibrational spectroscopic studies have concentrated on a few of them, less than 10 in all. The greatest interest has been in the homologous series of 4,4'-alkoxyazoxybenzenes (PAA and its homologs) and in the room temperature nematic (p-methoxybenzylidene)-p-N-butylaniline (MBBA). Some work has also been done on nematic molecules containing a CN end group.

For many years the main thing which vibrational spectroscopists did with their spectra was to assign the individual absorption or scattering bands to particular coordinates, generally internal coordinates such as bond stretches, angle deformations, etc. Such an activity is not too fruitful in the case of complex molecules, because the low symmetry and large number of atoms make mixing of internal coordinates in the normal vibrations inevitable. Nonetheless it is useful to obtain such information as can be extracted in this way for other studies of the temperature effects. The usual method for complex molecules is to study the spectra of progressively simpler components, assuming that a group frequency approach holds. In some cases, additional information can be obtained on liquid crystals from polarization measurements in the infrared. The prospect of doing even a restricted normal coordinate internal mode calculation for such complex molecules is a bit horrifying, but one has been carried out.

The Raman spectrum of PAA was first reported by Freymann and Servant [6]. Those data were very preliminary by today's standards and are of little interest except as history. An extensive study of the infrared spectra of PAA and its higher homologs was carried out by Maier and Englert [7]. Their work included infrared dichroism in the nematic phase of materials aligned homogeneously by rubbing NaCl plates with cat's fur. In addition, these workers investigated a large number of analogous compounds for the purpose of making assignments.

In later work, Amer and Shen [8] measured the Raman spectra of four members of the series, namely the methyl, ethyl, pentyl, and hexyl compounds. Not surprisingly, because of the low symmetry, the infrared and Raman spectra are almost identical. Table 1, taken from the paper of Amer and Shen, shows the Raman data of those workers, as well as the infrared and Raman spectra of earlier investigators. The assignments proposed by Maier and Englert are also included. The main contribution of the later Raman work is the addition of internal modes seen in the region from 100-500 cm^{-1}, as well as the external mode region for three of the compounds. The external modes are discussed separately below.

TABLE 1. Frequencies, Intensities, and Assignments of the Studied Compounds[a]

PAA				PAΦ		PAB		HAB		Assignment
Ref. 8	Ref. 6	Ref. 15	Ref. 7	Ref. 8	Ref. 7	Ref. 8	Ref. 7	Ref. 8	Ref. 7	Ref. 7
21(6)										
						37^B(6)				
40(7)										
52(9)				54^B(14)						
72(18)										
210^B(?)	213									
234^B(?)										
317(5)	313			318^B(6)		317^B(7)				
360^B(?)	365									
417(5)										
474(?)				467^B(4)						
494(5)						494^B(1?)				
536(2)		543								
611^B(2?)		618								
629^B(3)		637		628^B(6)		628^B(7)		629^B(7)		
670(6)		678	667		669	671^B(3)	665		663	Γ_1
			717		718		718^{sh}		718	Γ_3

TABLE 1.continued.

PAA				PAΦ		PAB		HAB		Assignment
Ref. 8	Ref. 6	Ref. 15	Ref. 7	Ref. 8	Ref. 7	Ref. 8	Ref. 7	Ref. 8	Ref. 7	Ref. 7
725(5)		733	723	724(?)	726	724^{B}(3)	724		?	
					813		731^{B}		728	$\rho(CH_2)$
									753	
							778^{sh}			
			753							
			756^{sh}		770		787		788	ω_{11} } $\rho(CH_2)$
797(4)		806	805				800		799	
					837^{sh}		810		807	
832^{B}(5)			836		844	833^{B}(2)	831		835	γ
									843	
848^{B}(6)						850^{B}(2)	853			
							892		897	$\rho(CH_3)$?
911(20)	914		908	911^{B}(8)	914	910(7)	910	910(11)	910	ω_1
					922		942			ω_c
			943		938		948		935	γ
			959		966		958		?	
							978			

PAA				PAΦ		PAB		HAB		Assignment	
Ref. 8	Ref. 6	Ref. 15	Ref. 7	Ref. 8	Ref. 7	Ref. 8	Ref. 7	Ref. 8	Ref. 7	Ref. 7	
									988		
			1006		1008				1008	δ_5	ω_c
							1007^{sh}				
		1013									
			1020				1017				
		1028							1028	ω_c-0	
					1045						
									1058		
							1060		1068		
		1078									
							1088^{sh}				
1095(36)	1094	1103	1091	1095(27)	1099	1095(17)	1097	1095(31)	1096		
			1109^{sh}		1110^{sh}				1111	?δ_4	
1114	1112	1122	1115	1114^{B}(11)	1122	1115^{B}(1)	1118	1113^{B}(2)	1118		
		1143		1140^{B}(7)		1142^{B}(14)	1129	1141(100)	1129	$\tau(CH_2)$	
1157^{sh}(6)			1154	1159^{B}(4)	1160	1159^{B}(2)	1157	1156^{B}(3)	1157	δ_3	
		1165	1167								

TABLE 1.continued.

PAA				PAΦ		PAB		HAB		Assignment	
Ref. 8	Ref. 6	Ref. 15	Ref. 7	Ref. 8	Ref. 7	Ref. 8	Ref. 7	Ref. 8	Ref. 7	Ref. 7	
1171(48)	1174	1180	1181	1178(18)	1174^{sh}	1178(18)	1171^{sh}	1176(25)	?	$\gamma(CH_3)$?	$\tau(CH_2)$
$1186^{sh}(4)$		1195				1188(2)		1192(25)	1203^{sh}		
		1203									
$1220^{sh}(11)$		1225	1218^{sh}	1220(4)	1217	1221(4)	1216^{sh}	$1219^{B}(?)$	1221^{sh}?		
1246(85)	1247	1253	1250	1246(32)			1253^{B}	1246(47)	1251^{B}	ω_4	
1252(19)			1260	1253(7?)	1258^{B}			1261(19?)			
1261(67)				1262(8?)		$1260^{sh}(28)^{b}$		1268(75)			
				1271(14)							
1276(100)	1276	1284	1277^{sh}	1280(82)	1278	1271(100)	1270^{sh}	1278(50)	1281		
						$1281^{sh}(11)$				ω_3	$\gamma(CH_2)$
$1301^{sh}(5)$		1308	1300	1300(?)	1300	1300(?)	1302	$1300^{B}(12)$	1296		
									1303		
$1319^{B,sh}(50)$		1328	1311	$1319^{sh}(28)$	1320	$1320^{B,sh}(28)$	1320	$1320^{B}(41)$	1318	ω_9, δ_1?	
1333(87)	1334	1342	1331	1332(100)	1332	1332(44)	1336	1333(47)	1336	ω(N.N.o)	
							1367^{sh}		1348^{sh}		
			1370				1380		1382^{sh}		
					1393		1394		1397		

PAA				PAΦ		PAB		HAB		Assignment
Ref. 8	Ref. 6	Ref. 15	Ref. 7	Ref. 8	Ref. 7	Ref. 8	Ref. 7	Ref. 8	Ref. 7	Ref. 7
1410(42)	1408	1418	1414	1411(33)	1414	1411(26)	1417	1410(93)	1417	ω_9 or δ_1?
			1425^{sh}		1422^{sh}		1427^{sh}		1425^{sh}	$\delta(CH_3)$
$1438^{B}(4)$		1446	1446	$1439^{B}(4)$	1449^{sh}		1440^{sh}		1438^{sh}	ω_5
1454(36)		1461	1456	$1454^{B,sh}(6)$	1457^{sh}	1454(40)		1454(50)		ω(N.N.o)
$1465^{b,B,sh}(37)$	1460	1466	1464	1464(73)	1463	$1466^{sh}(2)$	1464	1466(65)	1462^{sh}	$\delta(CH_2)$
		1477	1473				1472^{sh}		1470	
					1477		1481^{sh}			
$1501^{B}(18)$	1496	1501	1500	1499(18)	1502	1500(8)	1503	1500(22)	1502	ω_6
					1547		1540^{sh}			
$1570^{B}(28)$	1572		1567	1570(5)	1564		1566		1564	
$1582^{sh}(3)$				$1580^{B}(4)$				$1580^{B}(10)$		
1596(87)				1595(47)		$1595^{sh}(7)$	1595	$1595^{sh}(6)$	1594	
$1604^{sh}(11)$	1603	1605	1600	$1604^{B,sh}(19)$	1603	1604(21)	1603	1605(69)	1607^{sh}	ω_7, ω_8
							1626^{sh}		1629^{sh}	
			1648^{sh}		1650^{sh}		1640^{sh}			

[a]The accuracy of our frequency determination is ± 2 cm^{-1}.
[b]Normalized intensity given between parentheses.
[B]is broad line
[sh]is shoulder
[b]Broad band.

Examination of the assignments illustrates well the problems inherent in trying to deduce information about molecular interactions from an internal coordinate approach. Except for the CH stretches (not shown in the table) all the stretching and bending vibrations are concentrated below the 1,700-cm^{-1} region. The benzene ring stretching vibrations for para-substituted benzenes occur in the 1,400- to 1,600-cm^{-1} region. This is also the region of vibrations of the NNO group. In the higher homologs the CH_2 deformation modes will occur in this region as well. The latter may be quite well localized on the side chains, but the NNO and phenyl ring modes probably are very mixed in the potential energy distribution (PED). This is usually the case for conjugated, near-planar systems.

As one moves into the next lower frequency region, 900-1,400 cm^{-1}, C-O stretching modes, C-C alkyl stretches in the higher homologs, phenyl modes, and various CH_2 rocking motions all appear. This region is thus extremely complex from the viewpoint of the PED.

It is necessary to make the point about the breakdown of simple internal coordinate ideas in complex molecules quite strongly, because there has been some inclination to assign a definite motion to a particular absorption in these spectra, follow its intensity as a function of temperature, and draw broad conclusions about the molecular interactions which are occurring.

In the lower frequency internal mode region one may again have a tractable situation. Amer and Shen have not studied this region very much, but Schnur [9] has investigated it in some detail. This region is useful because one is now at energies below most of the vibrations of the phenyl rings and azoxy group, and vibrations of the hydrocarbon tail become most important. In particular, by studying the same homologous series, Schnur has identified the so-called accordion mode of the end chain. Actually, this is a band envelope which contains a number of these accordion modes, depending on the existing conformations of the chain. Based on the frequency dependence (275-325 cm^{-1}) of the envelope maximum as one proceeds from the n-butyl to n-heptyl compound, and calculation of this frequency for simple alkane models, one can see that the band is probably a relatively "pure" vibration in terms of the PED. Its temperature dependence will be discussed further below.

The only other nematogenic molecule which has been extensively investigated from the viewpoint of internal vibrational assignments is MBBA. Three papers have appeared on this assignment [10-12], but the work of Vergoten [12] is the most extensive. In the first report by Heger and Mercier [10], the Raman spectrum of MBBA was compared with those of the two compounds which form the Schiff base, p-n-butyl aniline and p-methoxy benzaldehyde. An infrared spectrum of p-n-butylaniline is also shown. The assignments made are pretty much those of a group frequency approach using the extensive infrared data of Bellamy's [13] book. Some are curiously out of place, however, for example, the assignment of a band at 1,270 cm^{-1} to a C=N stretch.

Bulkin and coworkers [11] have examined the infrared spectrum of MBBA using homogeneous and homeotropically aligned samples. These alignments are quite useful for doing vibrational assignments. Thus, modes which are composed of dipole changes along the long axis will be more intense in the homogeneous than in the homeotropic alignment. Those which are in the plane perpendicular to the long axis have the opposite behavior. Again, one finds that the mixing in the PED as well as intermediate values for the order parameter lead to less striking changes than one might expect initially (Fig. 2). Still, on the basis of these spectra a number of Heger and Mercier's assignments were clearly untenable.

The most extensive work is that of Vergoten [12], who carried out a normal coordinate calculation on benzylidene aniline, the Schiff base skeleton of MBBA. The resulting assignments are shown in Table 2. These assignments are in general agreement with those of Bulkin et al. More detailed information about these calculations is available in Vergoten's thesis [14]. Such calculations do, of course, give information about the PED, so that the internal coordinate assignments can be viewed with more perspective. However, it should be recognized that in these calculations, the amount of data available is insufficient to determine the internal coordinate force constants, and many of the interaction constants between internal coordinates are arbitrarily fixed at zero.

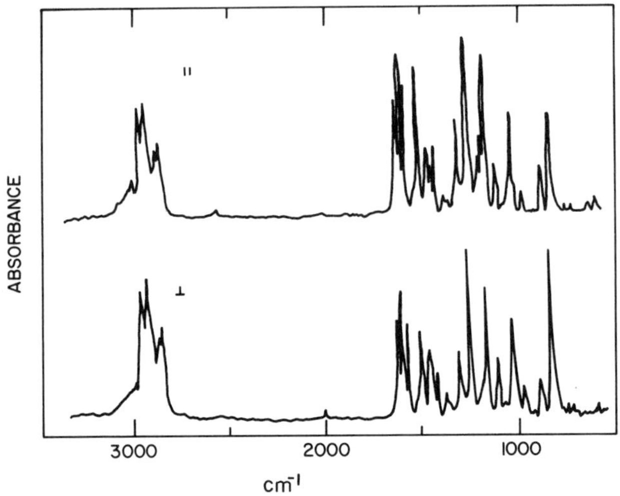

FIG. 2. Infrared spectrum of MBBA oriented homogeneously (labeled parallel) and homeotropically (labeled perpendicular). (Redrawn with permission from Ref. 11.)

B. The Crystal-Nematic Transition: Internal Modes

Maier and Englert observed that there were considerable changes in the infrared spectrum of alkoxyazoxybenzenes at the crystal-nematic phase transition. This has also been observed by Suschinski and coworkers [15]. This is not surprising, in that the large ΔH of this phase transition implies that most of the change in motional freedom occurs here, rather than at the nematic-isotropic transition.

Bulkin et al. [16] examined these changes in more detail as the phase transition is approached from the crystal side. They considered the origin of the modes which were present in the crystal but not observed in the nematic phase. A first possibility is that these bands are simply becoming very broad but not actually decreasing in integrated intensity. Such a possibility would be reasonable for a simple internal vibration where the degree of motional freedom was markedly altered during the phase transition. Amer and Shen [8] have reasoned in this manner. However, this did not appear to be the case in the measurements of Bulkin et al. [16], and more recent work of Nikolov and Simova [17] also shows that

TABLE 2. Coordinate Calculations of MBBA and Benzylidene Aniline (BA)

MBBA			BA
Frequency (cm^{-1})	Relative Intensity	Assignments (Wilson's notation)	Observed Frequency
180	6		
315	1		250
340	2		
410	4	16 a	405
			540
630	5	16 b	620
			650
720	2		
760	2		750
775	4		770
790	4		
825	3		825
886	3	δ(C-H)*	875
934	1	17 b	920
975	7	δ5	970
			1000
1014	1	18 a (I and II)	1025
1105	4		
1164	70	9 a (I and II)	1165
1182	37	$\nu(\phi$-N)	1190
1245	5	$\nu(\phi$-C)	1240
1305	5		1320
1370	4	δ(C-H)*	1370
1422	12	19 b (I and II)	1450
1503	9	19 a (I and II)	1480
1575	53	8 a (II)	1575
1596	100	8 a (I)	1590
1626	41	ν(C=N)	1630

the integrated intensity decreases as the phase transition is approached. Thus, it appears that Amer and Shen's discussion of the changes in the spectrum is probably not correct. It should be noted that these authors did not study the spectrum as a detailed function of temperature but only report the data in the various phases.

A second possibility which needs to be considered is the idea of coupling between molecules in the unit cell, giving rise to a factor group splitting of the vibrations. While this is a possibility, it was ruled out because the bands did not collapse to a central maximum as the phase transition was traversed, but rather a sharp band on the side of a stronger mode disappeared. If two energy levels of the isolated molecules were interacting in the unit cell to split apart, then we could expect the decoupled (nematic) energy levels to be approximately in the center of the split levels. Such was not the case.

This leaves a possibility which is not often observed in molecular crystals but is seen in ionic crystals frequently, i.e., sum and difference modes between lattice vibrations and internal vibrations. We would expect to see these in nematogenic materials because the intermolecular forces along certain directions are strong. This seems to be the correct explanation for the observed effect.

When the temperature dependence of the intensity of these "disappearing bands" is carefully monitored, they disappear gradually rather than abruptly at the phase transition [16]. The result for PAA is shown in Fig. 3. There are thus marked pretransition effects observed in this so-called soft solid region.

In some cases, frequency shifts of internal modes as the phase transition is approached have also been observed. Amer and Shen [8] reported such shifts in several cases, as did Itoh et al. [18]. The results of the latter group for PAA are shown in Fig. 4. The curve resembles that for PAA shown in Fig. 3.

Riste and Pynn [19] have made neutron diffraction measurements with a sample of deuterated PAA. Their results, shown in Fig. 5, show the same sort of pretransition effects as those discussed above. It thus appears that this effect is a quite reproducible one between laboratories and between techniques.

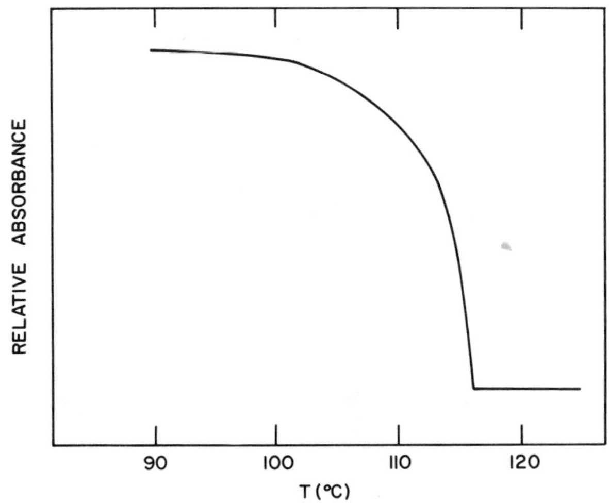

FIG. 3. Relative absorbance of an infrared band in the spectrum of PAA as a function of temperature in the region of the crystal-nematic phase transition. (Redrawn with permission from Ref. 16.)

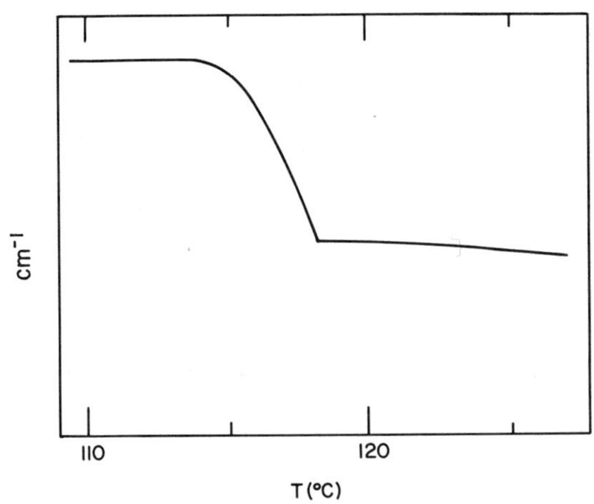

FIG. 4. Frequency shift of a band in the PAA spectrum as a function of temperature in the region of the crystal-nematic transition. (Redrawn with permission from Ref. 18.)

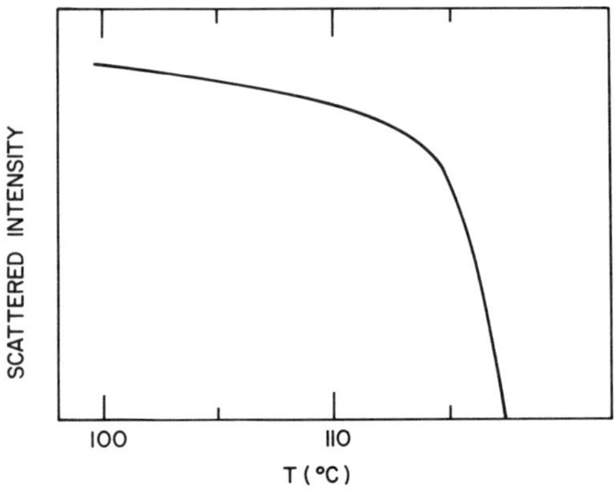

FIG. 5. Variation of integrated Bragg intensity with temperature for PAA in the crystal-nematic transition region. (Redrawn with permission from Ref. 19.)

An interesting alternative explanation to that of the above groups (pretransition effects) is a proposal by Schnur [9] who examined the integrated intensity of the accordion mode (340 cm^{-1}) for 4,4'-dibutoxy-azoxybenzene near the crystal-nematic transition. His results are shown in Fig. 6. He again observes that same gradually decreasing intensity as T_m is approached. However, he has interpreted this as being due to the presence of a second solid phase. He has stated that this transition is also observed calorimetrically and optically.

The particular point to be made about these observations is that Schnur has focused his attention on the accordion mode, and hence bases his interpretation of the results as being due to an increasing population of different alkyl chain conformers as the higher temperature phases are entered. This may indeed be the case, but the data need be considered in combination with all other data as one approaches this phase transition. Thus, the rapidly declining intensity in Schnur's "Solid II" phase region is seen for bands other than the accordion mode.

It is interesting to point out that the x-ray crystallographic studies of some PAA crystals [20] show smear lines along the long axis direction. These smear lines are indicative of the presence of disorder or large motional amplitudes along this direction. Although this has not been systematically investigated, not all crystals show such smear lines; they are probably a function of growth conditions.

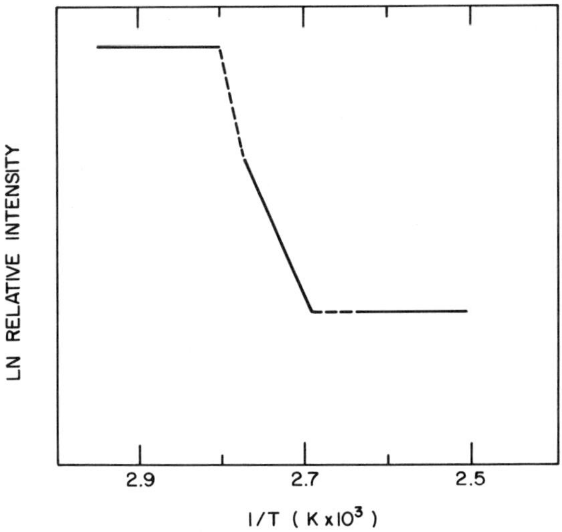

FIG. 6. Relative intensity of the accordion made of 4,4'-dibutoxyazoxybenzene as a function of temperature. (Redrawn with permission from Ref. 30.)

It would appear that the infrared, Raman, and neutron diffraction data all give indication that for the alkoxyazoxybenzenes there are considerable pretransition effects at the crystal-nematic phase transition. Either these are smoothly varying effects, or they involve a second phase in which there is progressive disorder. The difference between these explanations is more than semantic as the first implies a disordering of the crystalline structure while the second only speaks of a disorder in the internal molecular conformation.

Somewhat related to these studies is the work of Janik and coworkers [21] on MBBA. They have found that there are two solid forms of MBBA. One of them, the "metastable form," can be prepared by rapid cooling of the nematic phase. This is probably the form that Sorai et al. [22] has referred to as the glassy phase. If the metastable form is heated to temperatures about 20° below T_m, it is transformed to the stable form in a time-dependent annealing process.

The infrared spectra of the two forms are interesting, in that both show bands of comparable width, though varying in number, frequency, and relative intensity. In neither phase is there any indication of a glassy state, which would probably be characterized by broader bands.

The strangest aspect of the spectra of the two solid forms of MBBA is that the spectrum of the nematic and isotropic phases resembles most closely that of the "stable" form of the solid. This is most apparent when one views the spectrum in the 100-400 cm^{-1} region, where there are several striking changes. As this region also contains several modes associated with the end chain, one is tempted to follow Schnur's reasoning again regarding alkyl tail conformational changes near the crystal-nematic transition. Janik has also suggested that there are pretransition effects at the crystal-nematic transition for MBBA.

It seems possible that there is an alternative view of these data to that suggested by the authors, namely, that the metastable and stable phases found by them should actually be interchanged in designation. This would then correspond more closely to the findings of Schnur and be completely consistent with the thermodynamic data. To understand this, we can view the evidence regarding the similarity of the spectra of the higher temperature solid phase to that of the nematic phase as being indicative of the greater disorder in that solid. Further, one may view the metastable phase as being a monotropic solid, a type of solid phase found by Schnur for some of the alkoxyazoxybenzenes. Another way of looking at the transformation between the two solids is to say that once the higher temperature, disordered solid is formed it should spontaneously revert to the more ordered form on cooling. However, the disorder has been frozen out, and there is too great an energy barrier for the reversion to occur.

C. The Crystal-Nematic Transition: External Modes

In the external mode region we find good evidence relating to the ideas discussed in the previous section. If we are, indeed, interested in the vibrations of the crystal, then the lattice vibration region (in both the infrared and Raman spectra) should give the most direct information.

Most of the attention to date has been focused on PAA [8,23-25], although spectra have been reported for MBBA [26-28] and other alkoxyazoxybenzene [8,9] homologs as well. Schnur and Fontana [29] have also reported the spectrum of terephthal-bis-(4-n-butylaniline (TBBA), which shows a nematic phase but no crystal-nematic transition, as it has several smectic phases. This work will be discussed in more detail later.

The Raman spectrum of crystalline PAA between 10 and 150 cm^{-1} was first reported by Bulkin and Grunbaum [23], and subsequently was also published by Amer et al. [25]. The latter spectrum suffered from a very high background scatter. Schnur [9] has pointed out the possible existence of a problem of impurity in Amer's samples, and this may have been responsible for the high scattering observed. A more detailed presentation of the Raman data including oriented single crystal studies and temperature dependence near the crystal-nematic transition was presented by Bulkin and Prochaska [24]. Their single crystal spectrum is shown in Fig. 7. A subsequent spectrum published by Amer and Shen [8] was in complete agreement with these data, which were also reproduced in a paper by Schnur et al. [30].

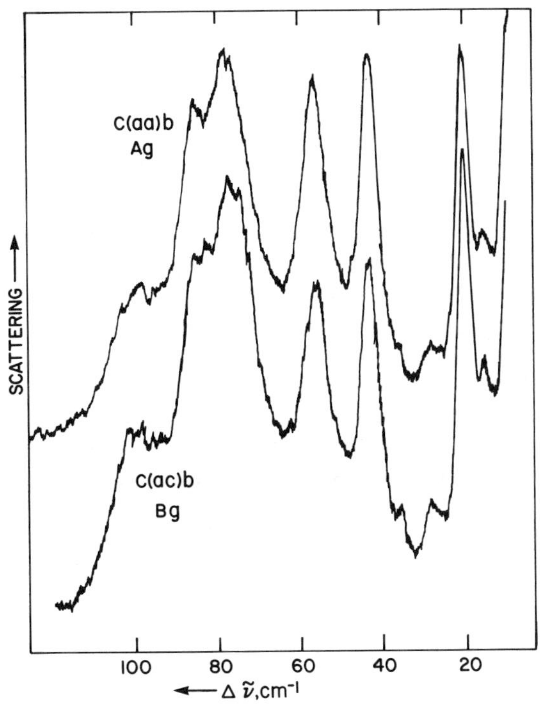

FIG. 7. Raman spectrum of two polarizations for an oriented single crystal of PAA in the lattice vibration region. (Redrawn with permission from Ref. 24.)

The far infrared region of PAA has also been reported by Bulkin and Lok [26]. Their data are shown in Fig. 8 for a polycrystalline sample. It is interesting to note (Table 3) that the far infrared absorption of PAA occurs at higher frequency than does the Raman scattering. We have previously noted that in the internal mode region the infrared and Raman spectra are identical, consistent with the expectation for a molecule obeying C_1 selection rules. Now, in the lattice vibration region we see the selection rules of the crystalline unit cell in operation. For the case of PAA, this is known to be C_{2h}^5, with four molecules per unit cell. For this group, there is a center of symmetry, so no mode should be both infrared and Raman active. This is indeed true to some extent. As we shall see below, the coincidences between the two spectra which do exist are of some interest.

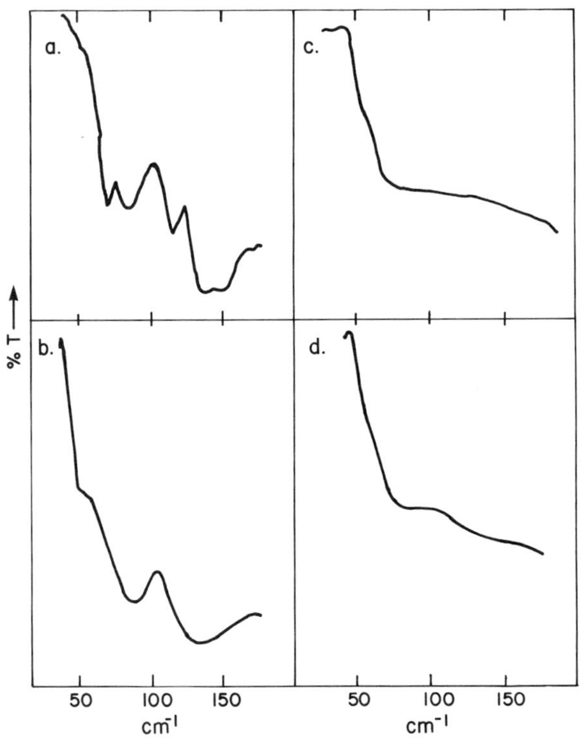

FIG. 8. Far-infrared spectrum of PAA as a function of temperature in the crystalline pahse: (a) 25, (b) 90, (c) 116, and (d) 117°C. (Redrawn with permission from Ref. 26.)

TABLE 3. Comparison of Far-infrared and Raman Frequencies (cm^{-1}) of Polycrystalline PAA[a]

Far infrared	Raman	Far Infrared	Raman
--	16	84	94
--	30	115	--
--	38	135	--
50	52	150	--
70	72 (broad)	--	--

[a]Single-crystal values for Raman bands available in Ref. 24. All values taken at 25°C.

Before considering the PAA spectrum further, we will make mention of the Raman spectra of its homologs which have been published. These have been studied by both Schnur [8] and Amer and Shen [9]. There seems to be some disagreement as to the results for these compounds. None of them have as strong or as well defined a Raman spectrum in the low-frequency region as PAA does. For p-azoxydiphenetole, Amer and Shen observe one band at 54 cm^{-1}. Schnur's spectrum also seems to show one band on the side of the Rayleigh wing about the same frequency.

For the butyl homolog, Amer and Shen report only one band at 37 cm^{-1}. Schnur has reported spectra of two solid forms, one of which has bands at 47, 80, and 93 cm^{-1}, and the other of which has bands at 41 and 78 cm^{-1}. Clearly there is some unresolved experimental difficulty here.

In the hexyl homolog, Amer and Shen did not find any bands in the lattice vibration region, although Schnur has found a rather intense band at 78 cm^{-1} for one solid phase, which shifted to 80 cm^{-1} in a second solid phase. Again, further work is necessary if these discrepancies are to be resolved.

PAA has been studied by Bulkin and Prochaska [24] and by Bulkin and Lok [26] in the crystal-nematic phase transition region. The results for the Raman spectrum are shown in Fig. 9. These results indicate that there are drastic changes in the spectrum in the crystalline phase as the

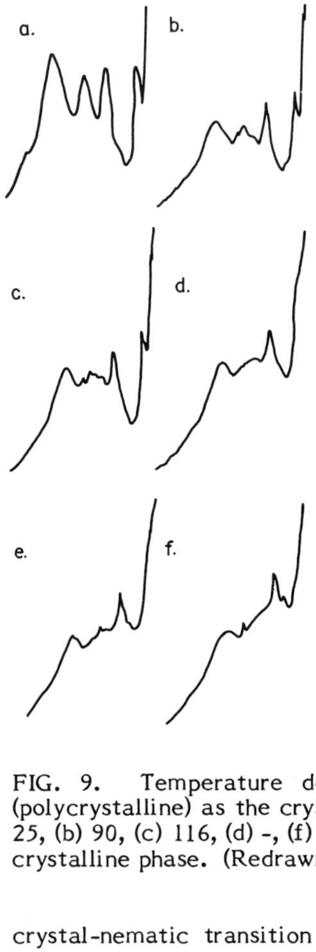

FIG. 9. Temperature dependence of the Raman spectrum of PAA (polycrystalline) as the crystal-nematic phase transition is approached: (a) 25, (b) 90, (c) 116, (d) -, (f) ≈ 0.3 °C temperature increments above (c) in the crystalline phase. (Redrawn with permission from Ref. 24.)

crystal-nematic transition is approached. They are consistent with the internal mode data reported earlier. The authors have interpreted the Raman data as being indicative of the presence of a soft mode as the transition is approached. Such a mode was predicted by the work of Kobayashi [31]. Ford [32] has also made a similar prediction.

The far infrared data did not indicate the presence of a soft mode, only a severe broadening of the absorption bands as the transition was approached. Both far infrared and Raman data, however, are consistent with a decrease in order in the crystalline phase as the crystal-nematic transition is approached.

Bulkin and Grunbaum [33] have carried out a calculation of the lattice vibration frequencies of PAA using an assumed potential. The molecules are fixed at the equilibrium crystal geometry, and an atom-atom potential function is constructed for the crystal using parameters transferred from other molecular crystals. It should be stressed that in this calculation the observed frequencies are not used at all to refine the calculated data.

The results of the calculation are shown in Table 4, where they are compared to the experimental data. We begin by noting a few things about the table. The experimental Raman frequencies can be divided according to irreducible representations of the factor group based on the single-crystal data. The infrared frequencies are arbitrarily assigned to the closest calculated value independent of representation. In addition, fewer bands are observed in the far-infrared region than are expected. Either this is due to the fact that some are falling too low in frequency for measurement, or to an overlap of the modes which is unresolved in the polycrystalline sample. We tend to believe it is the latter, as the calculations and our general experience indicate that lattice vibrations will be found in this spectral range. Thus, in some cases the same observed frequency maximum is assigned to more than one fundamental external vibration.

In setting up a calculation of external vibrations, one can consider motional coordinates analogous to the bond stretches and torsions of internal vibrations. These are the translations along three directions (x, y, z) and rotations about three axes (see Ref. 33 for a description of the rotational coordinate system). These motions are then combined under the operations of the factor group. In the case of PAA, with four molecules per unit cell, each motion will yield four lattice modes or external vibrations of the unit cell. We assume that all unit cells vibrate in phase. (Of the translations, three have zero frequency, representing the "acoustic modes" or translations of the entire lattice.)

If the story were no more complicated than this, it would be simple to identify each of the calculated force constants and frequencies with motions in the lattice. However, we find that the force constants which

TABLE 4. Observed and Calculated Lattice Vibration Frequencies of PPA[a]

Symmetry species			Translatory[b]		Rotatory[b]	
Ag						
observed	30	52	70	16	74	91
calculated	28	55	62	20	69	94
Bg						
observed	30,37	52	70	16	74,90	95
calculated	37	60	69	25	82	101
Au						
observed	50	70	--	135,150	84	50?
calculated	51	69	0[c]	130	89	55
Bu						
observed	--	--	70?	115	84?	50?
calculated	0[3]	0[3]	67	119	78	52

[a]Raman data (Ag, Bg) from Ref. 24, single-crystal frequencies at -90°C. Far-infrared (Au, Bu) data from Ref. 26, polycrystalline sample at 25°C.
[b]Modes are primarily rotatory or translatory in nature; however, the translation-rotation interaction force constants are not negligible.
[c]Acoustical modes.

involve interaction between different coordinates are substantial, and thus no motion is a pure translation of molecules along x, etc.

With these cautions in mind, we can view Table 4 with more understanding of its meaning. In the table, we have grouped the vibrations in rows according to representation and in columns according to motional coordinate. The first three columns are hindered rotatory modes. If the calculated frequencies in a particular column are similar, then this means that the coupling between molecules is small. The actual frequencies are determined by the molecular mass (translations) and moments of inertia (rotations), but the splitting between frequencies is determined solely by the potential function of intermolecular interaction.

We see that this splitting varies greatly for different coordinates. Of particular interest to our discussion in this section is the observation that modes in the vicinity of 70-80 cm^{-1} are <u>not</u> strongly coupled. It is in just this region that Bulkin and Prochaska [24] have proposed the existence of a soft mode. Clearly, this is where the requirement for a soft mode is most definitely met. It should be a hindered motion which can easily become free at the phase transition. Although one instinctively might expect the lowest frequency mode to be the soft mode, this is shown by the calculation not to be the case. The lowest and highest observed external vibrations are due to the strongest coupling. These are probably most preserved in the nematic phase.

We now turn our attention to the external vibrations of MBBA, which is the only other nematogenic material to be studied in both the Raman and far infrared. The Raman spectrum of solid (polycrystalline) MBBA was first reported by Billard et al. [34] and has since been discussed by Borer et al. [35] and Vergoten [12]. The spectrum of Billard and coworkers is shown in Fig. 10. Here we see the spectrum at two temperatures in the solid state. These workers have noted the abnormally great broadening of the external modes as the crystal-nematic transition is approached. No soft mode was observed, however.

Far-infrared spectra of MBBA have been studied by several groups [26-28]. The most accurate polycrystalline data appear to be those of Evans et al. [27], who studied the spectrum at 283 K between 40 and 200 cm^{-1}. Sciesinska et al. [28] have also studied the spectrum of the solid to 80 cm^{-1}, while Bulkin and Lok [26] have focused on the spectrum of the nematic phase. All the data are in agreement.

The spectrum of Evans et al. is shown in Fig. 11. Again we make the observation first noted by Bulkin and Lok [26] that the external modes seem to be most intense at higher frequencies in the infrared than in the Raman spectra. Sciesinska et al. [28] have questioned the attribution of all the intensity below 170 cm^{-1} to external modes, as was done by Evans et al. and Bulkin and Lok. This group feels that this absorption is due to internal modes only, and that any solvent, phase, and temperature dependences observed are due to mixtures of the two solid forms discussed above. Both Sciesinska and Bulkin have commented on the pretransition effects observed.

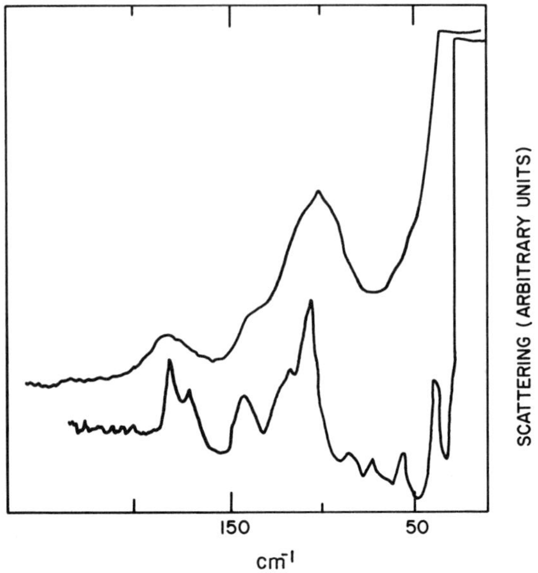

FIG. 10. Raman spectrum of solid MBBA in the lattice vibration region. Upper trace: 288 °K; lower trace: 100 °K. (Redrawn with permisison from Ref. 34.)

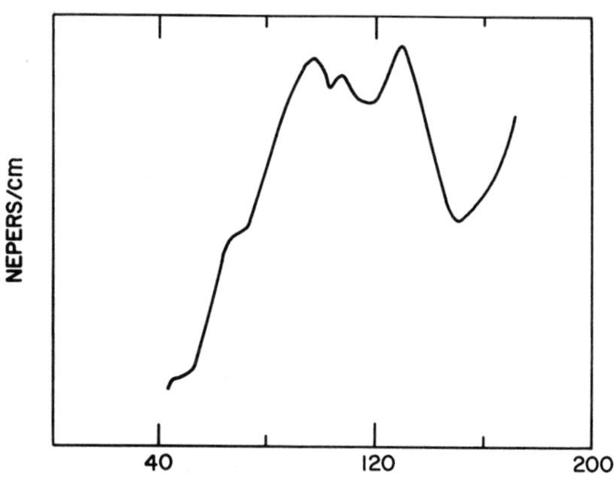

FIG. 11. Far-infrared absorption spectrum of solid MBBA. (Redrawn with permission from Ref. 27.)

Unfortunately, detailed interpretation of the far-infrared and Raman spectra of crystalline MBBA must await determination of its crystal structure using x-ray techniques. This is also the case for all other nematogenic materials. For p-azoxydiphenetole some x-ray data does exist, and some data on TBBA, to be discussed under the smectic liquid crystals below, has been obtained by Doucet et al. [36].

D. External Vibrations of the Nematic and Isotropic Phases

In principle, if there is strong intermolecular coupling in the nematic phase such that rotatory and/or translatory motions are hindered, it may be possible to observe either infrared absorption or Raman scattering associated with the intermolecular mode. This has been of some interest, because such an observation would constitute some direct evidence concerning intermolecular forces in mesophases. Further, there has been considerable discussion in the literature of the possible modes which give rise to the elastic constants of the nematic phase. Some workers have speculated that such modes might be observed in the external vibration region of the infrared or Raman spectra.

That such modes should be observable in fluid phases seems likely, at least in the infrared region. There have been numerous studies showing that polar liquids have a strong absorption in the 10- to 150-cm^{-1} region which is associated with intermolecular processes [37-39]. Such an absorption has been variously interpreted as arising from dipolar complexes [37], rotatory oscillations [38], and various relaxation processes [39]. The differences between the explanations offered are probably more semantic than real. All clearly establish the intermolecular origin. However, it should be noted that this is considerably easier in a small molecule where all fundamental vibrations can be accounted for, than it is in a large nematogenic material. In the latter case, proof of intermolecular origin of a vibrational band generally rests on the temperature or solution dependence of the intensity. Thus, if a band persists in the isotropic phase, but is nonetheless thought to be an external mode, it can be examined in dilute solution using longer pathlengths (in the infrared). The observation of external modes in the Raman effect is not so expected, as polar liquids do not show any definite band on the Rayleigh wing.

Let us now summarize what has been observed for nematogenic materials. For the alkoxyazoxybenzenes, several authors [8,9,24,30] have observed that there is broad Raman scattering below 100 cm^{-1} which can be distinctly seen on the Rayleigh wing. Not surprisingly, the various papers have differed somewhat in their choice of the positions of the frequency maximum. Such positions are almost impossible to determine for broad bands on an intense, sloping background. For PAA there seems to be a band near 40 cm^{-1}, and probably another band at higher frequency. Bulkin and Prochaska [24] reported a band near 70 cm^{-1}, Amer and Shen [8] at 52 cm^{-1}. The second band does not seem to be present in the spectrum of Sakamoto et al. [40], although they state that their spectrum is similar to that of Bulkin and Prochaska. In the isotropic liquid these bands disappear, and no distinct maximum can be seen. It seems that the band may disappear gradually as the nematic-isotropic transition is approached or, put another way, it may gradually become broader as the transition is approached. If this is the case, it would be important evidence regarding the order of the nematic-isotropic transition. More work is needed on these spectra to establish this point.

MBBA spectra offer some further evidence. Vergoten has observed a band in the Raman spectrum of nematic MBBA at 35 cm^{-1} at 25°C. This band seems to be shifted to lower frequency (23 cm^{-1}) in the isotropic phase just above the n - l transition, then disappear at higher temperatures. On the other, for TBBA, Schnur and Fontana [29] have reported that a band at 80 cm^{-1} in the nematic phase disappears discontinuously at the n - l transition.

Turning to the far-infrared region, we find that the observations are not the same as those made in the Raman spectrum. Again, as in the crystal phase, we can interpret this as evidence of the strong intermolecular origin of these vibrations. To reiterate the point made previously, for the molecules themselves, with C_1 symmetry, all vibrations are allowed in both the infrared and Raman spectra, and, in fact, in the 200- to 4,000-cm^{-1} region these spectra are identical. When the molecules are coupled other selection rules may apply, particularly to the external modes. Thus, for both PAA and MBBA, where detailed study has been done in the far infrared, the nematic phase shows a strong absorption band in the 100-to 150-cm^{-1} region. This band is quite broad, and in both cases it persists in the isotropic phase. Both PAA and MBBA have been reported by Bulkin and Lok [26], while the MBBA spectrum has been considered in some detail by Evans et al. [27] and by Sciesinska et al. [28]. The latter work is only extended as far as 80 cm^{-1}, while the two previous papers report the spectra to about 40 cm^{-1}.

The interpretation of these data by Bulkin and Lok and by Evans et al. was essentially the same, namely, the modes are due to hindered rotatory motions of the molecules in the nematic or isotropic phases. In both cases the solution dependence of the intensity was cited as evidence for this interpretation. Evans et al. have also carried out a rather extensive calculation using the Brot-Larkin model to show that a rotation about the long molecular axis is reasonable for this frequency region. This is consistent with measurements of the spectrum of MBBA aligned homogeneously and homeotropically reported by Bulkin and Lok. It should be mentioned that there is also a pronounced shoulder on the low-frequency side of the absorption band of MBBA in the nematic phase, probably centered near 80 cm^{-1}.

Sciesinska et al. have taken issue with this interpretation of the data, expressing the view that the absorption in this region is primarily if not completely of intramolecular origin. They attribute it to the internal torsions of the various carbon-carbon and carbon-oxygen single bonds in the molecules. They have carried out a rather simple calculation based on the moment of intertia only, which purports to show that the vibration about the long axis would fall in the region of 10-20 cm^{-1}. This calculation probably does not fully appreciate the depth and steepness of the potential well in which an MBBA molecule is located in the fluid phases. Further, the explanation is untenable on the grounds cited above, namely that the same mode is not found in the Raman spectrum, and it disappears in dilute (less than 10%) solution in nonpolar solvents.

The calculation of a barrier height for the hindered rotation by Evans et al. is interesting in that they have used a similar potential function to that used by Bulkin and Grunbaum [33] for crystalline PAA discussed above. They make the assumption that the barrier to rotation will be primarily that found by the interactions between the phenyl rings. While this qualitative calculation is not so well in agreement with experiment as is the Brot model, one can understand the reasons to some extent. The results are discussed in detail in Ref. 27.

In concluding this section, we wish to point out the importance of the observation of the intermolecular modes in the far-infrared spectrum of the isotropic phase for PAA and MBBA, but not in the Raman spectrum. This is exactly consistent with the observations for polar liquids. Further, the results have shown that different modes are observed in the intermolecular infrared and Raman spectra, indicating that there is perhaps sufficient unit cell symmetry preserved on the short time scale of molecular vibrations for some selection rules to be operative. More systems must be studied for this to be generalized. From the point of view of symmetry, we would then say that when this symmetry is partially broken at the nematic-isotropic transition, the Raman modes disappear and the infrared bands persist. Vergoten [12] has already begun to make such an argument for the case of MBBA in the Raman spectrum. Undoubtedly, we need to explore more deeply the application of group theoretical considerations to these spectra.

E. Polarization Measurements in the Nematic Phase

One of the earliest applications of infrared spectroscopy to the study of liquid crystals was made by Neff et al. [41], who used infrared dichroism measurements to determine the order parameter S of the nematic material. There seems to be little problem in carrying out such measurements if good infrared polarizers are available. Recently, some related measurements on solutions in (EBBA) were reported by Bulkin et al [11]. The S values thus obtained seem to be in fairly good agreement with those found from other methods on samples which had been similarly aligned (using fields or rubbing techniques).

Bulkin et al. have also pointed out that the same S value can be derived from measurements of infrared intensity of molecules which have been homogeneously and homeotropically aligned, without the use of a polarizer. In this way results which are again comparable to those from standard dichroism measurements using a polarizer on a homogeneously aligned sample are obtained.

Caesar and coworkers [42] have exploited the polarization measurements on solutes aligned in nematic phases to carry out vibrational assignments of solutes. This is similar to the technique discussed previously for the vibrational assignments of MBBA [11].

Of considerably greater difficulty is the measurement of polarization of Raman bands in the nematic phase. Some of the initial attempts in this direction, for example by Wang [43], observed a depolarization ratio of one in the nematic phase, which dropped to some other value, depending upon the band in question, in the isotropic phase. This is, of course, what one expects to obtain in an unaligned nematic, because the refractive index discontinuities in the nematic phase effectively scramble the polarization information. Indeed, this scrambling can also affect the relative intensity of the Raman bands in the absence of a polarization measurement, as shown by Bulkin et al. [44]. Thus, it clearly is not useful to attempt to make Raman measurements involving polarization ratios in unaligned or poorly aligned samples. This problem is much more severe than in the infrared, because, in general, infrared transmission measurements are made on very thin samples, for which excellent alignment may be produced, but Raman measurements generally require thicker samples. Another attempt by Lavrencic and Lugomer [45] also seems to be plagued by multiple scattering problems, which are at least partially responsible for some anomalies in the depolarization ratio observed by them at the phase transitions.

The only really thorough study of the depolarization of Raman-scattered radiation for nematic materials is the work of Priestly et al. [46-48]. They have used a 180° scattering geometry, rather than the usual 90° configuration used in Raman measurements. This has the advantage that the laser beam enters, and the scattered light is collected, along the same nematic direction. This permits sample thickness to be considerably less than that which would be required in a 90° configuration, and also makes analysis of the polarization ratios a bit simpler.

Because the Raman effect has its origin in the vibrational modulation of polarizability, and the polarizability is a second-rank tensor quantity, the measurement of the Raman polarizations has been shown by Priestly et al. to yield not only the usual S value (S_2) but also the next term in the distribution function, S_4. Thus the average values of $\cos^2\theta$ and $\cos^4\theta$ are determined in this measurement.

The method has been demonstrated for the case of p-butoxy benzylidene-p-cyanoaniline (BBCA) dissolved in MBBA. The solute was chosen because it has a CN stretching mode along the long axis (or at an angle very close to the long axis). This CN stretching mode is ideal for analysis of this sort, because it is well separated from any other vibrational modes. Thus, we do not have problems of overlapping bands or mixing of the PED which would greatly complicate matters. The only alternative to approaching the problem in this way would be to use a high-symmetry solute, but it would have to be one which did not perturb the nematic order.

The results of these measurements are that the S_2 data obtained by this technique reproduce those of other techniques and are consistent with the predictions of Maier-Saupe theory. For the S_4 measurements, however, the general shape of the temperature dependence is the same as that for Maier-Saupe predictions, but the actual values are considerably lower. In addition, the value of S_4 is found to be negative at temperatures close to the nematic-isotropic transition. The authors [48] have shown that one can interpret the data as being indicative of a larger tilt from the director than is predicted by Maier-Saupe theory near the nematic-isotropic transition. They point out that either molecular flexibility or intermolecular interactions of a pairwise nature are probably responsible for this.

F. Shapes of Infrared Bands in the Nematic Phase

In Sec. II, we referred to the various types of data that one could collect from vibrational spectra, and the remaining case for the nematic phase, which we discuss in this section, is the shapes of the absorption or scattering bands.

To date, no work has been reported on the shapes of Raman bands in the nematic phase. This is probably due to the complications introduced by multiple scattering, as discussed above. It seems likely that such work is being pursued, however, as it can potentially give interesting information about internal and external motions in these phases.

There has not been very much work done on the shapes of infrared bands, either. Actually, the only work reported to date is a series of studies on the half band widths of infrared bands for the alkoxy azoxy benzene series by Kirov and Simova [49-51]. This is, of course, closely related to a study of the band shape, though it is clearly not such a study. To determine a band shape, one need not, in principle, determine the entire shape, but only certain points along the band. Thus, if a Lorentzian-Gaussian sum or product function is assumed, determination of the absorption maximum, the half band width, and an additional point (conveniently twice the half band width) is sufficient to fix the Lorentzian or Gaussian contribution to the function. For more accurate studies of vibrational motion, however, the intensity must be determined more accurately far out into the wings of the band.

Simova and Kirov have determined the half widths for a number of bands of the series, and from these, using the so-called Rakov method [49], essentially an Arrhenius approach, have determined an orientational potential barrier. This barrier was found to have the same alternation property as the alkyl chain lengths, i.e., the odd-even effect was again seen. As pointed out by the authors, these potential barriers are more sensitive to the various phase transitions than are the vibrational frequencies themselves. Thus, we can expect that a more detailed study of infrared and Raman band shapes, and perhaps a study of the same band shape in both techniques, will yield considerable information about the intermolecular forces.

IV. INFRARED AND RAMAN SPECTRA OF SMECTIC MATERIALS

A. Thermotropic Phases

There has not been much work done on smectic materials to date, especially as compared with the volume of work previously discussed on nematics. This is probably due to the rather heavy concentration of all studies on liquid crystals in the years prior to 1972 on nematic phases, and a shift at about that time to greater interest in smectic phases.

If the problem were not approached in the historical context, however, one would expect that smectic liquid crystals would be more suitable choices for Raman and infrared investigations than nematics, because there is considerably less molecular motion in these phases. Hence we expect to find, and do find, better defined intermolecular modes in the smectic cases than in the nematics.

A few compounds have been studied in detail. Schnur [52] has considered the material diheptyloxyazoxybenzene, the first homolog of that series which shows a smectic phase. He has reported that the low-frequency Raman spectrum shows a complex of overlapped lattice vibration bands which seems to have a maximum at 76 cm^{-1}. In the smectic phase, some of the complex of bands seems to have disappeared, leaving a single sharp mode at 81 cm^{-1} superimposed on the background. According to Schnur, this mode persists at 78 cm^{-1} in the nematic phase, but disappears in the isotropic liquid. Again, as in the nematic members of this series discussed in Sec. III, the accordion mode (at 271 cm^{-1}) seems to decrease in intensity in the smectic phase and cannot be seen in the nematic and isotropic phases.

Amer and Shen [53] have reported the spectra of diethyl azoxybenzoate and diethylazoxy cinnamate in the 5- to 100-cm^{-1} region. According to their data, both of these compounds show a single band in this region in the crystal phase, at 22 cm^{-1} for the benzoate and 26 cm^{-1} for the cinnamate. At the crystal-to-smectic transition, the mode shifts to 14 cm^{-1} for both compounds. In the isotropic liquid it disappears completely. The intensity of the intermolecular mode is apparently reduced in the smectic phase, although this may be just an increase in width. The changes appear to be quite abrupt.

A final report on external modes for thermotropic smectic phases has been published by Schnur and Fontana [29]. They have studied the molecule TBBA. This molecule is interesting in that it exhibits smectic B, C, and A phases, as well as a nematic phase. For this case the solid phase shows a single low-frequency band at 22 cm^{-1}. This mode is also present in the smectic B phase, perhaps shifted slightly to lower frequency but with no great change in relative intensity. At the smectic B-C transition however, the band disappears completely. Some higher frequency scattering in the external mode region, at about 80 cm^{-1}, does seem to persist in the smectic B and C phases, and can still be seen, albeit greatly weakened in intensity, in the smectic A and nematic phases. This disappears in the isotropic liquid.

A curious feature in the TBBA spectra occurs in the 1,500- to 1,600-cm^{-1} region, where Schnur and Fontana have observed that a single band in the crystal and smectic B phases changes to three bands in the smectic C phase, a spectrum which persists in all the higher temperature phases as well.

The similarity of the smectic B to the crystal spectrum is good evidence that this phase is very like the solid and that motional freedom is quite restricted. This is in agreement with x-ray data. The 19-cm^{-1} mode, observed in the smectic B phase, is assigned to a hindered rotation of the molecules, and consistent with other data discussed above, this may be a reasonable assignment for such a heavy molecule. It would be interesting to see comparable far-infrared data on this molecule.

Amer and Shen [53] have also assigned the band observed in the smectic A phase of their compounds to a hindered rotation. It is curious that, in view of the similarity of the frequencies of these modes, one should be able to observe the low-frequency band in the smectic A phase in two cases (Amer and Shen) but not in a third (Schnur and Fontana). Clearly, more cases need to be studied before any generalizations about what can be learned concerning motional freedom from these spectra can be made.

As to the presence of modes reported by Schnur and Fontana, in the more disordered, higher temperature phases which are not present in the more ordered phases, the authors have given no explanation. If we consider this from the point of view of symmetry and selection rules, it seems possible that this could occur if the selection rules of the ordered phase were stricter than those of the disordered phase. In this case, some bands which were forbidden become allowed as the temperature is raised. It is a bit surprising, if this is the explanation, that the phenomenon is not seen for more of the internal modes. Again, it would appear that a group theoretical analysis of the problem might be informative.

B. Lyotropic Phases

It is somewhat surprising that there has not been more Raman spectroscopy done on lyotropic liquid crystals, in view of the ability of the Raman technique to probe systems involving water. There has been some work, of a qualitative nature, over many years, on some of the gels and liquid crystalline phases, but none of this, with the exception of the work on phospholipid-water systems discussed below, has really been aimed at elucidation of the phase structure or the nature of the phase transition.

Larsson [54] has recorded Raman spectra of a variety of simple lipids in the presence of water at various concentrations and temperatures. This study has concentrated on the CH stretching vibrations, which appear to be quite sensitive to the phase transition between solid and the various liquid crystalline phases. This sensitivity arises in the relative intensity of the various CH stretching vibrations, which are localized on an individual carbon atom. In a sample of a complex lipid, the overall intensity observed in the CH stretching region comes from several modes, symmetric and asymmetric CH_2 and CH_3 stretches. In the sample, the intensities of these modes will be averaged in some way due to the conformations of the hydrocarbon chains. Thus, a set of Raman band intensities will be observed. To date, it is not possible to make a priori predictions of what this intensity distribution will be. Nonetheless, Larsson has shown that even in simple lipids the pattern of relative intensities is quite sensitive to the phase structure. This implies that the arrangements, fluidities, and conformations of the chains is different in the different phases.

Completely analogous results have been seen in the more complex systems of phospholipids with water. The systems have been investigated by Lippert and Peticolas [55], Bulkin and Krishnamachari [56], and by Brown et al. [57]. The various investigations span the infrared [56] and Raman spectra [55-57]. In all cases it is found that modes associated with the hydrocarbon tails of the phospholipids are quite sensitive to the phase transition gel-liquid crystal which is characteristic of these systems in the presence of 20% water. The relative intensities of the CH stretching vibrations change at this transition, the width of the CH_2 deformation vibration increases, and the number of modes observed in the C-C stretching region increases. Any of these can and have been used to estimate fluidity in the lipid bilayer as a function of temperature. The fluidity is an important quantity in the use of these systems as membrane models. Brown et al. have noted, in addition to the changes in hydrocarbon vibrations, a change in the PO_2 symmetric stretching frequency which is attributed to a change in the exposure of this group at the phase transition, as well as a change in the frequency of the hydrocarbon translational mode near 150 cm^{-1}, which can also be correlated with the chain extension.

To date, no bands have been seen in the low-frequency infrared or Raman spectra of these lyotropic phases which could be associated with intermolecular modes. Such modes may be observable in the future, particularly pseudointermolecular modes of the polar head groups vibrating against the water sheath, and vibrations of the water bound to the polar head. In addition, one might expect that in the case of ionic (not zwitterionic) lyotropic phases, vibrations of the bound ions such as have been observed for ion pairs in the far-infrared region by others [58] might be seen in these cases. Such studies have not yet been carried out, presumably because of the substantial interference from the liquid water absorption band, which is very intense and broad.

V. INFRARED AND RAMAN SPECTRA OF CHOLESTERIC MATERIALS

There have been no extensive studies done on the infrared spectra of cholesteric materials from the point of view of intermolecular interactions or vibrational assignments; what has been done has simply been to use this phase as a special anisotropic medium. Thus, there have been several reports of induced optical activity in the infrared region using cholesteric phases [59,60] as well as with an analogous twisted nematic phase [61]. In addition, Lephardt and Bulkin [62] have used a cholesteric phase as a stationary medium for fractionating sample from carrier gas in a gas chromatography-infrared on-the-fly system. The cholesteric phase is an inert solvent in this respect. A related study by Bulkin and Krishnan [63] showed that the vibrational spectra of molecules dissolved in cholesteric solvents are basically unchanged in frequency from the spectra of pure liquids. The study implied that the environment of most of the molecules of solute in the cholesteric phase is the same as that in the pure liquid, and hence implied further that clustering of solute molecules might be occurring. This has not been investigated further.

Only three studies have been reported using Raman scattering [64-66], and one of these [65] seems to be plagued by nonlasing emission lines, or other spurious emission. In any case, the spectrum of cholesterol reported in that study is not reproduced by other workers.

The only detailed study to be presented of the effect of the phase transition on the Raman spectrum is that of Bulkin and Krishnan [64]. They found that, in contrast to the nematic materials, the spectrum of cholesterics in the middle frequency region, 200-1,800 cm^{-1}, is not particularly affected by the transitions from crystal-cholesteric or from cholesteric-isotropic liquid. This implies that the coupling of the internal modes in cholesterics to the lattice is weaker than in nematics.

In the CH stretching region, however, changes were observed for a number of cholesterol esters at the crystal-cholesteric and cholesteric-isotropic phase transitions. These changes could be explained as having their origin in changes in the environment of the angular methyl groups on the steroid framework as the transitions were traversed. It was found that the environment of these angular methyl groups in the cholesteric phase was more similar to that in the crystal than in the isotropic liquid. It appears that these groups, which protrude rather far from the plane of the steroid framework, are a good probe of the neighboring layer order. In the isotropic phase this order is perturbed, and the relative intensities of these bands change. Some of the changes reported by Larsson for lyotropic systems are similar to these [54].

An interesting case discussed by Bulkin and Krishnan is the mixture of 60:40% cholesteryl nonanoate-cholesteryl chloride. For this case, there are again changes in the spectrum at the Ch-I transition. If a spectrum is constructed of an appropriately weighted average of the spectra of the pure chloride and nonanoate spectra, this averaged result is quite different from the observed spectrum in the cholesteric phase but identical with the observed spectrum in the isotropic liquid. This confirms the idea that in the mixed liquid crystals there is a completely homogeneous system, with a different environment from that of either of the pure materials. An attempt was made in this study to see if the Raman spectra were sensitive to pitch, but no such dependence was found.

Chang [66] studied the Raman spectrum of several cholesterol esters in the 5- to 100-cm^{-1} region. He found that all the samples showed two sharp bands at about 18 and 24 cm^{-1}, in the crystalline phase. While one might suspect that such a result indicates that the bands are spurious in origin, this was ruled out by Chang on the basis of other measurements. He has discussed the assignments of these bands to lattice vibrations of the crystal. It should be pointed out that one might expect some variation in the lattice vibration frequencies between the various esters, as these depend on molecular mass and moment of inertia. It seems possible that the bands observed might correspond to a vibration of a fairly large part of the steroid framework, a pseudolattice vibration. Dimic and Osredkar [67] observed a mode in several liquid crystals at 15 cm^{-1} which they feel may be related to Chang's observations. Further work is needed to clarify whether this is the case. It is also possible that these modes have an intramolecular origin.

REFERENCES

1. W.F. Edgell, J. Phys. Chem., 75:1343 (1971).
2. S. Bhagavantam and K. Venkatarayudu, Theory of Groups and Its Application to Physical Problems, Andhra Univ. Press, Waltair, India, 1962.
3. W. Fateley, F.R. Dollish, N.T. McDevitt, and F.F. Bentley, Infrared and Raman Selection Rules for Molecular and Lattice Vibrations, Wiley, New York, 1972.
4. G. Herzberg, Infrared and Raman Spectra of Polyatomic Molecules, Van Nostrand, Princeton, 1945.
5. K.D. Moller and W.G. Rothschild, Far Infrared Spectroscopy, Wiley, New York, 1971.
6. R. Freymann and R. Servant, Ann. Phys. (Paris), 20:131 (1945).
7. W. Maier and G. Englert, Z. Elektrochem., 62:1020 (1958).
8. N.M. Amer and Y.R. Shen, J. Chem. Phys., 56:2654 (1972).

9. J. Schnur, Mol. Cryst. Liq. Cryst., 23:155 (1973).
10. J.P. Heger and R. Mercier, Helv. Phys. Acta, 45:886 (1972).
11. B.J. Bulkin, T. Kennelly, and W.B. Lok, in Liquid Crystals and Ordered Fluids, Vol. 2 (J.F. Johnson and R. Porter, eds.), 1974, p. 85.
12. G. Vergoten, Adv. Raman Spectr., 1:219 (1972).
13. L.J. Bellamy, The Infrared Spectra of Complex Organic Molecules, Wiley, New York, 1958.
14. G. Vergoten, Thesis, Univ. des Sciences et Techniques, Lille, France, 1973.
15. A.S. Zhadnova, L.F. Morozova, G.V. Peregudov, and M.M. Sushchinskii, Opt. Spektrosk., 26:209 (1969).
16. B.J. Bulkin, D. Grunbaum, and A. Santoro, J. Chem. Phys., 51:1002 (1969).
17. N.K. Nikolov and P.D. Simova, Proc. Bulg. Acad. Sci., 26:1013 (1973).
18. H. Itoh, H. Nakatsuka, and M. Matsuoka, J. Phys. Soc. Jap., 34:841 (1973).
19. T. Riste and R. Pynn, Solid State Commun., 12:409 (1973).
20. C.H. Carlisle and C.H. Smith, Acta Crystallogr., 25:547 (1969).
21. J.M. Janik, J.A. Janik, and W. Witko, Acta Phys. Pol., A44:483 (1973).
22. M. Sorai, T. Nakamura, and S. Seki, Proceedings International Symposium Liquid Crystals (Raman Jubilee), Bangalore, India, 1974.
23. B.J. Bulkin and D. Grunbaum, in Liquid Crystals and Ordered Fluids, (R. Porter and J. Johnson, eds.), Plenum, New York, 1970, p. 303.
24. B.J. Bulkin and F.T. Prochaska, J. Chem. Phys., 54:635 (1971).
25. N.M. Amer, Y.R. Shen, and H. Rosen, Phys. Rev. Lett., 27:718 (1970).
26. B.J. Bulkin and W.B. Lok, J. Phys. Chem., 77:326 (1973).
27. M. Evans, M. Davies, and I. Larkin, J. Chem. Soc., Far. Trans. II, 69:1011 (1973).
28. E. Sciesinska, J. Sciesinski, J. Twardowski, and J.A. Janik, Report No. 847/PS, Institute of Nuclear Physics, Cracow, September, 1973.
29. J.M. Schnur and M. Fontana, J. Phys. (Paris), 35:L-53 (1974).
30. J.M. Schnur, M. Hass, and W.L. Adair, Phys. Lett., 41A:326 (1972).
31. K.K. Kobayashi, Mol. Cryst. Liq. Cryst., 13:137 (1971).
32. W.G.F. Ford, J. Chem. Phys., 56:6270 (1972).

33. B.J. Bulkin and D. Grunbaum, J. Phys. Chem., in press; see also, D. Grunbaum, Ph.D. Thesis, City Univ., N.Y., 1974.

34. J. Billard, M. Delhaye, J.C. Merlin, and G. Vergoten, C. R. Acad. Sci., Paris, Ser. B, 273:1105 (1971).

35. W.J. Borer, S.S. Mitra, and C.W. Brown, Phys. Rev. Lett., 27:379 (1971).

36. J. Doucet, A. Levelut, and M. Lambert, Mol. Cryst. Liq. Cryst., in press.

37. B.J. Bulkin, Helv. Chim. Acta, 52:1348 (1969).

38. S.G. Kroon and J. Van der Elsken, Chem. Phys. Lett., 1:185 (1967).

39. E. Constant, Y. Leroy, and L. Raczy, C. R. Acad. Sci., Paris, Ser. 261:4687 (1965),

40. A. Sakamoto, K. Yoshino, U. Kubo, and Y. Inuishi, Japan J. Appl. Phys., 13:359 (1974).

41. V.D. Neff, L.W. Gulrich, and G.H. Brown, in Liquid Crystals (G.H. Brown, G.J. Dienes, and M.M. Labes, eds.), Gordon and Breach, New York, 1966, pp. 21-35.

42. G.P. Caesar, R.A. Levenson, and H.B. Gray, J. Amer. Chem. Soc., 91:772 (1969).

43. C.H. Wang, J. Amer. Chem. Soc., 94:8605 (1972).

44. B.J. Bulkin, J.O. Lephardt, and K. Krishnan, Mol. Cryst. Liq. Cryst., 19:295 (1973).

45. B. Lavrencic and S. Lugomer, Adv. Raman Spectr., 1:215 (1972).

46. E.B. Priestley and P.S. Pershan, Mol. Cryst. Liq. Cryst., 23:369 (1973).

47. E.B. Priestley, P.S. Pershan, R.B. Meyer, and D.H. Dolphin, Vijnanda Parishad Anusandhan Patrika, 14:93 (1971).

48. S. Jen, N.A. Clark, P.S. Pershan, and E.B. Priestley, Phys. Rev. Lett., 31:1552 (1973).

49. N. Kirov and P. Simova, Phys. Lett., 37A:51 (1971).

50. N. Kirov and P. Simova, Spectrochim. Acta, 29A:55 (1973).

51. P. Simova and N. Kirov, Adv. Mol. Relaxation Proc., 5:233 (1973); see also Acta Phys. Acad. Sci. Hung., 35:183 (1974).

52. J.M. Schnur, Phys. Rev. Lett., 29:1141 (1972).

53. N.M. Amer and Y.R. Shen, Solid State Commun., 12:263 (1973).
54. K. Larsson, Chem. Phys. Lipids, 10:165 (1973).
55. J.L. Lippert and W.L. Peticolas, Proc. Natl. Acad. Sci. USA, 68:1752 (1971).
56. B.J. Bulkin and N. Krishnamachari, J. Amer. Chem. Soc., 94:1109 (1972).
57. K.G. Brown, W.L. Peticolas, and E. Brown, Biochem. Biophys. Res. Commun., 54:358 (1973).
58. W.F. Edgell and J. Lyford IV, J. Amer. Chem. Soc., 93:6407 (1971).
59. I. Chabay, Chem. Phys. Lett., 17:283 (1972).
60. B. Schrader and E.H. Korte, Angew. Chem. (Int. Ed.), 11:226 (1972).
61. B.J. Bulkin, D. Grunbaum, T. Kennelly, and W.B. Lok, in Proceedings International Symposium Liquid Crystals (Raman Jubillee) (S. Chandrasekhar, ed.),
62. J.O. Lephardt and B.J. Bulkin, Anal. Chem., 45:706 (1973).
63. B.J. Bulkin and K. Krishnan, presented at the IV International Conference Liquid Crystals, Kent, Ohio, 1972.
64. B.J. Bulkin and K. Krishnan, J. Amer. Chem. Soc., 93:5998 (1971).
65. J.S. Chaudry, G. Subba Rao, and R.K. Mishra, Ind. J. Med.,
66. R. Chang, Mol. Cryst. Liq. Cryst., 12:105 (1971).
67. V. Dimic and M. Osredkar, Mol. Cryst. Liq. Cryst., 19:189 (1973).

Chapter 11

CHARACTERIZATION OF THE VISCOUS CRYSTALLINE AND OTHER PHYSICAL PHASES OF POLY(DIETHYLSILOXANE)

John M. Pochan, Michael F. Frois,*
Andreas O. Goedde, Charles L. Beatty† and Darlyn F. Pochan

Xerox Corporation
Rochester, New York

I. INTRODUCTION

Polymers are considered to exist in two phases, amorphous and crystalline. While many polymers have been shown to be completely amorphous, 100% crystalline systems have only been attained with special crystallization procedures, and hence most polymers that crystallize exist in the

*Present Affiliation: Celanese Research Company, Summit, New Jersey.
†Present Affiliation: Department of Materials Science & Engineering, University of Florida, Gainesville, Florida 32611

semicrystalline form [1]. Some polymers which contain optically active sites (for example, polybenzyl-γ-L-glutamate) have been shown to exhibit liquid crystalline behavior in solution. This type of behavior is known as lyotropism [2]. The poly(biphenylacrylates) [3-5] which have special steric constraints form thermally induced mesophases in which the polymers are homogeneously ordered throughout [5]. Recently, a new ordered polymeric form has been discovered in which a thermal phase of poly(diethylsiloxane) (PDES) exists as two phases: a partially ordered one (possibly similar to a liquid crystal phase) and a totally amorphous phase. Because no suitable nomenclature for such a system has been established the phase was termed "viscous crystalline." As far as can be ascertained, the phase is unique. A variety of studies on PDES and the viscous crystalline phase have been published [6-10], but a unified account of the studies has not been presented. It is the purpose of this report to review those findings briefly and to suggest relations between the viscous crystalline phase and the crystallization process in polymers.

II. EXPERIMENTAL

The extent of this experimental section is to cover new experimental techniques used in the PDES research which have not been previously published and to present relevant data on the PDES polymer.

The PDES used in all the experiments was obtained from C.L. Lee, Dow-Corning Corp., Midland, Michigan. The original sample was precipitated from toluene into cold methanol ($\overline{M}_n$ = 112,000, $\overline{M}_w/\overline{M}_n$ = 2.5). The material was fluid and transparent at room temperature. This sample is noted as PDES-I. The sample was further precipitated and fractionated ($\overline{M}_w$ = 250,000, $\overline{M}_w/\overline{M}_n$ = 2.0) yielding a transluscent viscous material at room temperature. This is denoted as PDES-II.

Dielectric data were taken by standard techniques [7] and also with a scanning dielectric device which is fully described elsewhere [11]. The essential parts of the instrument are a General Radio 1683 Automatic R.L.C. Bridge, a DORIC digital thermocouple, a Delta Design environmental oven (-180 °C to +250° C) with variable rate linear programming, and a DATOS (Dallas, Texas) multiplex system which cues the RLC bridge and digital thermocouple for parallel readings and then sequentially sends the data to a Texas Instruments Silent 700 tape and teletype terminal in ASCII code. After acquisition, the data can be manipulated via computer and plotted (see below).

III. DISCUSSION

Few scientific studies on PDES have been published [12a, b]. One of these [12a] indicated a double glass transition for PDES. This double T_g was the impetus for the initiation of our studies. The heat capacity curve for PDES-I is shown in Fig. 1 where it is seen that at least three calorimetric

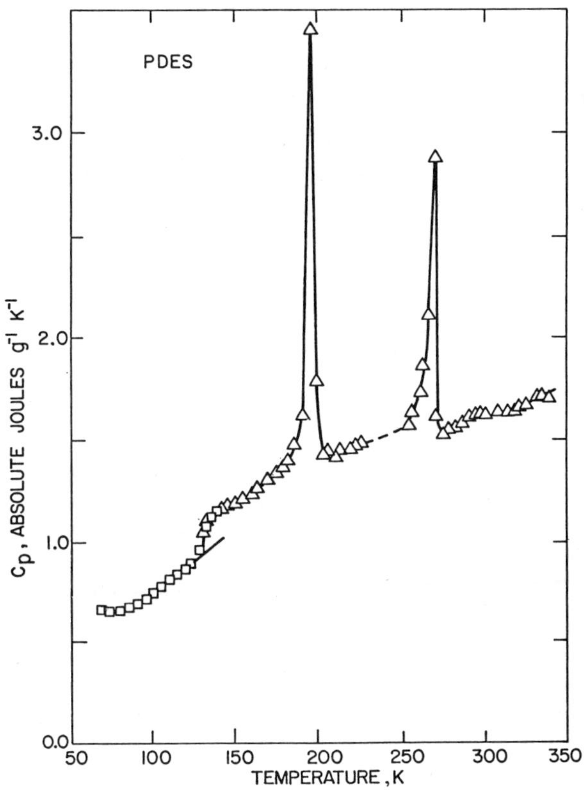

FIG. 1. Heat capacity of PDES as a function of temperature. Adiabatic calorimetry results.

events occur, Tg (≈ 140 K), a crystal-crystal transition (195 K) (as determined via x-ray diffraction [9]), and a crystal melt at 270 K. The apparent upturn in C_p at ≈ 70 K may also be indicative of another calorimetric event. Figure 2 shows a differential scanning calorimeter trace of PDES-I in the upper temperature region. The crystal-crystal and crystal melt are observed, and additionally an apparent weak transition at ≈ 295 K. In our original work, this additional peak was a curiosity. However, it became apparent, after reprecipitation, that the translucent phase and this calorimetric event were intimately related. Because of the translucence of the phase and its meltinglike behavior, our first hypothesis was that the phase was semicrystalline (similar to the lower temperature phases) but with much less crystallinity. These thoughts were dispelled when nuclear magnetic resonance (NMR), dielectric, x-ray, and optical microscopy experiments were done on the translucent phase. Rather than describe the results of these techniques together, the remainder of the discussion will be devoted to (1) a description of the results of the individual techniques, followed by (2) a general summation of the results, their interrelation, and what they tell of the viscous crystalline phase.

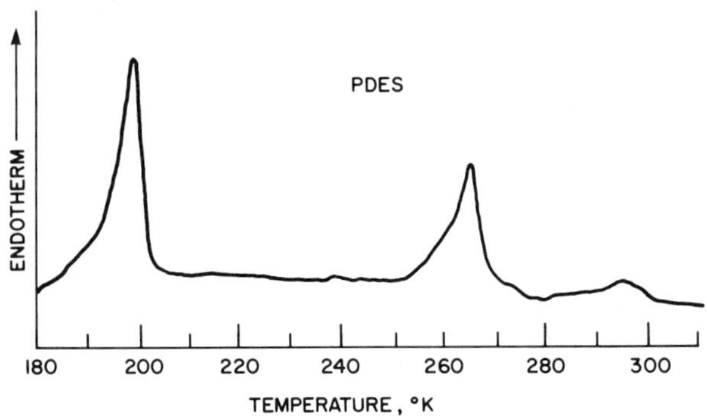

FIG. 2. Heat capacity of PDES as a function of temperature via differential scanning calorimetry.

A. Dielectric Results

Plots of tan δ and relative dielectric constant (ε′) for PDES-II are shown in Figs. 3 and 4. The peaks observed calorimetrically are also observed in one or the other dielectric scans. T_g, an apparent upturn in tan δ near room temperature signifying the crystal melt, and a small dip at slightly higher temperatures signifying the viscous crystalline melt are observed in the ε′ plot. Note is also made of a change in slope of relative ε′ that occurs at 255 K. This event is directly correlatable to an observed NMR phenomena discussed later. T_g and the crystal-crystal transition are easily discerned in the relative ε′ data. The relative ε′ data was not corrected for changes in density as a function of temperature. The abrupt increase in ε′ at the crystal-crystal transformation does indicate a more polar contribution to the dielectric properties. T_g was investigated dynamically [7] in the frequency range $10\text{-}10^5$ Hz and found to have an apparent activation energy of 14 kcal/mol, a value slightly less than that of poly(dimethylsiloxane). Thus, increasing the side-chain length of the siloxanes decreases the apparent activation energy for T_g, an effect noted earlier in other systems such as the poly(alkylmethacrylates) [13]. It should be noted that the sub-T_g relaxation observed calorimetrically was also observed dielectrically.

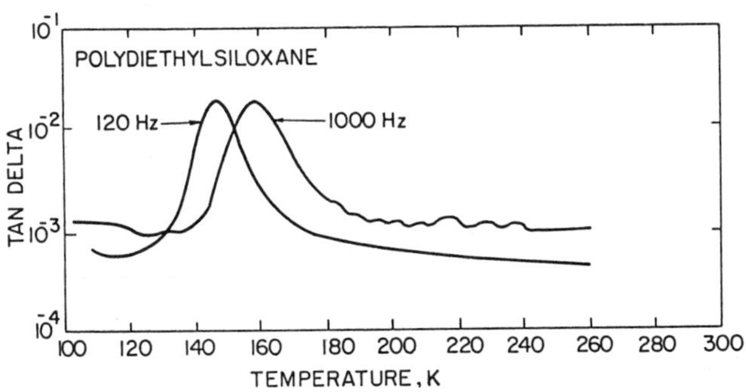

FIG. 3. Tan δ versus temperature for PDES-II at 120 Hz and 1 kHz).

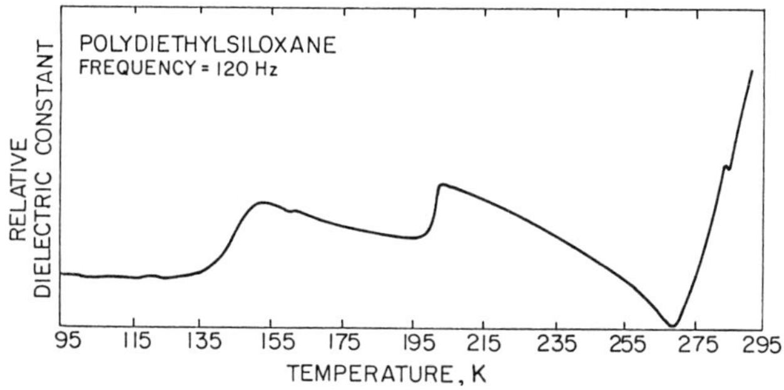

FIG. 4. Relative dielectric constant versus temperature for PDES-II at 120 Hz.

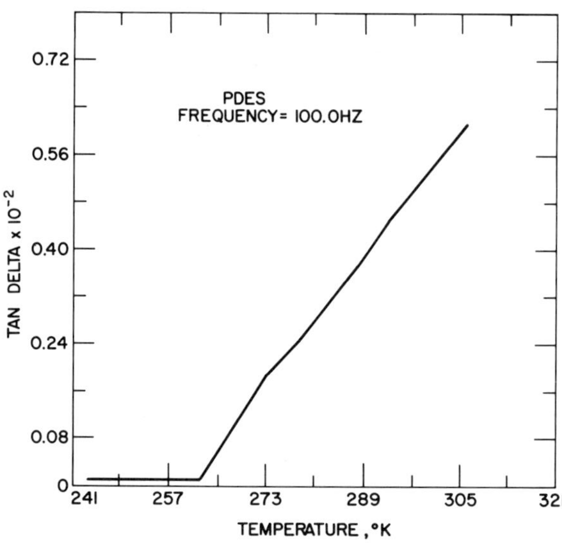

FIG. 5. Dielectric tan delta versus temperature for PDES.

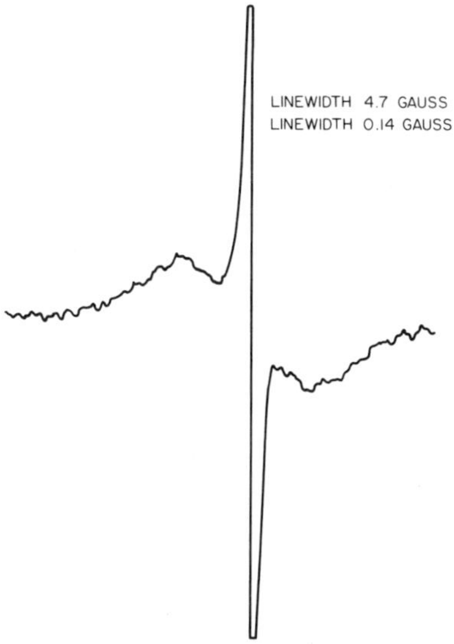

FIG. 6. Wide-line proton resonance of PDES-II at 258 K.

The gradual increase in tan δ when formation of the viscous crystalline phase occurs is the more important result. Data is shown in Fig. 5. This data initially led to the suggestion of a partially ordered phase, either homogeneous or heterogeneous. In the former case, increase in the molecular mobility with temperature would provide the increase. In the latter, a change in the heterogeneous structure with temperature would provide a more dielectrically mobile phase. The latter postulate was verified via NMR.

B. NMR: Wide-Line and Pulsed Results

Wide-line spectra of PDES-II are shown in Figs. 6 and 7. In Fig. 6, the spectra are shown at 258 K where the polymer is semicrystalline. At this temperature, the resonance line consists of two components: a broad component due to the crystalline region and a narrow component due to the amorphous region. The crystalline resonance is much narrower than one would expect for a rigid lattice. Above the crystal melt, the line width ascribed to the crystal lattice narrows (see Fig. 7) but does not become isotropic in nature, indicating that residual order exists in this temperature

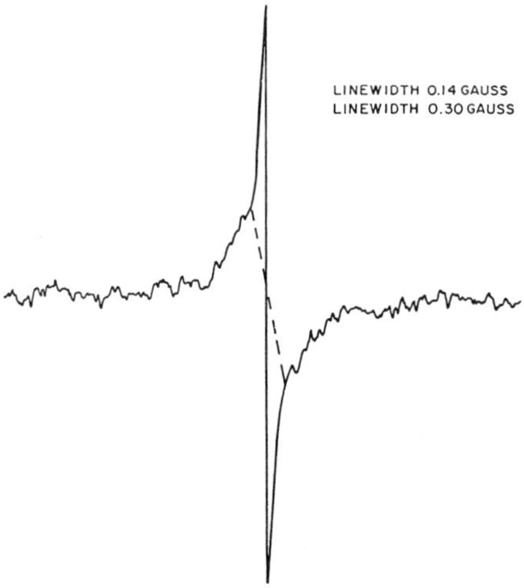

FIG. 7. Wide-line proton resonance of PDES-II at 283 K.

region. Comparing these data with the NMR relaxation time data (Figs. 8 and 9) reveals other features of the ordered phase. In the viscous crystalline region, the free induction decay (FID) consists of two components (see Fig. 9). In fact, prior to the crystal melt at ≈ 273 K, the amorphous phase exhibits increased mobility relative to the signals observed from the crystalline phase. This occurs at ≈ 258 K and as is seen in Fig. 9, where the spin-spin relaxation time T_{2a} (the subscript signifying the signal related to the amorphous component of the polymer) increases almost an order of magnitude and then levels prior to the crystal melt. At 270 K, crystal melt occurs, and T_{2c} (the spin-spin correlation time of the crystal component) increases until approximately 300 K where only one signal due to the totally amorphous polymer is observed. These effects are also manifest in the T_1 data. From the FID, the ratio of ordered to nonordered phase was found to decrease throughout the temperature range of the viscous crystalline region. The NMR data therefore demonstrates that the viscous crystalline region consists of regions of residual order in an amorphous phase and the relative ratio of these two phases (order:disorder) decreases as the isotropic melt is approached.

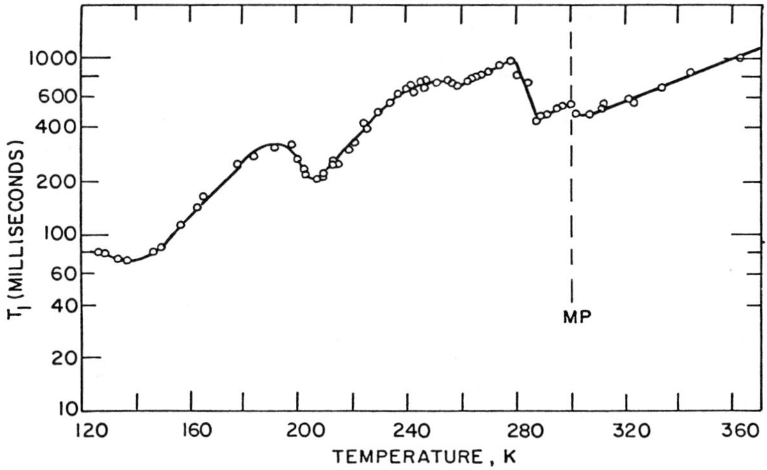

FIG. 8. Spin-lattice (T_1) relaxation times of PDES-II as a function of temperature.

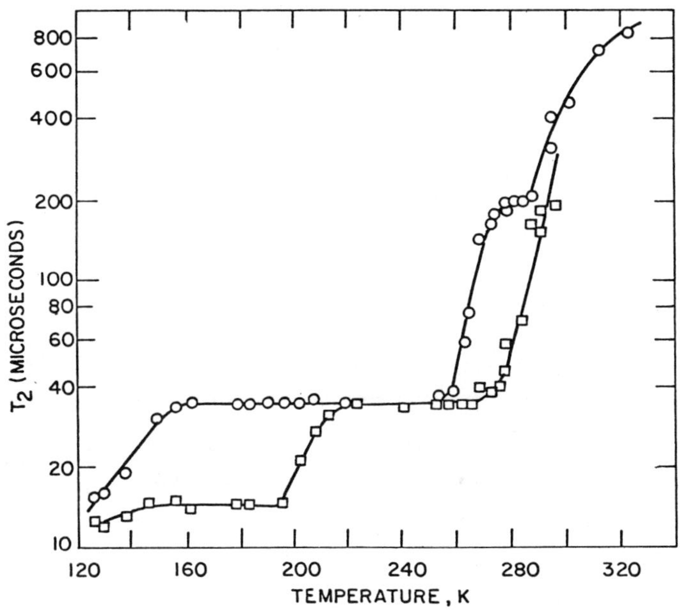

FIG. 9. Spin-spin relaxation times of PDES-II as a function of temperature (O, T_{2a}; □ T_{2c}).

When all the NMR relaxation data are considered, an interesting picture of PDES is obtained. All transitions or changes observed in T_2 are seen in the T_1 data: T_g at 138 K, the crystal-crystal transformation at ≈ 205 K, the change in the amorphous T_{2a} at ≈ 260 K, the crystal melt at ≈ 280 K, and the viscous crystalline melt at 300 K. More important is the T_2 data where relaxations from the amorphous and crystalline phases are observed. It is seen that (1) as T_{2a} changes at the glass transition a slight change in T_{2c} also occurs, (2) above the crystal-crystal transition $T_{2a} = T_{2c}$, and (3) T_{2a} does not increase monotonically with temperature above the glass transition but changes as certain relaxations (or phase changes) take place in the polymer. These facts indicate that the amorphous and crystalline components are in close association and that changes in one phase will many times affect the response of the other phase. Above the crystal-crystal transformation, both phases have identical T_2s over a 30-K range. This "coupled" relaxational behavior is usually not observed above T_g for the amorphous segments of a polymer, but instead T_{2a} would usually increase monotonically with temperature.

It must be noted that this technique does measure proton relaxation times, and, therefore, the results are due to ethyl or methyl group motion which is affected by backbone motion. Methyl group rotation usually takes place in polymeric systems at much lower temperatures, and it must be assumed that the observed relaxations are due to ethyl group relaxations or the effect of backbone motion on ethyl group relaxations.

C. X-Ray Diffraction

Since calorimetric and NMR data indicate ordering in the viscous crystalline region (270- → 295 K) of PDES-II, x-ray diffraction data was taken to see if the ordering was crystallike in nature. The wide-angle results are shown in Fig. 10 for temperatures in the amorphous region and the viscous crystalline region. The single diffraction maximum at 11.8 Å 2θ is similar to those observed and theoretically explained in nematic and smectic liquid crystals [14,15] and is indicative of one-or two-dimensional ordering. The value of 2θ corresponds to an interchain spacing of 8.7Å in the partially ordered state. The presence of the same diffraction peak at

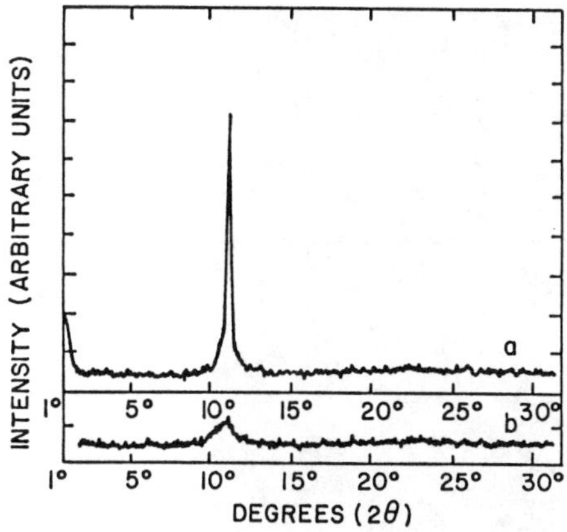

FIG. 10. Wide-angle X-ray diffraction traces of (a) the partially ordered phase and (b) the isotropic melt at 283 and 303 K, respectively, for PDES-II.

303 K (although of reduced intensity) indicates that (1) the spacings between chains in the melt are similar to the spacings on the partially ordered region, or (2) some ordered regions remain at the elevated temperature. In light of the NMR results [8] and the data in Fig. 10, the former model is better. The breadth of the transition in Fig. 10 is probably due to the poor heat transfer characteristics of the improvised apparatus used for scanning the temperature.

D. Optical Microscopy

The techniques cited above indicate that PDES-II exists as a two-phase System between 288 °C and 303 K, one partially ordered and the other amorphous. If the ordered phase is similar to a liquid crystal material, the material should be birefringent. Optical microscopy with polarized light is shown in Figs. 11, 12, and 13. Figure 11 shows a slightly birefringent phase with large areas (≈ 10μm), of birefringence. The birefringent pattern does not nucleate and grow in the classic sense of polymer crystallization [16,17] but develops and intensifies in time. More will be said of this later. Of great significance is the crystallization of the polymer from the viscous crystalline phase. If Figs. 11a and 11b are compared, the

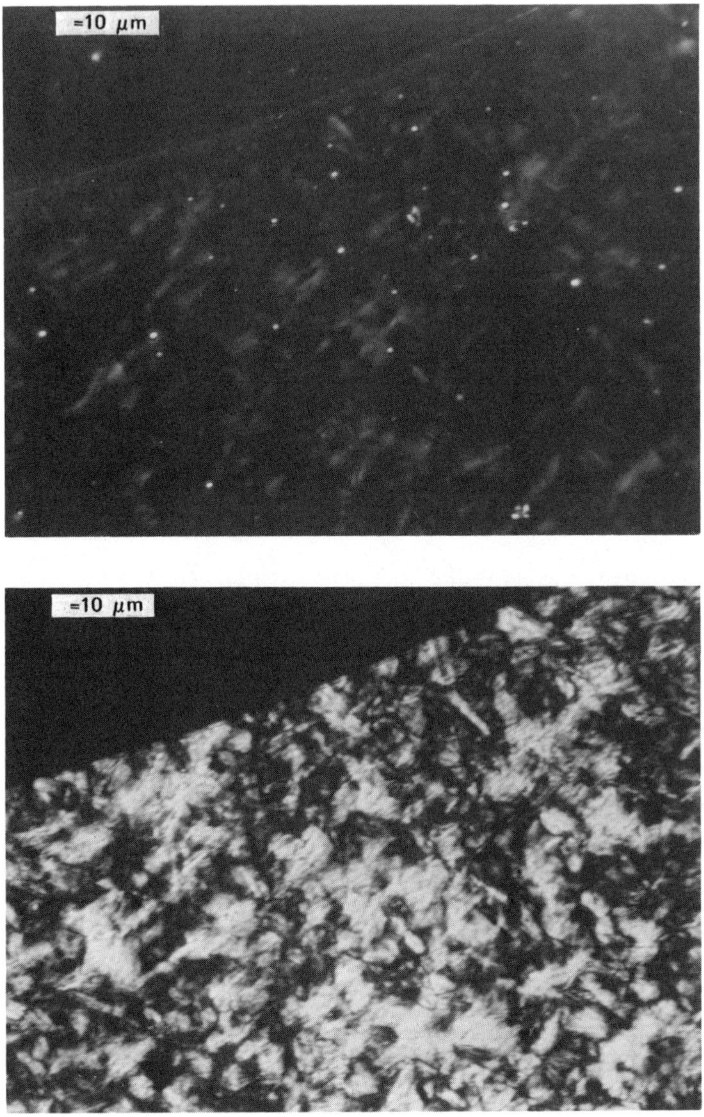

FIG. 11. (top) PDES-II at 273.2 K (crossed polarizers) at 200 X (viscous crystalline temperature); (bottom) PDES-II at 269.2 K (crossed polarizers) at 200 X (temperature of crystallization).

FIG. 12. (top) PDES-II at 269.2 K (crossed polarizers) at 200 X. Sample obtained by quenching to 269.2 K from 333 K; (bottom) PDES-II at 271.1 K (crossed polarizers) at 200 X. Sample formed by heating a to 271.2 K. Scale indicates 10 μm between markers.

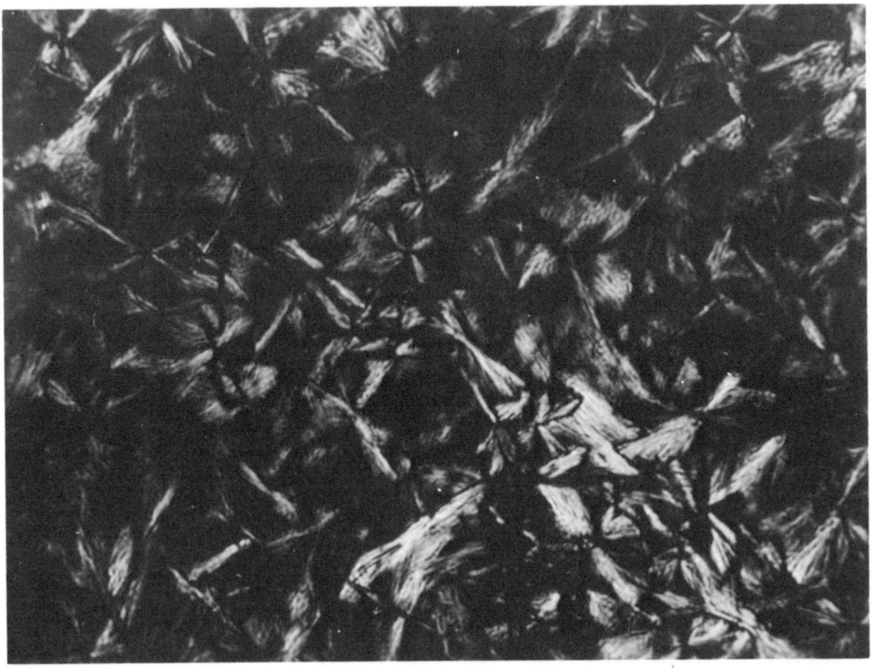

FIG. 13. PDES-II at 254.8 K (crossed polarizers) at 200 X. Sample quenched from 292.2 K. Same scale as Fig. 11.

birefringent areas of Fig. 11a give way to those of Fig. 11b. Likewise, if a sample is quenched to crystallinity from the amorphous phase, the crystal phase gives way to a birefringent viscous crystalline phase when heated above the melt (see Fig. 12). In the latter case, the birefringence of the viscous crystalline phase is greatly enhanced over that which is thermally formed by cooling the amorphous phase. Whether the formation of the phase via the alternative methods would produce identical birefringences at long times has not been investigated.

The results indicate that the viscous crystalline phase orders the polymer's structure prior to crystallization. In fact, when crystallization takes place from the viscous crystalline phase it occurs not as nucleation and growth, but rather as a development of the birefringence of the viscous crystalline phase. Similarly, crystal formation appears to preorder the viscous crystalline phase. Although it would appear that the phase is thermodynamically allowed, these results suggest that the kinetics governing viscous crystalline formation from the isotropic melt are slow.

Various morphologies of the crystalline material can be observed. This is seen by comparing Figs. 11 and 13. In Fig. 13, banding occurs which is characteristic of lamellar twisting during crystallization [1]. Also, if a free surface of the polymer is observed during viscous crystalline formation, surface ripples occur as the phase is formed. This has been taken as an indication of an increase in density, although density measurements have not been conducted.

E. Kinetics of Phase Formation

It is obvious from the microscopic studies that the viscous crystalline and crystalline structures of PDES are related. The following studies on the kinetics of crystalline formation and viscous crystalline formation were done in the hope that their results would shed light on this interrelation. As described in Ref. 10, two types of experiments were conducted. One monitored sample birefringence as a function of time and was done by monitoring the response of a phototube mounted in the eyepiece of a microscope (cross polarizers) during the quenching and crystallization process. The other experiment involved the use of sensitive photographic film to monitor the development of the viscous crystalline phase as a function of time after quenching from the amorphous state to a specific temperature. The former experiment monitored the crystallization phenomena (the birefringence developed by the viscous crystalline phase was too weak to monitor with the phototube), while the latter was used to observe the kinetics of viscous crystalline formation.

Typical phototube responses for crystallization are shown in Fig. 14. Two parameters were attainable from this date: (1) induction time (nucleation time) and (2) initial relative growth rate of the birefringence (initial slope of the change in transmission). Two other effects were noted during some of these experiments: a decrease in transmission near the initial nucleation (see Fig. 14a) and a change in the growth curve after an initial fast crystal growth. The former effect was attributed to turbidity and the latter to secondary nucleation phenomena.

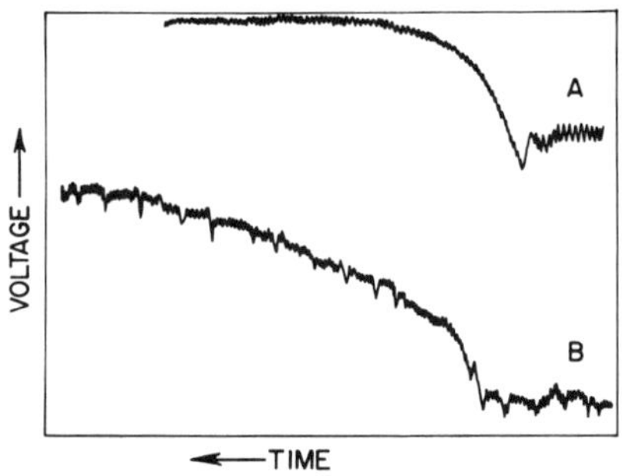

FIG. 14. Phototube response as a function of time for the crystallization of PDES-II after quenching at 266.2 K (a) and 261.2 K (b) from 323 K. Time scale for a is approximately 6 min. and that of b approximately 40 min.

Figure 15 is a plot of log induction time versus inverse temperature for PDES samples quenched from 323 K to the temperature noted. Data at temperatures below 269 C are not accurate because the quenching time of the sample is greater than the nucleation rate. The activation energy described by the linear position of the curve is quite high (> 500 kcal/mol) and is believed to represent only a small portion of the theoretical curve governing nucleation (and/or crystallization). Both effects are describable by equations of the form [16-19];

$$P = P_o \, EXP\left(\frac{E_d}{RT}\right) EXP \frac{\Delta F}{RT} \tag{1}$$

where P_o represents the rate at infinite temperature, E_d represents the viscosity barrier for molecular transport across the amorphous (or viscous crystalline)/crystalline interface, and ΔF represents the change in internal free energy upon recrystallization. ΔF is usually written in terms of a supercooling equation and in final form produces a plot similar to Fig. 16. It is obvious from the figure that there are two opposing effects during the nucleation (or crystal growth) phenomena. The thermodynamic driving force towards nucleation increases as the temperature decreases while the microviscosity of the system tends to retard the effect as the temperature decreases. We believe our data is representative of the far left of Fig. 16, near the crystal melting point.

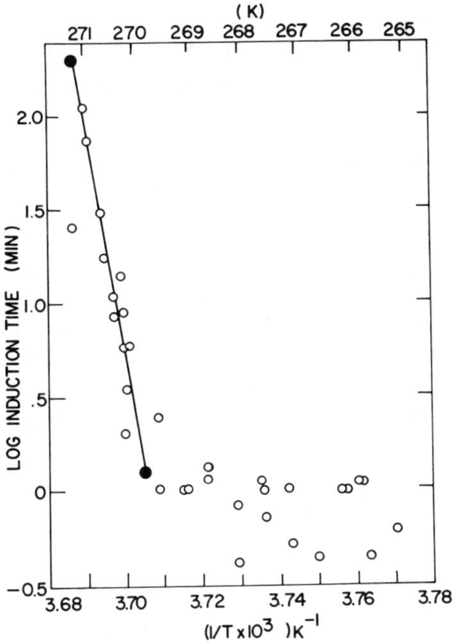

FIG. 15. Log induction time versus inverse temperature for PDES-II heated to 323 K and quenched to the specified temperature.

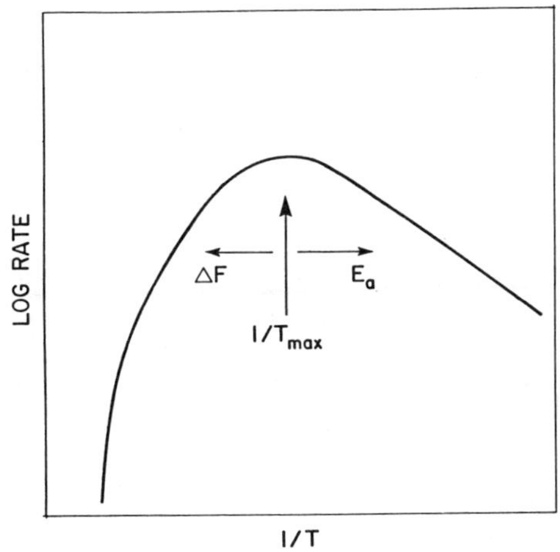

FIG. 16. Schematic representation of log nucleation rate or log growth rate versus inverse temperature.

It is also interesting to note that the time duration of these experiments did permit viscous crystalline formation prior to crystallization so that the process being monitored is more than likely the viscous crystal-to-crystal transformation.

A plot of the initial growth rate versus inverse temperature is shown in Fig. 17. The data coincide with the thermodynamically controlled portion of Fig. 16. The theory of Mandelkern can be applied [15-17] (with reservation) to determine an interfacial interaction parameter per molecule. The reservation considered is that the use of Eq. (1) and its derivatives involves the formalism of considering nucleation and growth in terms of critical nuclear sizes and growth boundaries, respectively. Crystallization from the viscous crystalline phase occurs as an intensification of the viscous crystalline areas. If the viscous-crystalline regions can be considered as nuclei, the application of Eq. (1) is directly understood. If, on the other hand, crystallization involves large-scale reorganizations throughout the preordered phase, where increased birefringence with time is due to increasing perfection and ordering, then the application of Eq. (1) must be made with caution. The observed phenomena could be explained in terms of microviscosity and free energy, as in Eq. (1), but the specifics, given below, could be in error. These reservations must be considered.

Mandelkern's conventional crystal growth theory involves plotting log (rate/T) versus $Tm^2/T\ \Delta T^2$ where Tm is the crystal melting point, T is the absolute temperature and ΔT is the degree of supercooling of the material below the crystal melt. A plot of this data is shown in Fig. 18. For a cylindrical nucleus, the slope of this line is equal to:

$$\frac{8\,\pi \sigma_\mu^{\,2} \sigma_e}{\Delta H^2 R \rho^2} \qquad (2)$$

where σ_μ is the lateral surface energy of a growing crystal boundary, σ_e is the end surface energy, ΔH is the heat of fusion of the crystal, ρ is the density, and R is the universal gas constant. The value of surface energy obtained from such an equation is the cubic $(\sigma_\mu^{\,2}\sigma_e)$. Using the data in Fig. 18 and a value of 10 J/cm as the heat of fusion [9], a value of 0.06 ergs/cm^2 is obtained for the average value of the molecular surface

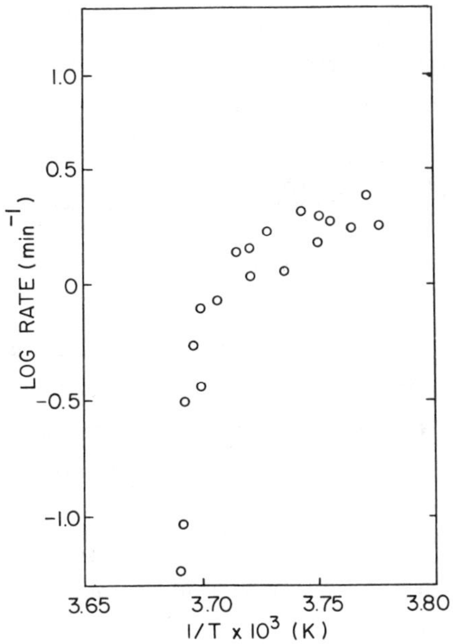

FIG. 17. Log (initial growth rate/T) versus 1/T (K^{-1}) for PDES-II heated to 50 °C and quenched to the specific temperature.

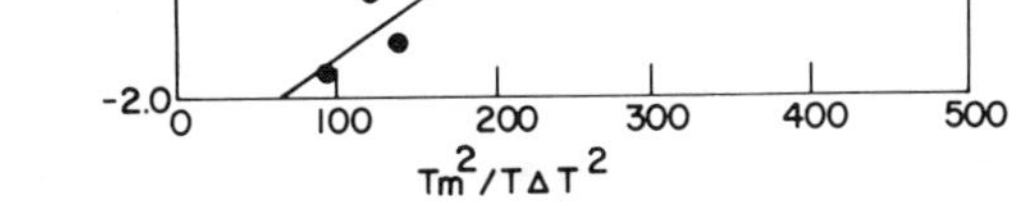

FIG. 18. Log (initial growth rate/T) versus $Tm^2/T\ \Delta T^2$ for PDES-II.

energy. This value could be in error because chain length has not been taken into account [17,20] and the dp (2,000) of PDES-II could affect the results. The value obtained for σ_{av} is much lower than those reported for other polymeric systems [17], and it is felt that this low value is due to the preordering by the viscous-crystalline phase. Such preordering would require less reorientational energy for the molecular chains forming the crystal structures from the viscous-crystalline structure and would provide less interactive free energy during the molecular transformation. These results are not unlike those obtained in liquid crystal systems [18,19] where preordering (the liquid crystal phase) facilitates crystallization and lowers the interfacial barrier.

Growth rate data for the viscous crystalline phase are shown in Fig. 19, a plot of arbitrary growth units versus inverse temperature. The data is Arrhenius activated, with $E_a = 47 \pm 8$ kcal/mol. The value of E_a is relatively large and may indicate that most ordering of the system takes place in the viscous crystalline region temperature region of the sample if enough time is permitted. Unfortunately, calorimetric measurements did not permit attainment of a "heat of fusion" for the viscous crystalline phase, and molecular theories could not be applied to obtain molecular information about the phase. Note that viscous crystalline formation is slow compared with crystal formation so that extreme quenching experiments can be done.

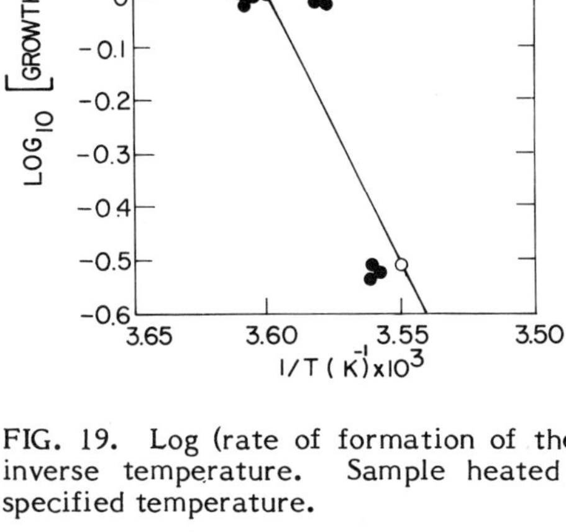

FIG. 19. Log (rate of formation of the viscous crystalline phase) versus inverse temperature. Sample heated to 323 K and quenched to the specified temperature.

F. Light Scattering

These experiments were done to ascertain, what, if any, effects the viscous-crystalline phase had on crystal structure. Two experiments were done. The samples were quenched from 50°C (totally amorphous region) to liquid nitrogen temperature or were cooled slowly through the viscous crystalline region (≈ 2 hr) and crystallized. The fast-quenched samples exhibited H_v patterns for unoriented spherulitic textures [21-23]. The patterns were similar to those observed by Stein and Mosra [21] in the early stages of crystallization of poly(ethyleneterephalate). The incomplete development was characterized as rodlike development of the crystals which transforms to spherulitic textures at longer times. The V_v pattern provided similar results with the experiment indicating incomplete crystallization as might be expected for a quenched sample.

The slow-cooled scattering patterns (which provide for formation of the viscous-crystalline phase prior to crystallization) provide quite different results. It is seen (Fig. 20) that the V_v pattern is isotropic and that the H_v pattern is almost isotropic in nature, but appears square in the intense scattering area, an indication of a possible small cloverleaf pattern. The small size of any cloverleaf pattern in the H_v scattering (as compared to the quenched sample's H_v scattering) would be indicative of very large spherulitic scattering. However, microscopic observation of the sample revealed a texture similar to that in Fig. 11 in which no spherulites were observed. The fact that H_v and V_v scattering are observed would indicate that there are correlated dipoles within the sample (crystallites) but that these crystallites are isotropically oriented. The differences between the quenched sample and the slowly cooled sample also indicate that the formation of the viscous-crystalline phase changes the morphological character of the PDES crystals.

G. Transition Map

A summary of the transitions observed in PDES is given in the transition map (Fig. 21). Data for T_g of poly(dimethylsiloxane) (PDMS) is also included in the figure. The T_g of PDES is very low, and when the entire map is considered, the activation energies of T_g for PDMS and PDES are

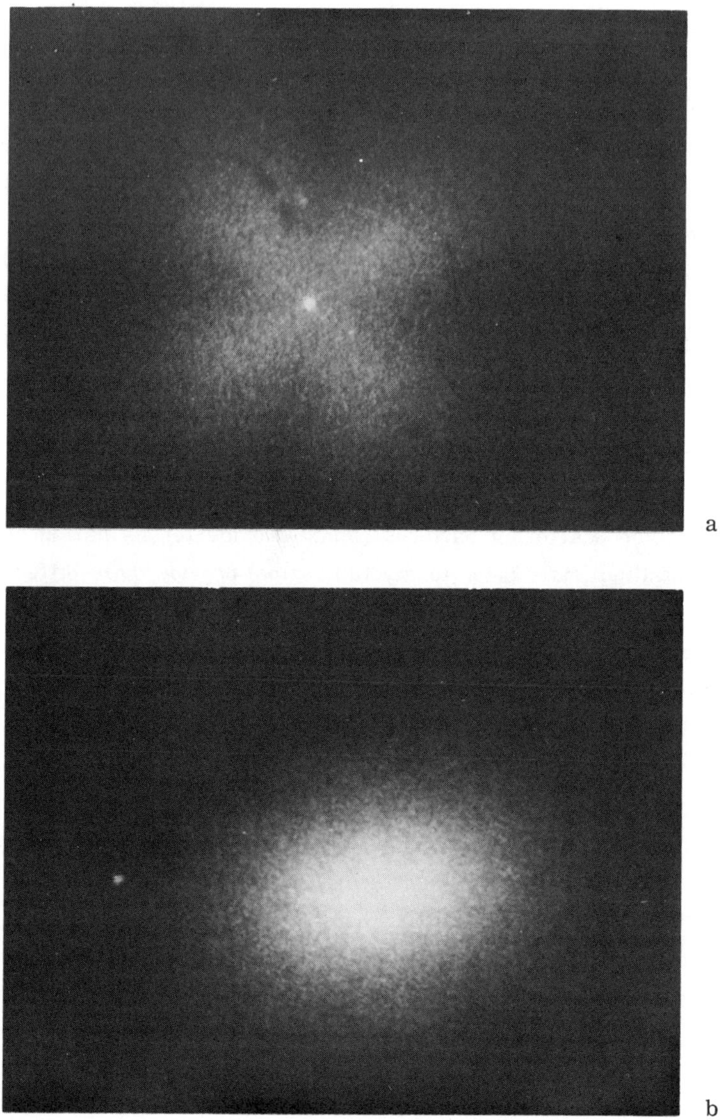

FIG. 20. (a) V_v light-scattering pattern for slowly cooled PDES-II. (b) H_v light-scattering pattern for slowly cooled PDES-II.

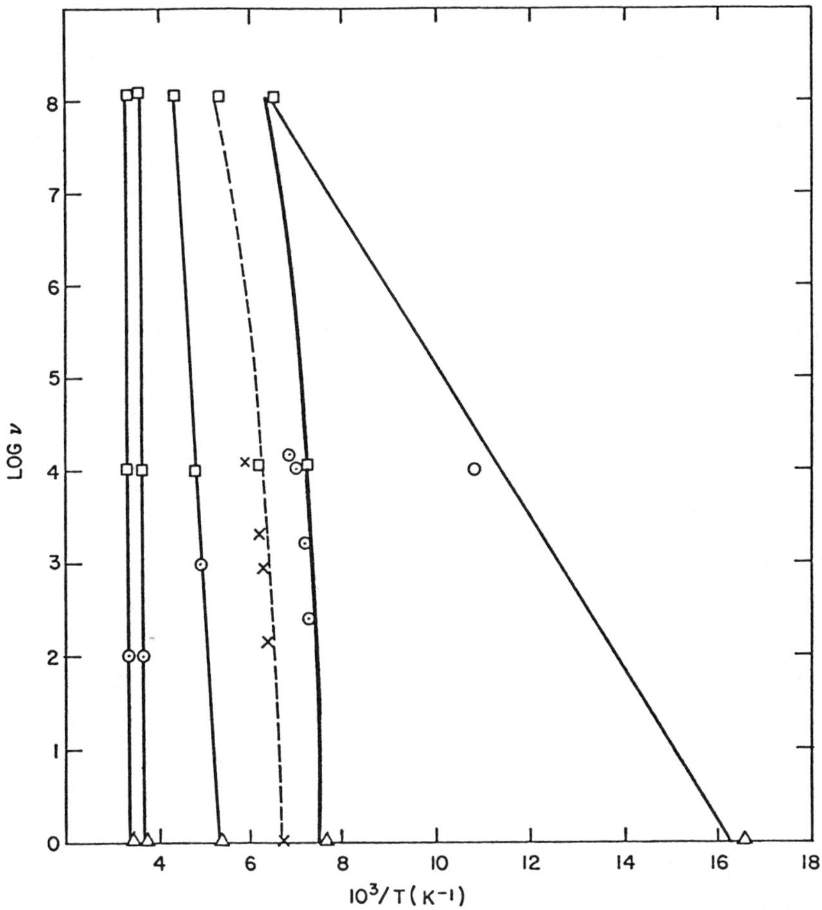

FIG. 21. Transition map for PDES.

similar. Since T_g curves are usually amenable to (WLF)-type analysis an activation energy over the entire frequency range cannot be considered. The data presented previously for a much narrower frequency range should be considered. In that case, E_a for PDES T_g was lower than that for PDMS, an expected result when other polymers are considered. Although not expected, the crystal-crystal transformation as observed via dielectric and NMR techniques is thermally activated with E_a of approximately 40 kcal/mol. This is thought to be due to a slight unwinding of the polymer helical structure [24]. The high-temperature crystal melt is first order, and the viscous crystalline to isotropic melt is also pseudofirst order in nature.

The sub-T_g relaxation is Arrhenius activated ($\approx$ 5 kcal/mol) and thought to be due to ethyl group reorientation which perturbed the chain backbone slightly.

IV. CONCLUSION

In many ways, PDES is similar to many polymers that have been studied. It is semicrystalline, exhibits at least two thermally stable crystal phases and a glass transition. As of this writing, the T_g of PDES is the lowest ever reported for an organic polymer. What makes this polymer unique is not only the viscous crystalline phase, but also the apparent interrelation between the amorphous, crystalline, and viscous crystalline phases. The NMR data show that the crystal and amorphous phases are in more direct contact than might be expected. The fact that T_{2a} does not increase monotonically above Tg indicates that the amorphous relaxations (motions) are perturbed directly by motions (or the structure) of the crystal phase. The converse is also true because T_{2c} changes slightly during the glass transition. In fact, it would be difficult to envision any volume of truly amorphous material, as it would lead to a third T_2 component above the glass transition. The fact that the amorphous and crystalline phases have identical T_2s above the crystal-crystal transformation would indicate that the phases restrict one another's motion. The dielectric data at the crystal-crystal transformation does indicate enhanced dipolar motion or else more contributing dipoles to the dielectric properties. The latter would indicate decreasing crystallinity at the transformation. An even more puzzling phenomena is the rise in T_{2a} at $\approx$ 255 K. This corresponds to a change in shape of the relative dielectric constant versus temperature which indicates the temperature coefficient of the density has increased. There are no calorimetric or dielectric events (other than the change in shape of the ε' versus 1/T) that indicate this event and yet, it is thermally stable and reproducible. The fact that T_{2a} levels above this transition and only increases dramatically above the viscous crystalline melt, again, indicates an intimate relation between the phases.

This relationship is further strengthened when the optical microscopy, x-ray data, and light scattering data are considered. These studies indicate: (1) the average interchain distance in the viscous crystalline and amorphous phases are identical; (2) the viscous crystalline and crystalline phases can act as preordering agents for each other depending on whether temperature is raised or lowered; (3) the kinetics of viscous crystalline phase formation are slow enough to prevent direct quenching to a crystal phase formation are slow enough to permit direct quenching to a crystal phase; and (4) preordering in the viscous crystalline phase lowers any molecular interaction barriers to crystallization. These facts might suggest that the viscous crystalline phase is a phenomena associated with molecular ordering prior to crystallization. The phase in PDES is unique. It may be an expanded (time-) temperature scale that reflects crystallization in all semicrystalline polymers. Alternatively, it may be that the judicious choice of structure in this polysiloxane produces identical interchain spacings in the amorphous and crystalline phases and that once the viscous crystalline phase is formed crystallization produces uniform lattice registry (side-chain crystallization?). Alternatively, the viscous crystalline phase may be formed because of attempted side-chain crystallization of the ethyl groups in PDES. Since the side chains are not long enough to produce true crystal structure, partial ordering may occur. It would appear that chain-folded lamellae do not exist in the crystal phase unless the system is quenched. The picture of molecular structure derived from the above experiments is that: (1) the viscous crystalline phase consists of scattered intramolecular chain folds or short-term intermolecular associations, and (2) upon crystallizing, the singular chain folds attain a uniform chain (lattice) registry (see Fig. 22). This latter effect is confirmed by NMR, where translational motion of the polymer chains occurs upon melting.

Unfortunately, all the experiments which would help define the molecular structure of the viscous crystalline phase have not been done. Small-angle x-ray scattering could indicate suprastructure, and density measurements could help define the observed calorimetric phenomena. As of now, this two-phase system is unique and we feel will lead to increased understanding of the crystallization phenomena in polymeric system.

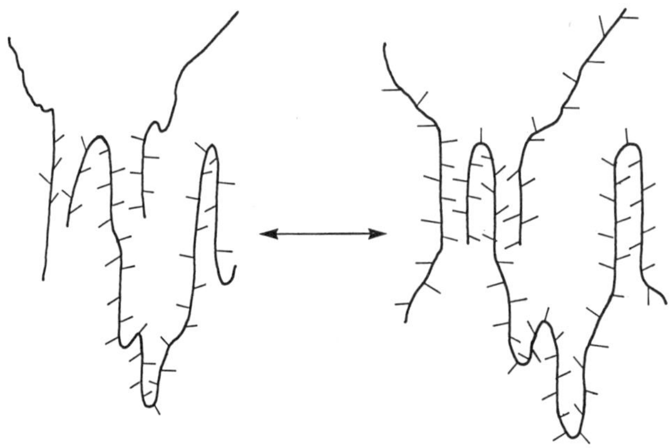

FIG. 22. Schematic of chain structure in the viscous crystalline and crystalline phases of PDES.

REFERENCES

1. F.W. Billmeyer, Textbook of Polymers Science, John Wiley and Sons, New York, 1962.
2. G.W. Gray, Molecular Structure and the Properties of Liquid Crystals, Academic, New York, 1962.
3. M. Baccaredda, P. Magagnini, G. Pizzirani, and P. Guiste, Polym. Lett., 9:303 (1971).
4. B.A. Newman, P.L. Magagnini, and V. Frosini, Adv. Polym. Sci. Eng. Proc. Symp., 21 (1972).
5. P.L. Magagnini, A. Marchetti, F. Materia, G. Pizzirani, and G. Turchi, Eur. Polym. J., 10:585 (1974).
6. C.L. Beatty, J.M. Pochan, M.F. Froix, and D.D. Hinman, Macromolecules, 8:547 (1975).
7. J.M. Pochan, C.L. Beatty, D.D. Hinman, and F. Karasz, J. Poly. Sci., 13:977 (1975).
8. M.F. Froix, C.L. Beatty, J.M. Pochan, and D.D. Hinman, J. Polym. Sci., 13:1269 (1975).
9. C.L. Beatty and F.E. Karasz, J. Polym. Sci., 13:971 (1975).

10. J.M. Pochan, D.F. Hinman, and M.F. Froix, Macromolecules, 9:611 (1976).

11. J.M. Pochan, D.F. Hinman, and S.R. Turner, J. Appl. Phys., 47:10, 4245 (1976).

12a. C.L. Lee, O.K. Johannson, O.L. Flanigan, and P. Hahn, Polym. Preprints, 10(2):1319 (1969).

12b. J.A. Turdakin and V.V. Tarasov, Tr. Mosk. Khim. Yekhno. Inst., 49:8 (1965).

13. N.G. McCrum, B.E. Read, and G. Williams, Anelastic and Dielectric Effects in Polymeric Solids, John Wiley and Sons, New York, 1967.

14. E.F. Westrum, Jr. and T.P. McCullough, in Physics and Chemistry of the Organic Solid State, Vol. 1 (D. Fox, M. Lakes, and A. Weissberger, eds.) Interscience, New York, 1963.

15. W.L. McMillan, Phys. Rev., 6:936 (1972).

16. L. Mandelkern, Crystallization of Polymers, McGraw-Hill, New York, 1964.

17. L. Mandelkern, J.G. Fatou, and C. Howard, J. Phys. Chem., 69:956 (1965).

18. J.M. Pochan and H.W. Gibson, J. Amer. Chem. Soc., 94:5573 (1972).

19. J.M. Pochan and H.W. Gibson, J. Amer. Chem. Soc., 93:1279 (1971).

20. L. Mandelkern, J.G. Fatou, and C. Howard, J. Phys. Chem., 68:3386 (1964).

21. R.S. Stein and A. Misra, J. Polym. Sci. Polym. Phys. Ed., 11:109 (1973).

22. R.J. Samuels, J. Polym. Sci., A-2, 9:2165 (1971).

23. R.J. Samuels, J. Polym. Sci. Polym. Phys. Ed., 12:1417 (1971).

24. M.F. Froix, J.M. Pochan, and D.F. Hinman, to be published.

Chapter 12

APPLICATIONS OF LIQUID CRYSTALS

Joseph A. Castellano*

Princeton Materials Science, Inc.
Princeton, New Jersey

I. INTRODUCTION

Liquid crystals were discovered in 1888 by F. Reinitzer and were first classified into various groups around the turn of the twentieth century. In the early days, the materials were primarily a laboratory curiosity having little or no known application. Although considerable basic research on the optical and electrical characteristics of these unique compounds was carried out during the first half of this century, it was not until the mid-

*Present Affiliation: Stanford Resources, Inc., San Jose, California.

fifties that extensive efforts to develop applications for these materials were made. During this latter period a whole spectrum of cholesteric liquid crystals which could be used as temperature sensors were discovered by workers at the Westinghouse Research Laboratories. As a result, applications to medical diagnostics, electronic component testing, and aerodynamic structure analysis became feasible.

Shortly after these discoveries, research at a number of industrial, university, and government laboratories began to focus on applications which exploited the electro- and/or magneto-optic characteristics of nematic- and cholesteric-type liquid crystals. A result of these experiments was the discovery of dynamic scattering at RCA Laboratories during the mid-sixties. At that time, it became evident that liquid crystals could be used in practical electronic display devices. Research conducted during the balance of that decade led to the identification of a number of other useful electro-optic effects. Now, during the seventies, we see the result of these research efforts. Liquid crystals are used in displays for watches, clocks, calculators, panel meters, and other digital displays, as well as point-of-purchase and point-of-sale displays.

In addition to these developments, liquid crystals were found to be useful laboratory tools for gas-liquid chromatography, absorption spectroscopy, and nuclear magnetic resonance spectroscopy. They have also been used in toys and artistic and decorative materials.

In this chapter we will deal with most of these applications in detail and attempt to provide an understanding of the means by which liquid crystals operate.

II. TEMPERATURE SENSORS

A. Thermal Mapping

The sharp changes in the color of cholesteric liquid crystals with small changes in temperature (e.g., Table 1) have resulted in a number of unique temperature-sensing applications. For a more detailed discussion of the nature of color (pitch) changes in cholesterics see Chap. 3 . One of these is the so-called thermal mapping of electronic components [1]. The microscopic size and high density of integrated and monolithic circuits

TABLE 1. Color changes as a function of temperature (°C) in a cholesteric liquid crystal

Human body temperature	36.9
Colorless → red	39.5
Red → orange	39.7
Orange → yellow	40.0
Yellow → green	40.2
Green → cyan	40.6
Cyan → blue	41.2
Blue → violet	42.0
Violet → colorless	43.0

make most conventional testing methods obsolete. However, thermal mapping with cholesteric compounds, which have a color-temperature response mechanism on the molecular scale, offers a new tool capable of high spatial resolution. The technique merely involves coating of the component with the liquid crystal. Thermal mapping is thus used to determine heat flow and temperature distribution patterns in operating devices as well as to reveal hidden structural characteristics by localization of heat sources or conduction paths.

Cholesteric liquid crystals can also be considered as a valuable tool for aerodynamic testing [1]. Scientists at the NASA-Ames Research Center have conducted a number of wind tunnel experiments with these materials coated on the surface of aircraft and rocket models. They have observed, for the first time, the finer details of boundary layer flow and the slight temperature rise produced by weak shock waves impinging on the surface of the models. They hope that this technique will provide a basis for determining heat transfer rate, an important parameter in aerodynamic design [1].

This technique has also been used to visualize two-dimensional temperature patterns developed during thermal nondestructive tests of other aerospace components and structures. These temperature-sensitive materials have been used to evaluate the efficiency of heat exchangers, to detect nonuniformities in electrically resistive coatings, and to observe the conversion of mechanical-to-thermal energy in tensile and fatigue test specimens. Discontinuities in adhesively bonded composite structures may be located by rapidly heating or cooling a liquid crystal-coated surface.

B. Medical Applications

The medical applications of cholesteric liquid crystals have also received a great deal of attention recently. Observation of skin temperature changes following blocks of the sympathetic nervous system enables the physician to determine if neurologic and vascular pathways are open [1]. Continuous monitoring of the skin temperature over an extended area with cholesteric liquid crystals provides the physician with a more detailed and more easily interpreted indication of circulatory improvements than point measurements with thermocouples or thermistors. It has also been possible to detect breast tumors by this technique [2]; since the temperature of the skin in the vicinity of a tumor is higher than that of the surrounding area.

Finally, a line of disposable thermometers which uses cholesteric liquid crystals on a thin film strip to detect changes in body temperature has become commercially available from several sources (e.g., Ashley-Butler, Inc., 208 Route 206 S., Somerville, New Jersey 08876).

III. ELECTRONIC DISPLAYS

The discovery of dynamic scattering in nematic liquid crystals [3] opened a whole new area of electronic display technology. Today, liquid crystals which exhibit either this mode of operation or one of the field-effect modes are used in displays for watches, clocks, calculators, and various digital panel meters. In this section an attempt will be made to describe the mechanism of operation of some of these effects and to review some of their important applications.

A. Dynamic Scattering Mode

In any discussion of electro-optic effects in liquid crystals, one must include a description of the basic fabrication techniques used in the construction of display devices. The basic cell consists of a parallel plate capacitor with liquid crystalline material acting as the dielectric. The plates are two pieces of glass, each having a thin conductive coating, such as tin or indium-tin oxide. The thickness of active area, which is generally in the range 6-25 μm, may be controlled by the use of appropriate polymeric spacers or glass frit materials.

Application of a dc or low-frequency ac signal across the plates changes the material into a milky white liquid. This appearance is due to the creation of scattering centers within the liquid and not to a chemical reaction. When the voltage is removed, the panel returns to its transparent state.

The initial studies on the electrical behavior of nematic compounds were conducted with p-azoxyanisole (1) and other azoxy compounds at elevated temperatures [4-7]. These compounds have strong negative dielectric anisotropy ($\varepsilon_{||} - \varepsilon_{\perp} < 0$) due to the dipolar nitrogen-oxygen group in the center of the molecule. Williams [5] discovered that thin layers of these materials form regular patterns or domains at low voltages. At higher voltages, these stationary patterns gave way to a turbulent condition. Similar observations have subsequently been reported in other papers [6-8].

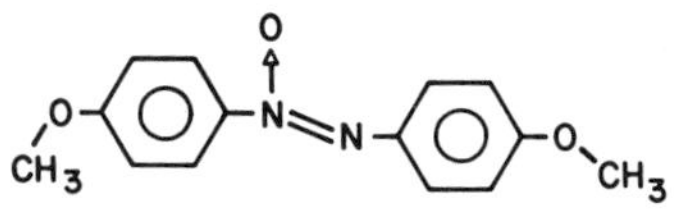

I

(NEMATIC RANGE 115-132°)

Although nematic mesomorphism has been observed in a variety of molecular structures, the large majority of compounds that exhibit the phase are aromatic Schiff-base derivatives. The discovery that certain of these compounds exhibit dynamic scattering [3] prompted studies directed toward the synthesis of materials with low melting points.

By using mixtures of selected Schiff-base compounds it was possible to prepare materials which exhibited, for the first time, stable nematic behavior at ambient temperatures [9]. The first example of a single Schiff base compound to exhibit nematic behavior at ambient temperatures was 2, prepared by Kelker and Scheurle [10]. This compound has since become a model in a number of theoretical studies of the dynamic scattering effect [11].

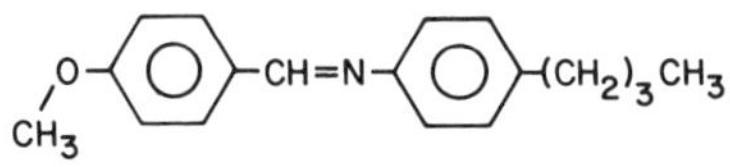

2 (MBBA)
(NEMATIC RANGE 22-47°)

Dynamic scattering is due to electrohydrodynamic instability. The instability can be produced by dc or ac excitation, and it is now believed that the dc and ac instability mechanisms are different. The mechanism proposed by Felici [12] is thought to be dominant with dc, while the ac regime appears to follow the Carr-Helfrich model [13].

Since most applications today involve the ac regime, let us briefly examine the conduction-induced hydrodynamic instability for this case. The presence of space charge is necessary for instability, and in present applications the charge carriers are introduced into the liquid crystal in the form of organic ionic "dopants" (e.g., a quaternary ammonium salt). Due to the anisotropy of the conductivity, the space charge accumulates at distortion maxima when a field is applied. The electric force acting on the space charges induces hydrodynamic flow which couples to the molecular alignment via viscous friction and tends to increase the amplitude of the initial perturbation. This increases with increasing field, and when the applied electric field exceeds a threshold value a turbulent instability results (Fig. 1). Although this turbulence is complex and not well understood, it is believed to be associated with the creation of disclination loops [13].

In current applications of the dynamic scattering mode, digital displays are made by photoetching a seven-segment pattern onto one or both of the indium-tin oxide coated surfaces. Reflective displays have one of the surfaces coated with a metal such as aluminum or a dielectric material such as thorium fluoride. Transmissive displays have etched patterns on both surfaces, and in some applications auxiliary lighting is provided.

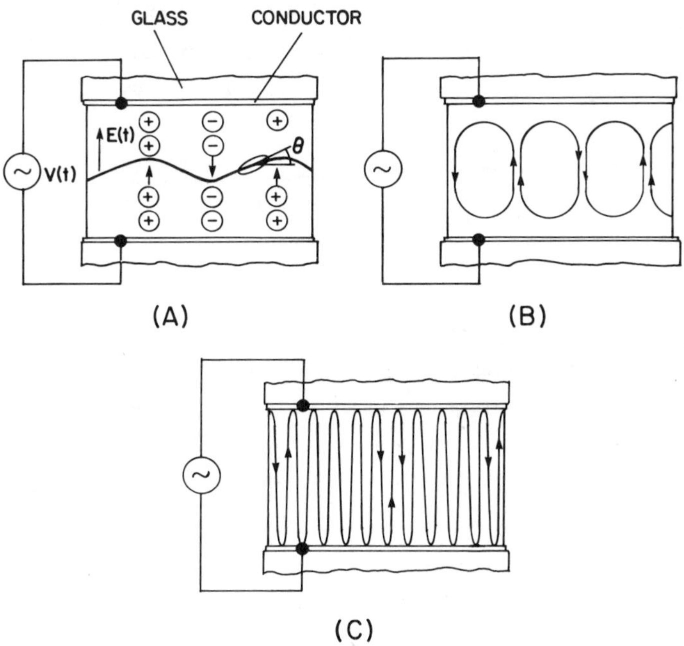

FIG. 1. Build-up of dynamic scattering in an ac-driven nematic: (a) below the instability threshold, (b) at the threshold, and (c) above the threshold.

Mixtures of nematic liquid crystals of the dynamic scattering type (such as 1) and cholesteric liquid crystals (almost all members of this class) produce another interesting electro-optical effect when they are used as the dielectric in the cell configuration described above [4]. The initially transparent panel becomes milky white, similar in appearance to the dynamic scattering mode when subjected to a dc or low-frequency ac signal (generally 25-30V for a 6-μm-thick active layer). In contrast to dynamic scattering, however, this highly scattering appearance remains after the electrical excitation is removed. The reflective contrast ratio fades gradually with time, reaching a value of about 7:1 in minutes, hours, or even days, depending on fabrication techniques and the specific materials used. The transparent state can be abruptly (milliseconds) restored by application of an ac signal in the kilohertz range. The mechanism of this process appears to involve the formation of scattering centers in which the helical axes of the cholesteric molecules have a distribution of angles with respect to the electrode surfaces. The high-frequency erasure field produces molecular alignment of the molecules so that the transparent, planar state is restored.

This effect, which can be operated either in a transmissive or reflective mode, provides a means for the optical storage of information requiring no maintaining power.

B. Field-Effect Modes

In discussing the field-effect modes of operation, we will assume that the device has the same parallel plate geometry described above. We will further assume that the dielectric anisotropy of the nematic material is positive (i.e. $\varepsilon_{\parallel} - \varepsilon_{\perp} > 0$) and the zero-field alignment is homogeneous. There are then two possible ways to construct the electro-optic device. The first approach involves molecular alignment of the nematic material parallel to both electrodes, while in the second method molecular alignment on one electrode surface is at 90° to the alignment on the other electrode surface. In both cases molecular alignment is achieved either by unidirectional rubbing of the plates or by angular evaporation of a dielectric film [14].

Figure 2 illustrates the action of an electric field on cells constructed using these approaches. When the field adds enough electrical energy to exceed the deformation potential of the molecular system (i.e., when a threshold is reached) the initial ordering becomes distorted and the molecules align with their long axes in the direction of the field.

For the first type of molecular alignment (i.e., alignment parallel to both electrodes), it was found [15] that this cooperative alignment could be used to orient dichroic dye molecules. The optical absorption spectrum of a dichroic dye molecule is a function of its molecular orientation with respect to the polarization of the incident light. Materials that exhibit dichroism are usually long, cylindrically shaped molecules containing chromophoric groups which form part of an extended aromatic system. Thus, if the dichroic molecule is oriented with its long axis parallel to the electric vector of the incident polarized light, absorption of light by the molecule (low-energy transition) occurs, and the characteristic color of the dye is observed.

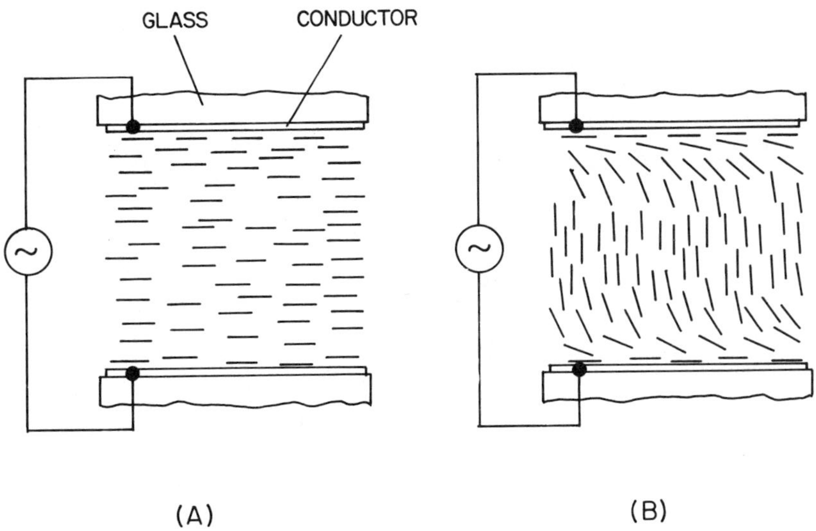

FIG. 2. Electric field ordering of a $\varepsilon \quad > \varepsilon$ nematic liquid: (a) below the dielectric reorientation threshold and (b) above the threshold. Spontaneous zero field ordering is homogeneous.

Conversely, orientation of the molecule with its long axis perpendicular to the electric vector results in little or no absorption by the visible transition, and the incident light is transmitted unchanged.

The effect of electric fields on mixtures of these dyes with nematic hosts is illustrated in Fig. 3. The very strong permanent dipole moment operating along the long molecular axis enables the molecules to align in the direction of the applied electric field and in turn to orient the dissolved dye molecules with their long axes perpendicular to the electric vector of the incident polarized light. This produces a large decrease in the optical density and hence in the disappearance of color.

The second type of zero-field molecular alignment was reported by Mauguin [22] and by Schadt and Helfrich [16]. They produced the molecular ordering by rubbing the top and bottom plates in orthogonal directions. This caused the ordering direction to twist 90° in going from one electrode to the other (hence the term "twisted nematic"). Application of an externally applied electric field then resulted in an "untwisting" of the structure and an alignment of the molecules in the direction of the

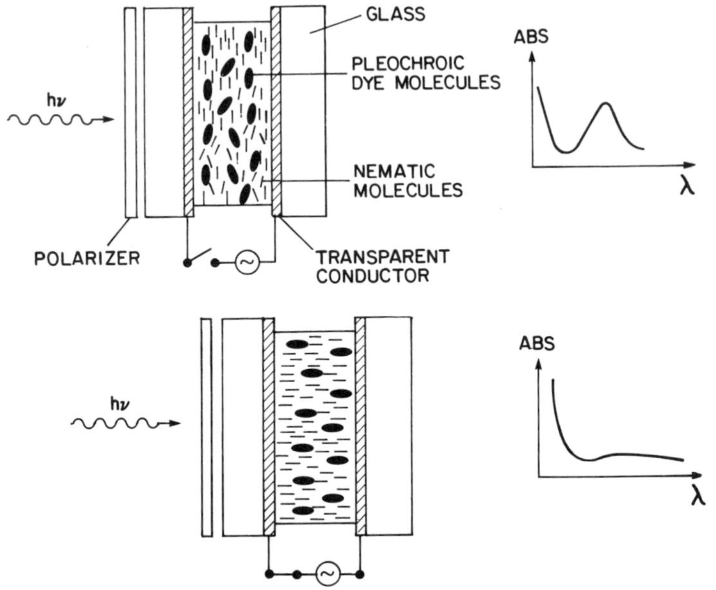

FIG. 3. Schematic diagrams of electro-optic cells exhibiting electronic color switching.

field. This electrical untwisting can be used to change the direction of polarization of light passing through the cell and thus to produce one of the most useful low-power electro-optic light valves known.

In Fig. 4 a twisted-nematic cell is shown with a linear polarizer and analyzer crossed with respect to their polarization directions. The polarized light entering the cell is thus rotated 90° and passes through the analyzer essentially unchanged (there are very small changes in the absorption of light by the cell). When the twist is destroyed by application of an electric field across the liquid crystal cell, however, the linearly polarized light is not rotated and becomes absorbed by the analyzer. If a seven-segment pattern is photoetched into the electrode surfaces, a digital display with dark numbers on a light background will result (see Fig. 6). The contrast of the display will vary depending upon the efficiency and color of the polarizer and/or analyzer.

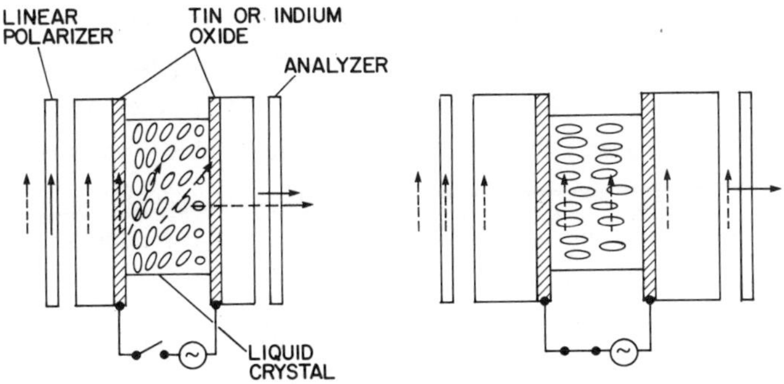

FIG. 4. Field-induced phase change in twisted nematic liquids.

The last field-effect mode we will consider deals with the effect of electric fields on mixtures of cholesteric compounds and nematic compounds (such as 3) which have their dipole moments along the molecular axis. When subjected to electric fields of 200 kV/cm, cells prepared in the manner described above, which initially exhibited a bright yellow iridescence, became transparent. This effect is the opposite of the dynamic scattering mode. The "cholesteric" material, which in this case is not a derivative of cholesterol, has a structure 3 very similar to that of 2. The asymmetric carbon atom (asterisk) permits the molecules of 3 to assume the helical conformation characteristic of the cholesteric mesophase. For a further discussion of cholesterics of the nonsterol type see Chap. . Application of an electric field causes disruption of the helical chains and thus a change from the cholesteric structure to an oriented nematic structure.

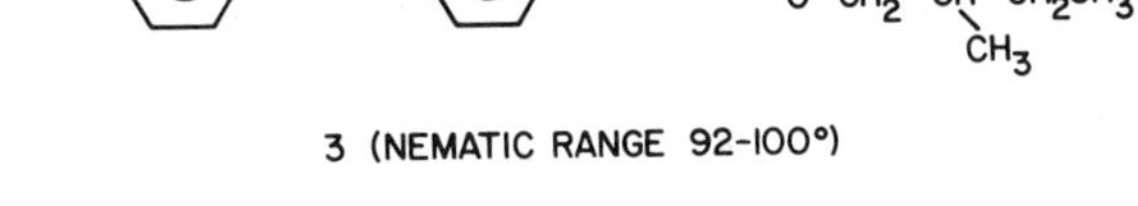

3 (NEMATIC RANGE 92-100°)

Electric field-induced structure changes are potentially useful display effects since no specular reflecting-back electrode is needed to maximize the contrast ratio (> 10:1 if a black background is used).

Numeric displays based on the field effect modes possess several important advantages over other electro-optic effects in liquid crystals. For example, the high resistivity of the material used in the displays enables one to have an extremely low power consumption and low operating voltages. In addition, it is possible to use both a transmissive and reflective mode of operation and to obtain various colors by the addition of dyes and/or selection of polarizer materials.

C. Applications of Liquid Crystal Electronic Displays

The main reason why liquid crystal displays have been utilized in digital display applications has been because of their ultralow power requirements and the associated advantages. However, other advantages of liquid crystal displays, including cost and the design flexibility that is possible in appearance and format of the display, will undoubtedly become increasingly significant for these and other commercial applications. The displays for the devices described here all utilize the standard seven-segment digital format. However, other format types are possible [17].

1. Watches

Figure 5 shows the basic schematic of a liquid crystal watch for displaying hours and minutes. The quartz crystal accurately controls an oscillator circuit. The oscillator frequency, typically 32,768 Hz, is divided down to generate "second" pulses. The latter are then counted to produce "minute" and "hour" pulses. Each of these time pulses is decoded to provide outputs of the type needed for seven-segment digital displays. The decoder outputs are converted in appropriate driver circuits to suitable ac signals for activating the liquid crystal display segments.

The dividers, counters, decoders, and driver circuits, typically containing a total of about 1,500 transistors, can all be produced on a single C-MOS chip having an active silicon area of about 0.25 x 0.40 cm. The power for the watch is provided by a 1.5-V silver oxide battery of capacity in the range of 100-240 mA• hr. Most of the electronic circuitry operates at 1.5 V, with the exception of the display drivers which operate at between 10 and 15 V for dynamic scattering displays and between 1.5 and 5 V for field-effect displays. The higher voltage is produced by conventional

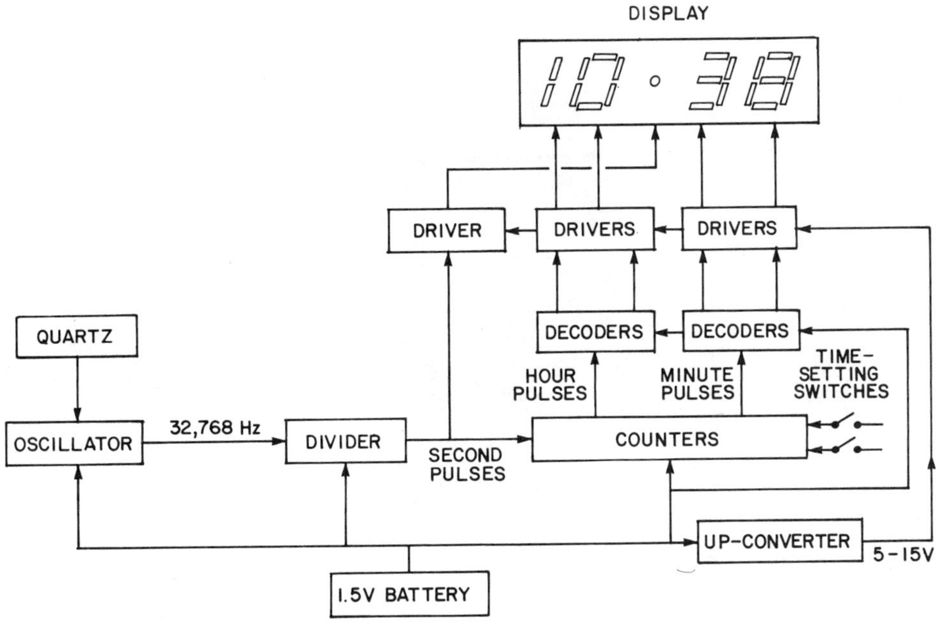

FIG. 5. Block diagram of a liquid crystal watch.

"up-converter" techniques, which if properly designed are capable of 85% efficiency. A dynamic scattering liquid crystal watch can operate for about 1 year on a 150 mA•hr battery. In the case of a field-effect liquid crystal watch, much less power is used in the display, drivers, and up-converter. As a result the watch can operate for much longer than a year, or alternatively a smaller battery can be used.

1973 was the first year of real production of the liquid crystal watches. During that year about 100,000 units were manufactured, of which about 50% had dynamic scattering displays and about 50% field-effect displays. An example of a field-effect liquid crystal display is shown in Fig. 6.

2. Clocks

A solid-state digital electronic clock is basically the same as a watch except that the display is much larger. Liquid crystal displays can be built with areas as large as about 400 cm^2. This size limit is determined by the

FIG. 6. Digital watch with field-effect liquid crystal display.

achievable flatness of the containing glass surfaces. The cost of a large liquid crystal clock display is low, and the power consumption is still sufficiently small to permit long operation from flashlight-sized batteries. As time goes by and manufacturing costs decrease (particularly those of the driving integrated circuits) liquid crystal clocks are expected to become an important factor in the digital clock market.

3. Calculators

A calculator consists of three parts: a keyboard, electronic circuitry, and a display. The electronics codes into binary pulses the information punched into the keyboard by the user. These pulses are then operated on, according to the mathematical function required (i.e., add, subtract, multiply, divide, etc.) in an arithmetic unit. This unit may or may not contain memory depending on the sophistication of the functions the calculator is designed to perform. The outputs from the arithmetic unit are decoded into digital form and applied through suitable driving circuits to the segments of the display digits. The electronics for coding, performing the arithmetic functions, decoding, and driving the display can all be realized on one or two integrated circuit chips.

In 1975 more than 100,000 calculators with liquid crystal displays were produced and sold. Most of these calculators were of the small desk-top variety and had individually driven (i.e. nonmultiplexed) 8-12 display digits. The liquid crystal displays were operated in a transmissive dynamic scattering mode with an auxiliary lamp provided for back lighting.

The consumer reaction to liquid crystal calculators has been very good. The advent of field-effect displays in commercial quantities further enhances the "readability" advantage that liquid crystal calculators already appear to enjoy over those that use other display technologies.

IV. LABORATORY AND OTHER USES

Nematic liquid crystals differ from normal isotropic liquids only by a parallel orientation of their rodlike molecules. This parallel orientation produces a high degree of macroscopic anisotropy. As a result, the molecules can be homogeneously oriented by a magnetic field such that their optic axis turns parallel to the field. Thus, it is possible to use these materials as anisotropic liquid solvents for nuclear magnetic resonance measurements. Information on the anisotropy of the chemical shift and molecular geometry may be obtained using this method.

Highly resolved spectra of partially oriented molecules such as mono-, di-, and trisubstituted benzenes have been obtained in nematic solutions [18]. The observed spectra were very nearly identical with those obtained through computer calculation.

Because of the ordered arrangement of their molecules, liquid crystals show some unusual solvent properties. Different-position isomers of disubstituted benzenes in particular show different solubilities in liquid crystals. The linear isomers dissolve more easily, their molecules resembling more closely the rod-shaped molecules of the solvent. This property makes liquid crystals useful as stationary phases in the gas-liquid chromatography of benzene-position isomers [19]. Both smectic and nematic compounds were evaluated for this application, and they were found to give better separation of meta and para isomers than conventional silicone liquid phases.

Cholesteric liquid crystals have also been used for novelty items such as toys and decorative materials. In these applications, however, the films must be protected from physical damage by microencapsulation [1], a process developed by the National Cash Register Company. In this process, the cholesteric compounds are emulsified into a warm aqueous solution of gelatin and gum arabic. As the solution is cooled, the polymers deposit on the liquid crystal forming individual microcapsules ranging from 5-50 μm in diameter.

Films of microencapsulated liquid crystals can be prepared by spraying, draw-bar coating, or silk screening. After evaporation of the water, the films are dry to the touch and are relatively free from damage by abrasion. The encapsulation process does not alter the color-temperature response, and for most applications the microcapsules can be directly substituted for the unencapsulated liquid crystals. Sheets of these materials are now being sold under various trade names.

A novel family of cholesteric liquid crystals have been developed by Du Pont for the decorative market. These materials, which are still in the experimental stage, may be used to provide attractive decorative effects on opaque, translucent, and transparent surfaces. In contrast to most commercial liquid crystals, however, these materials exist as solid glasses at ambient temperatures [20]. Thus, it is possible to lay down coatings that exhibit the coloration effects of liquid crystals but also possess rigidity and usable surface characteristics. Thermal aftertreatment of the coatings serve to enhance these colors and assist in the development of specific desired shades. By application of pressure or shearing action, it is possible to produce textured, iridescent patterns in the films. It is anticipated that these materials will be used to produce decorative patterns on metals, fabrics, leather, ceramics, and paper.

V. FUTURE OUTLOOK

enjoyed enormous growth over the period between 1965 and 1975 and that the future looks very bright indeed. During the second half of this decade, we expected the markets for watches, clocks, calculators, and digital panel readouts using liquid crystal displays to expand rapidly. We also expected to see these displays in automobiles, boats, airplanes, kitchen appliances, computers, and cash registers.

I am sure that before this century ends the number of applications will have trebled. Since I was fortunate to have been involved in this field during this period of rapid growth, it will be exciting and personally gratifying to see these developments unfold.

RECENT DEVELOPMENTS

During the past 4 years, liquid crystal applications to electronic display devices and systems have made major strides. In addition, a dramatic shift from dynamic scattering to field-effect-type displays has occurred. This was primarily due to the improved legibility of the field-effect displays and to their lower power consumption. In 1979, for example, we see field-effect liquid crystal displays in tens of millions of digital watches being sold world-wide. Because of improvements in integrated circuit technology, packaging, and liquid crystal display manufacturing, these watches are slimmer, far more attractive, and offer many more functions than those produced in 1973-1974.

Digital clocks with liquid crystal displays are also widely in use. Some versions still use the dynamic scattering effect. A major potential market, the OEM automotive business, has not yet been penetrated by liquid crystal because of the extremely difficult environmental requirements.

Field-effect liquid crystal displays now appear in virtually all hand-held calculators. The extremely low power requirements of the display, coupled with further improvements in complimentary symmetry-metal oxide semiconductor (C-MOS) integrated circuits has made possible the development of the "shirt-pocket" computer (e.g., the Sharp Electronic EL-5805). These ultrathin calculators operated by tiny silver oxide batteries are capable of performing trigonometric and logarithmic computations in addition to simple arithmetic.

Liquid crystal displays are gaining acceptance in the digital instrument field, replacing the power-hungry light-emitting displays in applications where portability is of prime interest.

The development of multiline, multicharacter, alphanumeric liquid crystal displays for application to data terminal readouts has recently been announced by Hitachi [21]. More activity in this area is expected.

Finally, the use of liquid crystal materials in the development of flat-panel television displays has recently received attention from Hitachi, Westinghouse, and Hughes Aircraft Co. [21]. Each company has developed its own unique electronic addressing schemes and fabrication techniques to demonstrate a full 500 x 500 line television picture. Although these are far from commercialization, it is conceivable that flat-panel, liquid crystal television displays could appear on the market before 1985.

REFERENCES

1. G.H. Brown (ed.), Liquid Crystals, Vol. 2, Part I, Gordon and Breach, New York, 1969, Synopsis of Panel Discussion on Applications, pp. 217-232.
2. J.L. Fergason, work conducted at Kent State University, Kent, Ohio.
3. G.H. Heilmeier, L.A. Barton, and L.A. Zanoni, Proc. IEEE, 56:1162 (1968).
4. J.A. Castellano, Electronics, McGraw-Hill, New York, 1970, p. 64.
5. R. Williams, J. Chem. Phys., 39:384 (1963).
6. A.P. Kapustin and L.S. Larinova, Sov. Phys. Cryst., 9:235 (1964).
7. G. Elliott and J.G. Gibson, Nature (London), 205:995 (1965).
8. R. Williams and G.H. Heilmeier, J. Chem. Phys., 44:638 (1966).
9. J.A. Castellano and J.E. Goldmacher, U.S. Patent 3,540,796, 1970, applied for June 9, 1966.

10. H. Kelker and B. Scheurle, Angew. Chem., 81:903 (1969).
11. Abstracts of Papers presented III Liquid Crystal Conference, Berlin, 1970. I. Haller and J.D. Litster, p. 26; Orsay Liquid Crystal Group, p. 98; W. Haas, J. Adams, and J.B. Flannery, p. 109; H. Koelmans and A.M. van Bostel, p. 113; R. Borel and F. Tobazeon, p. 117.
12. N. Felici, Rev. Gen. Electr. 78:717 (1969).
13. E.F. Carr, Mol. Cryst., 7:253 (1969); W. Helfrich, J. Chem. Phys., 51:4092 (1969).
14. J.L. Janning, Appl. Phys. Lett., 21:173 (1972).
15. G.H. Heilmeier, J.A. Castellano, and L.A. Zanoni, Mol. Cryst., 8:293 (1969).
16. M. Schadt and W. Helfrich, Appl. Phys. Lett., 18:127 (1971).
17. G.W. Taylor and I. Lefkowitz, Proc. IEEE, 61:487-489, 1973; Opt. Commun., 8:426-429, 1973.
18. A. Saupe, in Liquid Crystals (Brown, Dienes, and Labes, eds.), Gordon and Breach, New York, 1967, p. 207; L.C. Snyder, J. Chem. Phys., 43:4041 (1965); M. Pavar, W.D. Phillips, and J.C. Rowell, J. Chem. Phys., 43:3442 (1965).
19. H. Kelker, Z. Anal. Chem., 198:254 (1963); M.J. Dewar and J.P. Schroeder, J. Org. Chem. 30:3486 (1965).
20. W. Mahler and M. Panar, J. Amer. Chem. Soc., 94:7195 (1972); L.B. Leder, U.S. Patent 3,789,225 (1974).
21. Proceedings of the 1978 Society for Information Display Conference, San Francisco, Ca. April, 1978.
22. C. Mauguin, Bull. Soc. Fr. Min., 34:71, 1911.

AUTHOR INDEX

Numbers in brackets are reference numbers and indicate that an author's work is referred to although his name is not cited in the text. Underlined numbers give the page on which the complete reference is listed.

SUBJECT INDEX